MATHEMATICS RESEARCH DEVELOPMENTS

ESTIMATION OF STOCHASTIC PROCESSES WITH MISSING OBSERVATIONS

Mathematics Research Developments

Additional books and e-books in this series can be found on Nova's website under the Series tab.

MATHEMATICS RESEARCH DEVELOPMENTS

ESTIMATION OF STOCHASTIC PROCESSES WITH MISSING OBSERVATIONS

MIKHAIL MOKLYACHUK
MARIA SIDEI
AND
OLEKSANDR MASYUTKA

nova
science publishers
New York

Library of Congress Cataloging-in-Publication Data

ISBN: 978-1-53615-890-8

Published by Nova Science Publishers, Inc. † New York

Contents

Introduction

The problem of estimation of unobserved values of stochastic processes is of constant interest in the theory and applications of stochastic processes. The problem of forecasting future values of economic and physical processes, the problem of restoring lost information, cleaning signals or other data from observations with noise, is magnified in an information-laden word. Therefore, the development of estimation methods is one of the main tasks of the modern theory of stochastic processes. Formulation of the interpolation, extrapolation and filtering problems for stationary stochastic sequences with known spectral densities and reducing these estimation problems to the corresponding problems of the theory of functions belongs to A. N. Kolmogorov [1]. The main properties of such sequences were obtained based on the H. Cramér [2] and M. Stone [3] results. Based on the H. Wold [4] results, A. N. Kolmogorov proposed a decomposition of the stationary sequences into a sum of regular and singular stationary sequences. He also derived a number of properties of stationary sequences and received characterization of regular stationary sequences. He introduced the concept of minimality, and was the first one who proved the minimality condition $\int_{-\pi}^{\pi} f^{-1}(\lambda)d\lambda < \infty$ for the spectral density function $f(\lambda)$. These results were used for solving the extrapolation and interpolation problems for stationary sequences $x(k)$, $k \in \mathbb{Z}$. Effective methods of solution of the estimation problems for stationary stochastic sequences and processes were developed by N. Wiener [5], N. Wiener and P. Masani [6], [7], A. M. Yaglom [8, 9]. Further results are presented in the books by Yu. A. Rozanov [10] and E. J. Hannan [11], P. J. Brockwell and R. A. Davis [12].

The basic techniques of statistics of stochastic processes are summarized in the books by D. Z. Arov and H. Dym [13], L. Aggoun and R. J. Elliott [14],

I. V. Basawa and B. L. S. Prakasa Rao [15], S. Cohen and R. J. Elliott [16], M. S. Grewal and A. P. Andrews [17], G. Kallianpur [18], Yu. A. Kutoyants [19], [20], R. S. Liptser and A. N. Shiryaev [21], [22], B. L. S. Prakasa Rao [23, 24], B. L. Rozovsky and S. V. Lototsky, [25], M. B. Rajarshi [26], W. A. Woodward, H. L. Gray and A. C. Elliott [27]. Spectral theory of random fields is described in the monograph by M. I. Yadrenko [28]. A detailed description of different time series models is given in the classical book by G. E. P. Box and G. M. Jenkins et al [29]. The resent investigations on this topic have been made, for example, by S. Arvanitis and T. Magdalinos [30], W. Bell and D. Martin [31], M. J. Chambers and M. A. Thornton [32], M. J. Chambers [33], D. Findley and D. Martin [34], B. Spangl and R. Dutter [35], T. Proietti et al. [36], B. Seong et al. [37], M. Zhong [38].

For the recent results on signal extraction problem, we refer to S. Koopman and A. Harvey [39], T. McElroy and A. Sutcliffe [40], T. McElroy [41], M. Wildi [42]. The prediction problem for the stationary processes based on observation of a finite segment was introduced by M. G. Krein [43] (a translation of the original work published in 1954) and developed in the article [44] and the book [45] by H. Dym and H. P. McKean dedicated to extrapolation and interpolation of Gaussian stationary processes. The estimation problems occur in different studies. We refer to D. V. Koroliouk [46], who investigated the difference stochastic equation $\Delta\alpha_{t+1} = -V_0\alpha_t + \sigma_0\Delta W_{t+1}$, $t \geq 0$, which determines a sequence α_t, $t \geq 0$, for the stochastic component ΔW_{t+1}, $t \geq 0$. D. V. Koroliouk and V. S. Koroliuk [47] studied the problem of filtration of stationary Gaussian statistical experiments considered for the solution α_t, $t \geq 0$, of the indicated equation.

The crucial assumption of most of the methods developed for estimating unobserved values of stochastic processes is that the spectral densities of the involved stochastic processes are exactly known. However, in practice complete information on the spectral densities is impossible in most cases. In this situation one finds parametric or nonparametric estimate of the unknown spectral density and then apply one of the traditional estimation methods provided that the selected density is the true one. This procedure can result in a significant increasing of the value of the estimation error as K. S. Vastola and H. V. Poor [48] have demonstrated with the help of some examples. To avoid this effect one can search the estimates which are optimal for all densities from a certain class of admissible spectral densities. These estimates are called minimax since they minimize the maximal values of the errors of estimates for all densities from a given class of admissible spectral densities.

The paper by Ulf Grenander [49] was the first one where this approach to extrapolation problem for stationary processes was proposed. The author investigated the stationary stochastic process $\{x(t), t \in \mathbb{R}\}$ with continuous time. He described a game of two players X and P, where the player X chooses a process $x(t)$ and the player P builds a forecast px of the linear functional $Ax = \int_0^1 a(t)x(t)dt$ from observations of the process $x(t)$ at points $t \leq 0$. The first player may choose a stationary process $x(t)$ from the class of stationary processes satisfying conditions $Ex(t) = 0$ and $Ex^2(t) = 1$. The player X aims to maximize the value $E(Ax - px)^2$ and the player P aims to minimize this value. The game price and the value px are determined by a given function $a(t), t \in (0,1)$. Similar problems for two person games in the case of stationary sequences and processes were investigated in the works of M. P. Moklyachuk [50, 51, 52, 53, 54] (see also the book [55] and the review article [56]).

Another type of uncertainty was analysed by P. J. Huber [57]. He investigated the normal distribution with ε–mixture and introduced an M-estimator to solve the problem. Later on, P. J. Huber [58] found the robustness conditions (insensitivity to small deviations) of the censored likelihood ratio test and proposed a method of checking if a spectral density is the least favourable in a given class. These results were developed by S. A. Kassam and H. V. Poor in their work [59]. L. Breyman [60] used the minimax method of filtering of stationary sequences in the case of convex sets of spectral distributions of signals.

Extrapolation and interpolation problems for stationary sequences in the case of ε-mixture models were investigated by Y. Hosoya [61] and M. Taniguchi [62]. S. A. Kassam [63], [64] summarized the results for the interpolation problem for stationary sequences. To find the least favourable spectral densities he used the statistical hypotheses testing approach.

The minimax estimation of stationary sequence from observations with noise was studied by J. Franke [65], [66], J. Franke and H. V. Poor [67]. Among others, they managed to find solutions of the extrapolation and interpolation problems in the case of one missing value using the convex optimization methods. The minimax filtering problem was studied by H. V. Poor [68]. He applied the obtained results to the problem of signal processing [69]. Then S. K. Vastola and H. V. Poor [70] considered the linear filtering problem for a functional in the case of spectral uncertainty. We should also mention the work by S. Verdu and H. V. Poor [71] dedicated to robust methods of estimation and the works by G. K. Golubev and M. S. Pinsker [72], [73] who studied signals with bounded average energy.

The main results of the theory and applications of the robust estimation techniques for the stationary processes published up to 1985 were summarized in the review article by S. A. Kassam and H. V. Poor [59]. We can also mention the book by O. M. Kurkin, Yu. B. Korobochkin, and S. D. Shatalov [74], dedicated to the minimax methods of information processing. Further results on minimax extrapolation, interpolation and filtering problems for linear functionals can be found in the papers by M. P. Moklyachuk [50] – [54], [56, 75, 76, 77, 78, 79, 80]. In these works the author investigated the estimation problem for stationary sequences and processes and found the classical mean square optimal and the minimax estimates based on observations with a noise.

Robust approach to estimation problem is widely applied in different areas of stochastic process researches, see for example the works by D. Bertsimas et al.[81], P. S. Bhattacharya and D. D. Thomakos [82], E. Carrizosaa et al. [83], E. Delage and Y. Ye [84], J. Ishihara et al.[85], T. L. Molloy and J. J. Ford [86], P. Ruckdeschel et al.[87], S. Song [88], B. Spangl and R. Dutter [35]. We also refer to the book by Yu. Kharin [89] dedicated to the robust methods in statistical forecasting.

The estimation problems for functionals which depend on unobserved values of multidimensional stationary stochastic processes is the aim of the papers by M. P. Moklyachuk and O. Yu. Masyutka [90, 91, 92, 93, 94, 95, 96]. Extrapolation, interpolation and filtering problems for linear functionals depending on periodically correlated sequences and processes have been set up and solved by M. P. Moklyachuk and I. I. Golichenko(Dubovets'ka) [97, 97, 98, 99, 100, 101, 102, 103, 104]. Harmonizable stochastic sequences are being studied by M. P. Moklyachuk and V. I. Ostapenko [105, 106, 107, 108, 109, 110, 111]. In particular, they found the minimax solutions of the interpolation problem for these sequences in the case of observations with and without noise. Extrapolation, interpolation and filtering problems for linear functionals depending on unobserved values of stochastic sequences and processes with stationary nth increments are investigated by M. M. Luz and M. P. Moklyachuk

[112, 113, 114, 115, 116, 117, 118, 119, 120, 121, 122, 123, 124, 125, 126]. Estimation of functionals constructed from random sequences with periodically stationary increments is the aim of the paper by P. S. Kozak and M. P. Moklyachuk [127].

The prediction problem for stationary sequences with missing observations is investigated in the papers by P. Bondon [128], [129], R. Cheng, A. G. Miamee and M. Pourahmadi [130], R. Cheng and M. Pourahmadi [131], Y. Kasahara, M. Pourahmadi and A. Inoue [132], [133]. The detailed analysis of the estimation problems with missing observations are presented in the paper by B. Abraham[134], books by M. J. Daniels and J. W. Hogan [135], C. K. Enders [136], R. J. A. Little and D. B. Rubin [137], P. E. McKnight et al.[138], M. M. Pelagatti [139].

In the papers by M. P. Moklyachuk and M. I. Sidei [140, 141, 142, 143, 144, 145, 146] results of investigations of the interpolation, extrapolation and filtering problems for stationary stochastic sequences and processes with missing observations are proposed. The results of the study of the extrapolation, interpolation and filtering problems for linear functionals constructed from unobserved values of multidimensional stochastic sequences and processes are presented in the papers by O. Yu. Masyutka, M. P. Moklyachuk and M. I. Sidei [147, 148, 149, 150, 151]. See also the book by M. P. Moklyachuk, O. Yu. Masyutka and I. I. Golichenko [152] where results of the investigation of the problem of mean square optimal estimation (forecasting, interpolation, and filtering) of linear functionals constructed from unobserved values of periodically correlated isotropic random fields are described.

In this book, we propose results of the investigation of the problem of mean square optimal estimation (forecasting, interpolation, and filtering) of linear functionals constructed from unobserved values of stationary stochastic processes. Estimates are based on observations of the processes with additive stationary noise. We develop methods for finding the optimal estimates of the functionals in the case where some observations are missing. Formulas for computing values of the mean-square errors and the spectral characteristics of the optimal linear estimates of the functionals are derived in the case of spectral certainty, where the spectral densities of the processes involved are exactly known. Formulas that determine the least favourable spectral densities and the minimax robust spectral characteristics of the optimal estimates of the functionals are derived in the case of spectral uncertainty, where the spectral densities of the processes are not exactly known while some classes of admissible spectral densities are specified.

Keywords: stationary stochastic process, optimal estimate, minimax robust estimate, mean square error, least favourable spectral density, minimax spectral characteristic.

2010 Mathematics Subject Classification: Primary: 60G10, 60G25, 60G35, Secondary: 62M20, 93E10, 93E11 93E10, 93E11

Chapter 1

Estimation of Functionals from Stationary Stochastic Sequences with Missing Observations

In this chapter, we propose results of the investigation of the problem of mean square optimal estimation of linear functionals constructed from unobserved values of a stationary stochastic sequence. Estimates are based on observations of the sequence with a stochastic noise sequence. We develop methods for finding optimal estimates of functionals in the case where some observations are missing. Formulas are derived for computing the mean-square errors and the spectral characteristics of the optimal linear estimates of the functionals in the case of spectral certainty, where the spectral densities of the sequences are exactly known. Formulas are proposed that determine the least favourable spectral densities and the minimax robust spectral characteristics of the optimal estimates of the functionals in the case of spectral uncertainty, where the spectral densities are not known exactly, while some classes of admissible spectral densities are given.

1. Stationary Sequences. Spectral Representation

In this section, we present the basic definitions and spectral properties of stationary stochastic sequences (for more details see books by I. I. Gikhman and A. V. Skorokhod [153], A. M. Yaglom [8], [9]).

Definition 1.1. *A sequence of complex-valued random variables $\xi(k), k \in \mathbb{Z}$ with finite first and second moments $\mathsf{E}|\xi(k)| < \infty$, $\mathsf{E}|\xi(k)|^2 < \infty$, $k \in \mathbb{Z}$ is called stationary (in wide sense) if for all $k \in \mathbb{Z}$*

$$\mathsf{E}\xi(k) = \mathsf{E}\xi(0) = m,$$

$$\mathsf{E}(\xi(j+k)\overline{\xi(j)}) = \mathsf{E}(\xi(k)\overline{\xi(0)}) = R(k), \quad j \in \mathbb{Z}.$$

Without loss of generality we will assume that $m = 0$.

Theorem 1.1. (Herglotz's theorem.) *Let $R(k)$ be the covariance function of a stationary sequence $\xi(k), k \in \mathbb{Z}$ with zero first moment. Then there exists a finite measure $F = F(B)$, $B \in \mathcal{B}([-\pi,\pi))$, on the interval $[-\pi,\pi)$ such that the covariance function can be represented in the form*

$$R(k) = \int_{-\pi}^{\pi} e^{i\lambda k} F(d\lambda), \quad k \in \mathbb{Z}. \tag{1.1}$$

This representation of the covariance function is called the spectral representation. The finite measure $F = F(B)$, $B \in \mathcal{B}([-\pi,\pi))$, is called the spectral measure of the stationary sequence $\xi(k)$, and the function $F(\lambda) = F([-\pi,\lambda))$ is called the spectral function of the stationary sequence with the covariance function $R(k)$.

In the case where the spectral measure of the stationary sequence $\xi(k), k \in \mathbb{Z}$ is absolutely continuous, the spectral function $F(\lambda)$ has a spectral density $f(\lambda)$. In this case, the covariance function can be represented as

$$R(k) = \frac{1}{2\pi}\int_{-\pi}^{\pi} e^{i\lambda k} f(\lambda) d\lambda.$$

Theorem 1.2. *Let $\xi(k), k \in \mathbb{Z}$ be a stationary stochastic sequence with the covariance function $R(k)$ which admits the spectral representation* (1.1). *Then* *there exists an orthogonal stochastic measure* $Z = Z(\Delta)$, $\Delta \in \mathcal{B}([-\pi, \pi))$, *such that the stationary sequence admits the spectral representation*

$$\xi(k) = \int_{-\pi}^{\pi} e^{i\lambda k} Z(d\lambda), \quad k \in \mathbb{Z}, \tag{1.2}$$

and $\mathsf{E}\left(Z(\Delta_1)\overline{Z(\Delta_2)}\right) = F(\Delta_1 \cap \Delta_2)$.

Denote by $H = L_2(\Omega, \mathcal{F}, P)$ the Hilbert space generated by the complex-valued random variables ξ with zero first moment, $\mathsf{E}\xi = 0$, finite second moment, $\mathsf{E}|\xi|^2 < \infty$, and the inner product $(\xi, \eta) = \mathsf{E}(\xi\overline{\eta})$.
Denote by $H(\xi)$ the closed linear subspace of the Hilbert space $H = L_2(\Omega, \mathcal{F}, P)$ generated by values $\{\xi(k) : k \in \mathbb{Z}\}$ of the stationary sequence $\xi(k), k \in \mathbb{Z}$.
Denote by $L_2(f)$ the space of complex-valued functions on $[-\pi, \pi)$ that are square-integrable with respect to a measure whose density is $f(\lambda)$.

Theorem 1.3. *Let* $\xi(k), k \in \mathbb{Z}$ *be a stationary stochastic sequence which admit the spectral representation* (1.2), *and let a random variable* $\eta \in H(\xi)$. *Then there exists a function* $\varphi \in L_2(f)$, *such that*

$$\eta = \int_{-\pi}^{\pi} \varphi(\lambda) Z(d\lambda).$$

We also consider stationary correlated sequences $\{\xi(j), j \in \mathbb{Z}\}$ and $\{\eta(j), j \in \mathbb{Z}\}$ with the covariance functions which admit the spectral representations

$$\begin{aligned}
R_\xi(k) &= \mathsf{E}\xi(j+k)\overline{\xi(j)} = \frac{1}{2\pi}\int_{-\pi}^{\pi} e^{ik\lambda} f(\lambda)d\lambda, \\
R_{\xi\eta}(k) &= \mathsf{E}\xi(j+k)\overline{\eta(j)} = \frac{1}{2\pi}\int_{-\pi}^{\pi} e^{ik\lambda} f_{\xi\eta}(\lambda)d\lambda, \\
R_{\eta\xi}(k) &= \mathsf{E}\eta(j+k)\overline{\xi(j)} = \frac{1}{2\pi}\int_{-\pi}^{\pi} e^{ik\lambda} f_{\eta\xi}(\lambda)d\lambda, \\
R_\eta(k) &= E\eta(j+k)\overline{\eta(j)} = \frac{1}{2\pi}\int_{-\pi}^{\pi} e^{ik\lambda} g(\lambda)d\lambda.
\end{aligned} \tag{1.3}$$

The orthogonal spectral measures $Z_\xi(\Delta)$, and $Z_\eta(\Delta)$, $\Delta \in \mathcal{B}([-\pi,\pi))$ of stationary correlated sequences satisfy the following relations

$$\begin{aligned} \mathsf{E}Z_\xi(\Delta_1)\overline{Z_\xi(\Delta_2)} &= \frac{1}{2\pi}\int_{\Delta_1\cap\Delta_2} f(\lambda)d\lambda, \\ \mathsf{E}Z_\xi(\Delta_1)\overline{Z_\eta}(\Delta_2) &= \frac{1}{2\pi}\int_{\Delta_1\cap\Delta_2} f_{\xi\eta}(\lambda)d\lambda, \\ \mathsf{E}Z_\eta(\Delta_1)\overline{Z_\xi}(\Delta_2) &= \frac{1}{2\pi}\int_{\Delta_1\cap\Delta_2} f_{\eta\xi}(\lambda)d\lambda, \\ \mathsf{E}Z_\eta(\Delta_1)\overline{Z_\eta}(\Delta_2) &= \frac{1}{2\pi}\int_{\Delta_1\cap\Delta_2} g(\lambda)d\lambda. \end{aligned} \tag{1.4}$$

The spectral properties of stationary stochastic sequences described above are used in the following sections to study the problems of estimating of unobservable values of stationary sequences.

2. The Problem of Interpolation of Stationary Sequences with Missing Observations

In this section, the problem of mean-square optimal linear estimation of the functional

$$A_s\xi = \sum_{l=0}^{s-1}\sum_{j=M_l}^{M_l+N_{l+1}} a(j)\xi(j),$$

constructed from unobserved values of a stationary stochastic sequence $\xi(j)$ is investigated. Estimates are based on observations of the sequence $\xi(j)$ at points $j \in \mathbb{Z}\setminus S$, where

$$S = \bigcup_{l=0}^{s-1}\{M_l, M_l+1, \ldots, M_l+N_{l+1}\},$$

$$M_l = \sum_{k=0}^{l}(N_k+K_k),\; N_0 = K_0 = 0.$$

The problem is investigated in the case of spectral certainty, as well as in the case of spectral uncertainty. Formulas for calculating the spectral characteristic and the mean square error of the optimal linear estimate of the functional are derived in the case of spectral certainty, where the spectral density of the stationary sequence $\xi(j)$ is exactly known. The minimax method of estimation is applied in the case of spectral uncertainty, where thc spectral density of the stationary sequence is not exactly known, while a set of admissible spectral densities is given. This method allows us to find estimates that minimize the maximal values of the mean-square errors of estimates for all spectral densities from a given class of admissible spectral densities. Formulas that determine the least favourable spectral densities and the minimax-robust spectral characteristics of the optimal estimates of the functional are derived for some specific classes of admissible spectral densities.

2.1. Hilbert Space Projection Method of Interpolation

Let $\xi(j), j \in \mathbb{Z}$ be a (wide sense) stationary stochastic sequence. We consider values $\xi(j)$ of the sequence as elements of the Hilbert space $H = L_2(\Omega, \mathcal{F}, P)$. We consider stationary stochastic sequences with absolutely continuous spectral measures and covariance functions of the form

$$R(k) = \frac{1}{2\pi} \int_{-\pi}^{\pi} e^{ik\lambda} f(\lambda) d\lambda,$$

where $f(\lambda)$ is the spectral density of the sequence $\xi(j)$ which satisfies the minimality condition

$$\int_{-\pi}^{\pi} f^{-1}(\lambda) d\lambda < \infty. \tag{1.5}$$

This condition is necessary and sufficient to make impossible the error-free interpolation of unobserved values of the sequence (see the book by Yu. A. Rozanov [10] for more details).

The stationary stochastic sequence $\xi(j)$ admits the spectral representation (1.2) with the spectral measure $Z(d\lambda)$ so to hold the first equality (1.4) true.

Consider the problem of mean-square optimal estimation of the linear functional

$$A_s\xi = \sum_{l=0}^{s-1} \sum_{j=M_l}^{M_l+N_{l+1}} a(j)\xi(j)$$

based on observations of the sequence $\xi(j)$ at points $j \in \mathbb{Z}\backslash S$,

$$S = \bigcup_{l=0}^{s-1} \{M_l, M_l + 1, \dots, M_l + N_{l+1}\},$$

$$M_l = \sum_{k=0}^{l} (N_k + K_k),\ N_0 = K_0 = 0.$$

It follows from the spectral representation (1.2) of the sequence $\xi(j)$ that we can represent the functional $A_s\xi$ in the form

$$A_s\xi = \int_{-\pi}^{\pi} A_s(e^{i\lambda}) Z(d\lambda), \tag{1.6}$$

$$A_s(e^{i\lambda}) = \sum_{l=0}^{s-1} \sum_{j=M_l}^{M_l+N_{l+1}} a(j) e^{ij\lambda}.$$

Denote by $H^s(\xi)$ the subspace of the Hilbert space $H = L_2(\Omega, \mathcal{F}, P)$, generated by elements $\{\xi(k) : k \in \mathbb{Z}\backslash S\}$.
Let $L_2^s(f)$ be the subspace of the space $L_2(f)$, generated by functions $\{e^{ij\lambda} : j \in \mathbb{Z}\backslash S\}$.

The mean-square optimal linear estimate $\hat{A}_s\xi$ of the functional $A_s\xi$ based on observations of the sequence $\xi(j)$ at points $j \in \mathbb{Z}\backslash S$ is an element of the subspace $H^s(\xi)$. It can be represented in the form

$$\hat{A}_s\xi = \int_{-\pi}^{\pi} h(e^{i\lambda}) Z(d\lambda), \tag{1.7}$$

where $h(e^{i\lambda}) \in L_2^s(f)$ is the spectral characteristic of the estimate $\hat{A}_s\xi$.

The mean-square error $\Delta(h; f)$ of the estimate $\hat{A}_s\xi$ is given by the formula

$$\Delta(h; f) = \mathsf{E}\left|A_s\xi - \hat{A}_s\xi\right|^2 = \frac{1}{2\pi}\int_{-\pi}^{\pi} \left|A_s(e^{i\lambda}) - h(e^{i\lambda})\right|^2 f(\lambda) d\lambda.$$

Based on the Hilbert space projection method proposed by A. N. Kolmogorov [1] we find the spectral characteristic $h(e^{i\lambda})$ and the mean-square error $\Delta(h; f)$ of the optimal linear estimate of the functional $A_s\xi$ in the case, where the spectral density $f(\lambda)$ of the sequence is exactly known and the minimality condition (1.5) is satisfied. The spectral characteristic can be found from the following conditions:

$$1) h(e^{i\lambda}) \in L_2^s(f),$$
$$2) A_s(e^{i\lambda}) - h(e^{i\lambda}) \perp L_2^s(f).$$

It follows from the second condition that for any $\eta \in H^s(\xi)$ the following equations should be satisfied

$$\left(A_s\xi - \hat{A}_s\xi, \eta\right) = \mathsf{E}[(A_s\xi - \hat{A}_s\xi)\overline{\eta}] = 0.$$

The last relation is equivalent to equations

$$\mathsf{E}[(A_s\xi - \hat{A}_s\xi)\,\overline{\xi}_k] = 0, \quad k \in \mathbb{Z}\backslash S.$$

By using representations (1.6), (1.7) and properties of the inner product in the space $H^s(\xi)$, we get

$$\mathsf{E}\left(\int_{-\pi}^{\pi}\left(A_s(e^{i\lambda}) - h(e^{i\lambda})\right)Z(d\lambda)\right)\left(\int_{-\pi}^{\pi} e^{-ik\lambda}\overline{Z(d\lambda)}\right) =$$

$$= \frac{1}{2\pi}\int_{-\pi}^{\pi}\left(A_s(e^{i\lambda}) - h(e^{i\lambda})\right)f(\lambda)e^{-ik\lambda}d\lambda = 0, \quad k \in \mathbb{Z}\backslash S.$$

It follows from this relation that the Fourier coefficients of the function

$$(A_s(e^{i\lambda}) - h(e^{i\lambda}))f(\lambda)$$

are equal to zero for $k \in \mathbb{Z}\backslash S$, and the function is of the form

$$(A_s(e^{i\lambda}) - h(e^{i\lambda}))f(\lambda) = C_s(e^{i\lambda}), \quad C_s(e^{i\lambda}) = \sum_{l=0}^{s-1}\sum_{j=M_l}^{M_l+N_{l+1}} c(j)e^{ij\lambda},$$

where $c(j), j \in S$ are unknown coefficients that must be found.

It follows from the last relation, that the spectral characteristic $h(e^{i\lambda})$ of the optimal linear estimate of the functional $\hat{A}_s$ is of the form

$$h(e^{i\lambda}) = A_s(e^{i\lambda}) - C_s(e^{i\lambda})f^{-1}(\lambda). \tag{1.8}$$

To find equations for determination the unknown coefficients $c(j), j \in S$, we use the decomposition of the function $f^{-1}(\lambda)$ into the Fourier series

$$f^{-1}(\lambda) = \sum_{m=-\infty}^{\infty} r_m e^{im\lambda}, \tag{1.9}$$

where r_m are the Fourier coefficients of the function $f^{-1}(\lambda)$.
Inserting the decomposition (1.9) of the function $f^{-1}(\lambda)$ into equation (1.8), we obtain the following representation of the spectral characteristic of the estimate $\hat{A}_s\xi$

$$\begin{aligned} h(e^{i\lambda}) = & \left(\sum_{l=0}^{s-1} \sum_{j=M_l}^{M_l+N_{l+1}} a(j)e^{ij\lambda} \right) - \\ & - \left(\sum_{l=0}^{s-1} \sum_{j=M_l}^{M_l+N_{l+1}} c(j)e^{ij\lambda} \right) \left(\sum_{m=-\infty}^{\infty} r_m e^{im\lambda} \right). \end{aligned} \tag{1.10}$$

It follows from the first condition, $h(e^{i\lambda}) \in L_2^s(f)$, that the Fourier coefficients of the function $h(e^{i\lambda})$ are equal to zero for $j \in S$, so that

$$\frac{1}{2\pi} \int_{-\pi}^{\pi} h(e^{i\lambda}) e^{-ij\lambda} d\lambda = 0, \quad j \in S.$$

Using the last relations and representation (1.10), we get the following system of equations that determine the unknown coefficients $c(j), j \in S$,

$$\begin{aligned} & a(M_{k-1}) - \sum_{l=0}^{s-1} \sum_{j=M_l}^{M_l+N_{l+1}} c(j) r_{M_{k-1}-j} = 0; \\ & a(M_{k-1}+1) - \sum_{l=0}^{s-1} \sum_{j=M_l}^{M_l+N_{l+1}} c(j) r_{M_{k-1}+1-j} = 0; \\ & \dots \\ & a(M_{k-1}+N_k) - \sum_{l=0}^{s-1} \sum_{j=M_l}^{M_l+N_{l+1}} c(j) r_{M_{k-1}+N_k-j} = 0, \quad k = 1, \dots, s. \end{aligned} \tag{1.11}$$

Denote by $\vec{\mathbf{a}}_s$ the vector

$$\vec{\mathbf{a}}_s = (\vec{a}_1, \vec{a}_2, \ldots, \vec{a}_s),$$

where $\vec{a}_k = (a(M_{k-1}), \ldots, a(M_{k-1}+N_k))$, $k = 1, \ldots, s$, and denote by

$$q = N_1 + N_2 + \ldots + N_s + s.$$

Next, denote by B_s the $q \times q$ matrix

$$B_s = \begin{pmatrix} B_{11} & B_{12} & \ldots & B_{1s} \\ B_{21} & B_{22} & \ldots & B_{2s} \\ \vdots & \vdots & \ddots & \vdots \\ B_{s1} & B_{s2} & \ldots & B_{ss} \end{pmatrix},$$

where B_{mn} are $(N_m+1) \times (N_n+1)$ matrices with elements that are Fourier coefficients of the function $f^{-1}(\lambda)$

$$B_{mn}(k,j) = \frac{1}{2\pi} \int_{-\pi}^{\pi} f^{-1}(\lambda) e^{-i(k-j)\lambda} d\lambda = r_{k-j}, \quad m,n = 1, \ldots, s,$$

$$k = M_{m-1}, \ldots, M_{m-1}+N_m, \quad j = M_{n-1}, \ldots, M_{n-1}+N_n.$$

Making use of the introduced notations we can write formulas (1.11) in the form of equation

$$\vec{\mathbf{a}}_s = B_s \vec{\mathbf{c}}_s, \tag{1.12}$$

where $\vec{\mathbf{c}}_s$ is the vector constructed from the unknown coefficients $c(j), j \in S$. It is of the form

$$\vec{\mathbf{c}}_s = (\vec{c}_1, \vec{c}_2, \ldots, \vec{c}_s),$$

where $\vec{c}_k = (c(M_{k-1}), \ldots, c(M_{k-1}+N_k))$, $k = 1, \ldots, s$.
Since the matrix B_s is invertible (see papers by M. Pourahmadi,[154], H. Salehi [155] for additional details), we get the formula

$$\vec{\mathbf{c}}_s = B_s^{-1} \vec{\mathbf{a}}_s,$$

and the unknown coefficients $c(j), j \in S$, are calculated by the formula

$$c(j) = \left(B_s^{-1} \vec{\mathbf{a}}_s\right)_j,$$

where $\left(B_s^{-1}\vec{\mathbf{a}}_s\right)_j$ is the j-th component of the vector $B_s^{-1}\vec{\mathbf{a}}_s$.

From the derived relations we can conclude that the spectral characteristic of the estimate $\hat{A}_s\xi$ is calculated by the formula

$$\begin{aligned} h(e^{i\lambda}) = &\left(\sum_{l=0}^{s-1}\sum_{j=M_l}^{M_l+N_{l+1}} a(j)e^{ij\lambda}\right) - \\ &-\left(\sum_{l=0}^{s-1}\sum_{j=M_l}^{M_l+N_{l+1}} \left(B_s^{-1}\vec{\mathbf{a}}_s\right)_j e^{ij\lambda}\right)\left(\sum_{m=-\infty}^{\infty} r_m e^{im\lambda}\right), \end{aligned} \tag{1.13}$$

and the mean-square error of estimate of the functional can be calculated by the formula

$$\Delta(h;f) = \frac{1}{2\pi}\int_{-\pi}^{\pi}\left|C_s(e^{i\lambda})\right|^2 f^{-1}(\lambda)d\lambda =$$

$$= \frac{1}{2\pi}\int_{-\pi}^{\pi}\left(\sum_{l=0}^{s-1}\sum_{k=M_l}^{M_l+N_{l+1}} c(k)e^{ik\lambda}\right)\left(\sum_{l=0}^{s-1}\sum_{j=M_l}^{M_l+N_{l+1}} \overline{c(j)}e^{-ij\lambda}\right)\sum_{m=-\infty}^{\infty} r_m e^{im\lambda}d\lambda$$

$$= \langle c_s, c_s B_s\rangle = \langle B_s^{-1}\vec{\mathbf{a}}_s, \vec{\mathbf{a}}_s\rangle, \tag{1.14}$$

where $\langle\cdot,\cdot\rangle$ is the inner product in the space C^q.

Let's summarize our reasoning and present them as a theorem.

Theorem 1.4. *Let $\xi(j)$ be a stationary stochastic sequence with the spectral density $f(\lambda)$, that satisfies the minimality condition (1.5). The mean-square error $\Delta(h,f)$ and the spectral characteristic $h(e^{i\lambda})$ of the optimal linear estimate $\hat{A}_s\xi$ of the functional $A_s\xi$ based on observations of the sequence $\xi(j)$ at points $j\in\mathbb{Z}\backslash S$, $S=\bigcup_{l=0}^{s-1}\{M_l,\dots,M_l+N_{l+1}\}$, are calculated by formulas (1.13), (1.14).*

Example 1.1. *As an example, let's investigate the problem of linear estimation of the functional*

$$A_2\xi = a(0)\xi(0) + a(1)\xi(1) + a(5)\xi(5)$$

constructed from unobserved values of the stationary sequence $\xi(j)$ based on observations of the sequence at points $j\in\mathbb{Z}\backslash S$, $S=\{0,1\}\cup\{5\}$. In this case *the spectral characteristic of the optimal linear estimate $\hat{A}_2\xi$ is calculated by the formula*

$$\begin{aligned} h(e^{i\lambda}) &= a(0)+a(1)e^{i\lambda}+a(5)e^{5i\lambda} \\ &\quad -(c(0)+c(1)e^{i\lambda}+c(5)e^{5i\lambda})\cdot f^{-1}(\lambda), \end{aligned} \tag{1.15}$$

where $f(\lambda)$ is the known spectral density of the sequence $\xi(j)$. The function $f^{-1}(\lambda)$ admits the decomposition into the Fourier series

$$f^{-1}(\lambda)=\sum_{m=-\infty}^{\infty} r_m e^{i\lambda m},$$

and the coefficients $c(0),c(1),c(5)$ are determined by the system of equations

$$\begin{aligned} a(0) &= c(0)r_0+c(1)r_{-1}+c(5)r_{-5}, \\ a(1) &= c(0)r_1+c(1)r_0+c(5)r_{-4}, \\ a(5) &= c(0)r_5+c(1)r_4+c(5)r_0. \end{aligned}$$

The matrix B_2 is of the form

$$B_2=\begin{pmatrix} B_{11} & B_{12} \\ B_{21} & B_{22} \end{pmatrix},$$

$$B_{11}=\begin{pmatrix} r_0 & r_{-1} \\ r_1 & r_0 \end{pmatrix}, B_{12}=\begin{pmatrix} r_{-5} \\ r_{-4} \end{pmatrix}, B_{21}=\begin{pmatrix} r_5 & r_4 \end{pmatrix}, B_{22}=\begin{pmatrix} r_0 \end{pmatrix}.$$

Let $\vec{\mathbf{a}}_2=(\vec{a}_1,\vec{a}_2)$, where $\vec{a}_1=(a(0),a(1))$, $\vec{a}_2=(a(5))$, and let $\vec{\mathbf{c}}_2=(\vec{c}_1,\vec{c}_2)$, where $\vec{c}_1=(c(0),c(1)),\vec{c}_2=(c(5))$. In this case equations (1.12) can be written in the form

$$\vec{\mathbf{a}}_2=B_2\vec{\mathbf{c}}_2.$$

From this equation we obtain

$$\vec{\mathbf{c}}_2=B_2^{-1}\vec{\mathbf{a}}_2.$$

We invert the matrix B_2 and denote by

$$D=\det(B_2)=r_0^3-r_0r_{-4}r_4-r_1r_{-1}r_0+r_1r_{-5}r_4+r_5r_{-1}r_{-4}-r_5r_{-5}r_0.$$

Thus formulas for calculating the unknown coefficients from (1.15) are as follows

$$c(0) = \left[(r_0^2 - r_{-4}r_4)a(0) - (r_{-1}r_0 - r_{-5}r_4)a(1) + (r_{-1}r_4 - r_{-5}r_0)a(5)\right] \cdot D^{-1},$$

$$c(1) = \left[(-r_1r_0 + r_{-4}r_5)a(0) + (r_0^2 - r_5r_{-5})a(1) - (r_0r_{-4} - r_{-5}r_1)a(5)\right] \cdot D^{-1},$$

$$c(5) = \left[(r_1r_4 - r_0r_5)a(0) - (r_0r_4 - r_5r_{-1})a(1) + (r_0^2 - r_{-1}r_1)a(5)\right] \cdot D^{-1}.$$

The mean-square error of the estimate is calculated by the formula

$$\Delta(f) = \langle B_2^{-1}\vec{\mathbf{a}}_2, \vec{\mathbf{a}}_2\rangle = D^{-1}\left[(a(0))^2r_0^2 - (a(0))^2r_{-4}r_4 - a(0)a(1)r_{-1}r_0\right.$$
$$+a(0)a(1)r_{-5}r_4 + a(0)a(5)r_{-1}r_{-4} - a(0)a(5)r_{-5}r_0 - a(0)a(1)r_0r_1$$
$$+a(0)a(1)r_{-4}r_5 + (a(1))^2r_0^2 - (a(1))^2r_5r_{-5} - a(1)a(5)r_0r_{-4} + a(1)a(5)r_{-5}r_1$$
$$+a(0)a(5)r_1r_4 - a(0)a(5)r_0r_5 - a(1)a(5)r_0r_4 + a(1)a(5)r_{-1}r_5 + (a(5))^2r_0^2$$
$$\left.-(a(5))^2r_1r_{-1}\right].$$

Consider this problem for the stationary sequence with the spectral density

$$f(\lambda) = \frac{1}{\left|1 - \alpha e^{-i\lambda}\right|^2}, \ |\alpha| < 1.$$

In this case the decomposition of the function $f^{-1}(\lambda)$ into the Fourier series is as follows

$$f^{-1}(\lambda) = \left|1 - \alpha e^{-i\lambda}\right|^2 = \left(1 + |\alpha|^2 - \bar{\alpha}e^{i\lambda} - \alpha e^{-i\lambda}\right).$$

The coefficients in formulas, which determine the spectral characteristic and the mean-square error of the functional $A_2\xi$, are the following

$$r_4 = r_5 = r_{-4} = r_{-5} = 0,$$

$$r_0 = \left(1 + |\alpha|^2\right), \ r_1 = -\bar{\alpha}, \ r_{-1} = -\alpha.$$

The spectral characteristic of the estimate is calculated by the formula

$$h(e^{i\lambda}) = \left(a(0) + a(1)e^{i\lambda} + a(5)e^{5i\lambda}\right) -$$
$$- \left(c(0) + c(1)e^{i\lambda} + c(5)e^{5i\lambda}\right)\left(1 + |\alpha|^2 - \bar{\alpha}e^{i\lambda} - \alpha e^{-i\lambda}\right),$$

and coefficients $c(0), c(1), c(5)$ *are calculated by formulas*

$$c(0) = \left[(1+|\alpha|^2)^2 a(0) + (\alpha(1+|\alpha|^2))a(1)\right] \cdot D^{-1},$$

$$c(1) = \left[(\bar{\alpha}(1+|\alpha|^2))a(0) + (1+|\alpha|^2)^2 a(1)\right] \cdot D^{-1},$$

$$c(5) = \left[((1+|\alpha|^2)^2 - |\alpha|^2)a(5)\right] \cdot D^{-1},$$

where

$$D = \left(1+|\alpha|^2\right)\left(1+|\alpha|^2+|\alpha|^4\right).$$

By inserting values $c(0), c(1), c(5)$ *into the formula for spectral characteristic, we obtain*

$$h(e^{i\lambda}) = \Bigg[\left(a(0)\alpha\left(1+|\alpha|^2\right)^2 + a(1)\alpha^2\left(1+|\alpha|^2\right)\right)e^{-i\lambda} +$$
$$+\left(a(0)\,(\bar{\alpha})^2\left(1+|\alpha|^2\right) + a(1)\bar{\alpha}\left(1+|\alpha|^2\right)^2\right)e^{2i\lambda} +$$
$$+a(5)\alpha\left(1+|\alpha|^2+|\alpha|^4\right)e^{4i\lambda} + a(5)\bar{\alpha}\left(1+|\alpha|^2+|\alpha|^4\right)e^{6i\lambda}\Bigg] \times$$
$$\left(\left(1+|\alpha|^2\right)\left(1+|\alpha|^2+|\alpha|^4\right)\right)^{-1}.$$

The mean-square error is of the form

$$\Delta(f) = D^{-1}\Big[(a(0))^2(1+|\alpha|^2)^2 + a(0)a(1)\alpha(1+|\alpha|^2) + a(0)a(1)(1+|\alpha|^2)\bar{\alpha}$$
$$+(a(1))^2(1+|\alpha|^2)^2 + (a(5))^2(1+|\alpha|^2)^2 - (a(5))^2|\alpha|^2\Big].$$

Next, consider the problem for different values of the coefficients $a(0), a(1), a(5)$.

a) Let $a(0)=1$, $a(1)=0$, $a(5)=1$. *We get the following formula for calculating the spectral characteristic*

$$h(e^{i\lambda}) = (1+e^{5i\lambda}) - (c(0)+c(1)e^{i\lambda}+c(5)e^{5i\lambda})\left(1+|\alpha|^2-\bar{\alpha}e^{i\lambda}-\alpha e^{-i\lambda}\right),$$

and coefficients $c(0), c(1), c(5)$ *are calculated by the formula*

$$c(0) = \frac{\left(1+|\alpha|^2\right)}{\left(1+|\alpha|^2+|\alpha|^4\right)}, c(1) = \frac{\bar{\alpha}}{\left(1+|\alpha|^2+|\alpha|^4\right)}, c(5) = \frac{1}{\left(1+|\alpha|^2\right)}.$$

Thus, the spectral characteristic of the estimate is of the form

$$h(e^{i\lambda}) = \Big[\alpha\left(1+|\alpha|^2\right)^2 e^{-i\lambda} + (\bar{\alpha})^2\left(1+|\alpha|^2\right)e^{2i\lambda} +$$
$$+\alpha\left(1+|\alpha|^2+|\alpha|^4\right)e^{4i\lambda} + \bar{\alpha}\left(1+|\alpha|^2+|\alpha|^4\right)e^{6i\lambda}\Big] \times$$
$$\left(\left(1+|\alpha|^2\right)\left(1+|\alpha|^2+|\alpha|^4\right)\right)^{-1}.$$

The mean-square error of the estimate is calculated by the formula

$$\Delta_1 = \Delta(f) = \frac{\left(2(1+|\alpha|^2)^2 - |\alpha|^2\right)}{\left(1+|\alpha|^2\right)\left(1+|\alpha|^2+|\alpha|^4\right)}.$$

b) Let $a(0) = 1, a(1) = 1, a(5) = 0.$
We get the following form of the spectral characteristic

$$h(e^{i\lambda}) = 1 + (c(0) + c(1)e^{i\lambda})\left(1+|\alpha|^2 - \bar{\alpha}e^{i\lambda} - \alpha e^{-i\lambda}\right),$$

where coefficients $c(0), c(1)$ *are calculated by the formula*

$$c(0) = \frac{\left(1+|\alpha|^2\right)}{\left(1+|\alpha|^2+|\alpha|^4\right)}, \quad c(1) = \frac{\bar{\alpha}}{\left(1+|\alpha|^2+|\alpha|^4\right)}.$$

The coefficient $c(5) = 0.$
The spectral characteristic of the estimate is of the form

$$h(e^{i\lambda}) = \frac{\left(\alpha\left(1+|\alpha|^2\right)^2 e^{-i\lambda} + (\bar{\alpha})^2\left(1+|\alpha|^2\right)e^{2i\lambda}\right)}{\left(1+|\alpha|^2+|\alpha|^4\right)}.$$

The mean-square error is calculated by the formula

$$\Delta_2 = \Delta(f) = \frac{\left(1+|\alpha|^2\right)}{\left(1+|\alpha|^2+|\alpha|^4\right)}.$$

It can be easily checked that $\Delta_1 > \Delta_2$.

Example 1.2. *Consider the problem of linear interpolation of the functional*

$$A_3\xi = a(-n)\xi(-n) + a(0)\xi(0) + a(m)\xi(m)$$

constructed from unobserved values of a stationary sequence $\xi(j)$, *based on observations of the sequence at points* $j \in \mathbb{Z}\backslash S$, *where* $S = \{-n\} \cup \{0\} \cup \{m\}$.

Let $f(\lambda)$ *be the spectral density of the sequence* $\xi(j)$ *and let the function* $f^{-1}(\lambda)$ *admit the decomposition* $f^{-1}(\lambda) = \sum\limits_{m=-\infty}^{\infty} r_m e^{i\lambda m}$.

The spectral characteristic of the estimate $\hat{A}_3\xi$ *is of the form*

$$\begin{aligned} h(e^{i\lambda}) =& (a(-n)e^{-ni\lambda} + a(0) + a(m)e^{mi\lambda}) - \\ & (c(-n)e^{-ni\lambda} + c(0) + c(m)e^{mi\lambda}) \cdot f^{-1}(\lambda), \end{aligned}$$

where coefficients $c(-n), c(0), c(m)$ *are determined by formulas*

$$\begin{aligned} a(-n) &= c(-n)r_0 + c(0)r_{-n} + c(m)r_{-n-m}, \\ a(0) &= c(-n)r_n + c(0)r_0 + c(m)r_{-m}, \\ a(m) &= c(-n)r_{n+m} + c(0)r_m + c(m)r_0. \end{aligned}$$

The matrix B_3 *is constructed from the Fourier coefficients of the function* $f^{-1}(\lambda)$. *It is of the form*

$$B_3 = \begin{pmatrix} B_{11} & B_{12} & B_{13} \\ B_{21} & B_{22} & B_{23} \\ B_{31} & B_{32} & B_{33} \end{pmatrix},$$

$$B_{11} = \left(r_0 \right), B_{12} = \left(r_{-n} \right), B_{13} = \left(r_{-n-m} \right),$$
$$B_{21} = \left(r_n \right), B_{22} = \left(r_0 \right), B_{23} = \left(r_{-m} \right),$$

$$B_{31} = \left(\; r_{n+m} \; \right), B_{32} = \left(\; r_m \; \right), B_{33} = \left(\; r_0 \; \right).$$

Let $\vec{\mathbf{a}}_3 = (\vec{a}_1, \vec{a}_2, \vec{a}_3)$, *where* $\vec{a}_1 = (a(-n))$, $\vec{a}_2 = (a(0))$, $\vec{a}_3 = (a(m))$, *and let* $\vec{\mathbf{c}}_3 = (\vec{c}_1, \vec{c}_2, \vec{c}_3)$, *where* $\vec{c}_1 = (c(-n))$, $\vec{c}_2 = (c(0))$, $\vec{c}_3 = (c(m))$. *In this case, equation (1.12) can be written as*

$$\vec{\mathbf{a}}_3 = B_3 \vec{\mathbf{c}}_3$$

and we get

$$\vec{\mathbf{c}}_3 = B_3^{-1} \vec{\mathbf{a}}_3.$$

Denote by

$$D = \det(B_3) = r_0^3 - r_0 r_{-m} r_m - r_n r_{-n} r_0 +$$

$$r_n r_{-n-m} r_m + r_{n+m} r_{-n} r_{-m} - r_{n+m} r_{-n-m} r_0.$$

The unknown coefficients are calculated by the formulas

$$\begin{aligned} c(-n) =& [(r_0^2 - r_{-m} r_m) a(-n) - (r_{-n} r_0 - r_{-n-m} r_m) a(0) + \\ & (r_{-n} r_m - r_{-n-m} r_0) a(m)] \cdot D^{-1}, \end{aligned}$$

$$\begin{aligned} c(0) =& [(-r_n r_0 + r_{-m} r_{n+m}) a(-n) + (r_0^2 - r_{n+m} r_{-n-m}) a(0) - \\ & (r_0 r_{-m} - r_{-n-m} r_n) a(m)] \cdot D^{-1}, \\ c(m) =& [(r_n r_m - r_0 r_{n+m}) a(-n) - (r_0 r_m - r_{n+m} r_{-n}) a(0) + \\ & (r_0^2 - r_{-n} r_n) a(m)] \cdot D^{-1}. \end{aligned}$$

The mean-square error is calculated by the formula

$$\begin{aligned} \Delta(f) = & \langle B_3^{-1} \vec{\mathbf{a}}_3, \vec{\mathbf{a}}_3 \rangle = D^{-1} \left[(a(-n))^2 r_0^2 - (a(-n))^2 r_{-m} r_m - \right. \\ & a(-n)a(0) r_{-n} r_0 + a(-n)a(0) r_{-n-m} r_m + a(-n)a(m) r_{-n} r_{-m} - \\ & a(-n)a(m) r_{-n-m} r_0 - a(-n)a(0) r_0 r_n + a(-n)a(0) r_{-m} r_{n+m} + (a(0))^2 r_0^2 - \\ & (a(0))^2 r_{n+m} r_{-n-m} - a(0)a(m) r_0 r_{-m} + a(0)a(m) r_{-n-m} r_n + \\ & a(-n)a(m) r_n r_m - a(-n)a(m) r_0 r_{n+m} - a(0)a(m) r_0 r_m + a(0)a(m) r_{-n} r_{n+m} \\ & \left. + (a(m))^2 r_0^2 - (a(m))^2 r_n r_{-n} \right]. \end{aligned}$$

Next, let's find formulas for calculating the spectral characteristic and the mean-square error of the functional

$$A_3(\xi) = \xi(-n) + \xi(0) + \xi(m)$$

when the spectral density of the sequence is of the form

$$f(\lambda)=\frac{1}{\left|1-\alpha e^{-i\lambda}\right|^2},\ |\alpha|<1.$$

It follows from the first example that

$$f^{-1}(\lambda)=\left(1+|\alpha|^2-\bar{\alpha}e^{i\lambda}-\alpha e^{-i\lambda}\right).$$

The Fourier coefficients of the function $f^{-1}(\lambda)$, *except* r_0, r_1, r_{-1}, *are equal to zero.*

Consider the following four particular cases:

$1)\, n\neq 1, m\neq 1;\ 2)\, n\neq 1, m=1;\ 3)\, n=1, m\neq 1;\ 4)\, n=1, m=1.$

The spectral characteristic and the mean-square error of the estimate $\hat{A}_3\xi$ *of the functional* $A_3(\xi)$ *are calculated by the following formulas in accordance with the indicated cases.*

In the first case

$$h(e^{i\lambda})=\left(\alpha e^{(-n-1)i\lambda}+\bar{\alpha}e^{(-n+1)i\lambda}+\alpha e^{-i\lambda}+\bar{\alpha}e^{i\lambda}+\alpha e^{(m-1)i\lambda}+\right.$$
$$\left.+\ \bar{\alpha}e^{(m+1)i\lambda}\right)(1+|\alpha|^2)^{-2},$$

$$\Delta(f)=\frac{3}{(1+|\alpha|^2)}.$$

In the second case

$$h(e^{i\lambda})=\Big[(\alpha+\alpha|\alpha|^2+\alpha^2)(1+|\alpha|^2)e^{-i\lambda}+(\bar{\alpha}^2+\bar{\alpha}+\bar{\alpha}|\alpha|^2)(1+|\alpha|^2)e^{2i\lambda}$$
$$+\bar{\alpha}(1+|\alpha|^2+|\alpha|^4)e^{(-n+1)i\lambda}+\alpha(1+|\alpha|^2+|\alpha|^4)e^{(-n-1)i\lambda}\Big]\times$$
$$\left((1+|\alpha|^2)(1+|\alpha|^2+|\alpha|^4)\right)^{-1},$$

$$\Delta(f)=\left(3+2|\alpha|^2+(1+|\alpha|^2)(\alpha+\bar{\alpha})\right)\left((1+|\alpha|^2)(1+|\alpha|^2+|\alpha|^4)\right)^{-1}.$$

In the third case

$$h(e^{i\lambda}) = \Big[(\bar{\alpha}^2 + \bar{\alpha}(1+|\alpha|^2))(1+|\alpha|^2)e^{i\lambda} +$$
$$+ (\alpha^2 + \alpha(1+|\alpha|^2))(1+|\alpha|^2)e^{-2i\lambda}$$
$$+\bar{\alpha}(1+|\alpha|^2+|\alpha|^4)e^{(m+1)i\lambda} + \alpha(1+|\alpha|^2+|\alpha|^4)e^{(m-1)i\lambda}\Big] \times$$
$$\Big((1+|\alpha|^2)(1+|\alpha|^2+|\alpha|^4)\Big)^{-1},$$
$$\Delta(f) = \Big(3+2|\alpha|^2+(1+|\alpha|^2)(\alpha+\bar{\alpha})\Big)\Big((1+|\alpha|^2)(1+|\alpha|^2+|\alpha|^4)\Big)^{-1}.$$

In the fourth case

$$h(e^{i\lambda}) = \Big[(\bar{\alpha}^2 + \bar{\alpha}(1+|\alpha|^2) + 1 + |\alpha|^2 + |\alpha|^4)\bar{\alpha}e^{2i\lambda} +$$
$$(1+|\alpha|^2+|\alpha|^4 + \alpha(1+|\alpha|^2) + |\alpha|^2)\alpha e^{-2i\lambda}\Big] \times$$
$$\Big(1+|\alpha|^2+|\alpha|^4+|\alpha|^6\Big)^{-1},$$
$$\Delta(f) = \Big(3+2|\alpha|^2+(1+|\alpha|^2)(\alpha+\bar{\alpha})+\alpha^2+\bar{\alpha}^2\Big)\cdot$$
$$\Big((1+|\alpha|^2)(1+|\alpha|^2+|\alpha|^4)\Big)^{-1}.$$

2.2. Minimax Robust Method of Interpolation

The mean square error and the spectral characteristic of the optimal estimate $\hat{A}_s\xi$ of the functional $A_s\xi$ based on observations of the sequence $\xi(j)$ at points $j \in \mathbb{Z}\backslash S$, where $S = \bigcup_{l=0}^{s-1}\{M_l,\ldots,M_l+N_{l+1}\}$, can be calculated by formulas (1.13), (1.14) respectively, provided the spectral density $f(\lambda)$ of the stochastic sequence $\xi(j)$ is exactly known. However, in practical cases we do not have complete information on the spectral density of the sequence. If in such cases a set D of admissible spectral densities is defined, the minimax robust method of estimation of the functionals may be applied. According to this method, we find an estimate which minimizes the maximal values of mean square errors for all spectral densities from a given class D of admissible spectral densities simultaneously.

For a description of the minimax method of estimation, we propose the following definitions.

Definition 1.2. *For a given class of spectral densities D, the spectral density $f_0(\lambda) \in D$ is called the least favourable in D for the optimal linear estimation of the functional $A_s\xi$ if the following relation is true*

$$\Delta(f_0) = \Delta(h(f_0); f_0) = \max_{f \in D} \Delta(h(f); f).$$

Definition 1.3. *For a given class of spectral densities D the spectral characteristic $h^0(e^{i\lambda})$ of the optimal linear estimate of the functional $A_s\xi$ is called minimax-robust if the following relations hold true*

$$h^0(e^{i\lambda}) \in H_D = \bigcap_{f \in D} L_2^s(f),$$

$$\min_{h \in H_D} \max_{f \in D} \Delta(h; f) = \sup_{f \in D} \Delta\left(h^0; f\right).$$

It follows from the introduced definitions and the obtained formulas, that the following statement is true.

Lemma 1.1. *The spectral density $f_0(\lambda) \in D$ is the least favourable in a class of admissible spectral densities D for the optimal linear estimation of the functional $A_s\xi$ if the Fourier coefficients of the function $f_0^{-1}(\lambda)$ define the matrix B_s^0 that determines a solution of the optimization problem*

$$\max_{f \in D} \langle B_s^{-1} \vec{\mathbf{a}}_s, \vec{\mathbf{a}}_s \rangle = \langle (B_s^0)^{-1} \vec{\mathbf{a}}_s, \vec{\mathbf{a}}_s \rangle. \tag{1.16}$$

The minimax spectral characteristic $h^0 = h(f_0)$ is calculated by formula (1.13) if $h(f_0) \in H_D$.

For more detailed analysis of properties of the least favourable spectral densities and the minimax-robust spectral characteristics we observe that the minimax spectral characteristic h^0 and the least favourable spectral density f^0 form a saddle point of the function $\Delta(h; f)$ on the set $H_D \times D$. The saddle point inequalities

$$\Delta(h; f_0) \geq \Delta\left(h^0; f_0\right) \geq \Delta\left(h^0; f\right) \quad \forall f \in D, \forall h \in H_D$$

hold true if $h^0 = h(f_0)$ and $h(f_0) \in H_D$, where f_0 is a solution of the constrained optimization problem

$$\tilde{\Delta}(f) = -\Delta\left(h^0; f\right) \to \inf, \quad f(\lambda) \in D. \tag{1.17}$$

The functional $\Delta(h^0; f)$ is calculated by the formula

$$\Delta\left(h^0; f\right) = \frac{1}{2\pi} \int_{-\pi}^{\pi} \frac{\left|C_s^0(e^{i\lambda})\right|^2}{f_0^2(\lambda)} f(\lambda) d\lambda,$$

$$C_s^0(e^{i\lambda}) = \sum_{l=0}^{s-1} \sum_{j=M_l}^{M_l+N_{l+1}} \left((B_s^0)^{-1} \vec{\mathbf{a}}_s\right)_j e^{ij\lambda}.$$

The constrained optimization problem (1.17) is equivalent to the unconstrained optimization problem (see B. N. Pshenichnyj [156])

$$\Delta_D(f) = \tilde{\Delta}(f) + \delta(f\,|D) \to \inf,$$

where $\delta(f\,|D)$ is the indicator function of the set D, namely $\delta(f|D) = 0$ if $f \in D$ and $\delta(f|D) = +\infty$ if $f \notin \mathcal{D}$. A solution f_0 of this problem is characterized by the condition $0 \in \partial\Delta_D(f_0)$, where $\partial\Delta_D(f_0)$ is the subdifferential of the convex functional $\Delta_D(f)$, which is the necessary and sufficient condition for the point f^0 to belong to the set of minima of the convex functional $\Delta_D(f)$ (see M. P. Moklyachuk [55], B. N. Pshenichnyj [156], J. Franke [66]).
By $\partial\Delta_D(f^0)$ we denote the subdifferential of the functional $\Delta_D(f)$ at point $f = f^0$, namely, the set of all continuous linear functionals Λ in the space L_1 satisfying the inequality

$$\Delta_D(f) - \Delta_D(f^0) \geq \Lambda(f - f^0), \quad f \in D.$$

This condition makes it possible to find the least favourable spectral densities in some special classes of spectral densities D.

Note, that the form of the functional $\Delta\left(h^0; f\right)$ is convenient for application of the Lagrange method of indefinite multipliers for finding a solution of the optimization problem (1.17). By using the Lagrange method and the form of subdifferentials of the indicator functions we describe relations that determine the least favourable spectral densities in some special classes of spectral densities.

2.3. Least Favourable Spectral Densities in the Class $D = D_0^-$

Consider the problem of minimax estimation of the functional $A_s\xi$, based on observations of the sequence $\xi(j)$ at points $j \in \mathbb{Z}\backslash S$, in the case where we do not have complete information on the spectral density of the sequence, while we know that the density is from the class

$$D_0^- = \left\{ f(\lambda) \left| \frac{1}{2\pi} \int_{-\pi}^{\pi} f^{-1}(\lambda) d\lambda \geq P \right. \right\}.$$

Let the sequence $a(k), k \in S$, that determines the functional $A_s\xi$, be strictly positive (see M. G. Krein and A. A. Nudelman [157]). To find solutions of the constrained optimization problem (1.17) we use the Lagrange method of indefinite multipliers. Making use of this method we get the equation

$$\frac{1}{2\pi} \int_{-\pi}^{\pi} \left[\frac{\left|C_s^0(e^{i\lambda})\right|^2}{f_0^2(\lambda)} - p_0^2 \frac{1}{f_0^2(\lambda)} \right] \rho(f(\lambda)) d\lambda = 0,$$

where p_0^2 is a constant (the Lagrange multiplier), and $\rho(f(\lambda))$ is a variation of the function $f(\lambda)$. From a generalization of the Lagrange lemma we get that the Fourier coefficients of the function f_0^{-1} satisfy the equation

$$\left| \sum_{l=0}^{s-1} \sum_{k=M_l}^{M_l+N_{l+1}} c(k) e^{ik\lambda} \right|^2 = p_0^2, \tag{1.18}$$

where $c(k), k \in S$ are components of the vector $\vec{\mathbf{c}}_s$, that satisfies the equation

$$B_s^0 \vec{\mathbf{c}}_s = \vec{\mathbf{a}}_s.$$

The matrix B_s^0 is constructed from matrices $B_{mn}^0(k, j)$, each of which is determined by the Fourier coefficients of the function $f_0^{-1}(\lambda)$

$$B_{mn}^0(k, j) = \frac{1}{2\pi} \int_{-\pi}^{\pi} f_0^{-1}(\lambda) e^{-i(k-j)\lambda} d\lambda = r_{k-j}^0, \quad m, n = 1, \ldots, s.$$

$$k = M_{m-1}, \ldots, M_{m-1} + N_m, \; j = M_{n-1}, \ldots, M_{n-1} + N_n.$$

The Fourier coefficients $r_k = r_{-k}, k \in S$, satisfy both equation (1.18) and equation

$$B_s^0 c_s = a_s.$$

These coefficients can be found from the equation

$$B_s \vec{\mathbf{p}}_s^0 = \vec{\mathbf{a}}_s,$$

where $\vec{\mathbf{p}}_s^0 = (p_0, 0, \ldots, 0)$. The last relation can be presented in the form of the system of equations

$$r_k p_0 = a(k), k \in S.$$

From the first equation of this system (for $k = 0$) we find the unknown value $p_0 = a(0)(r_0)^{-1}$. It follows from the extremum condition (1.16) and the restriction on spectral densities from the class D_0^- that the Fourier coefficients

$$r_0 = \frac{1}{2\pi} \int_{-\pi}^{\pi} f_0^{-1}(\lambda) d\lambda = P,$$

$$r_k = P a(k) a^{-1}(0).$$

We construct the following sequence of coefficients

$$r_k = r_{-k} = \begin{cases} P a(k) a^{-1}(0) & \text{if } k \in S, \\ 0 & \text{if } k \in \{0, \ldots, M_{s-1} + N_s\} \setminus S. \end{cases} \tag{1.19}$$

We can represent the function $f^{-1}(\lambda)$ in the form

$$f_0^{-1}(\lambda) = \sum_{k=-(M_{s-1}+N_s)}^{M_{s-1}+N_s} r_k e^{ik\lambda}.$$

Since the sequence $a(k), k \in S$ is strictly positive, the sequence $r_k, k = 0, 1, \ldots, M_{s-1} + N_s$ is also strictly positive and the function $f_0^{-1}(\lambda)$ is positive, so it can be represented in the form (see M. G. Krein and A. A. Nudelman [157])

$$f_0^{-1}(\lambda) = \left| \sum_{k=0}^{M_{s-1}+N_s} \gamma_k e^{-ik\lambda} \right|^2, \quad \lambda \in [-\pi, \pi],$$

where $\gamma_k = 0$, $k \in \{0,\dots,M_{s-1}+N_s\}\setminus S$. So that, $f_0(\lambda)$ is the spectral density of an autoregressive stochastic sequence of the order $M_{s-1}+N_s$, determined by the equation

$$\sum_{k=0}^{M_{s-1}+N_s} \gamma_k \xi(n-k) = \sum_{l=0}^{s-1} \sum_{k=M_l}^{M_l+N_{l+1}} \gamma_k \xi(n-k) = \varepsilon_n, \tag{1.20}$$

where ε_n is a "white noise" sequence of uncorrelated random variables such that $E\varepsilon_n = 0$, $E|\varepsilon_n|^2 = 1$.
The minimax spectral characteristic $h(f_0)$ of the optimal linear estimate of the functional $A_s\xi$ can be calculated by the formula (1.8), where

$$C_N(e^{i\lambda}) = \sum_{l=0}^{s-1} \sum_{k=M_l}^{M_l+N_{l+1}} c(k)e^{ik\lambda} = p_0 = P^{-1}a(0),$$

$$\begin{aligned} h(f_0) &= \sum_{l=0}^{s-1} \sum_{k=M_l}^{M_l+N_{l+1}} a(k)e^{ik\lambda} - P^{-1}a(0) \sum_{k=-(M_{s-1}+N_s)}^{M_{s-1}+N_s} r_k e^{ik\lambda} \\ &= \sum_{k=1}^{N_1} a(k)e^{-ik\lambda} + \sum_{l=1}^{s-1} \sum_{k=M_l}^{M_l+N_{l+1}} a(k)e^{-ik\lambda}. \end{aligned} \tag{1.21}$$

Summing up our reasoning, we conclude that the following theorem is true.

Theorem 1.5. *The least favourable in the class D_0^- spectral density for the optimal linear estimation of the functional $A_s\xi$, determined by a strictly positive sequence $a(k), k \in S$, is the spectral density of the autoregressive sequence (1.20) of order $M_{s-1}+N_s$ whose Fourier coefficients are $r_k = r_{-k} = Pa(k)a^{-1}(0), k \in S$. The minimax spectral characteristic $h(f_0)$ is given by formula (1.21).*

Example 1.3. *Consider the problem of the optimal linear estimation of the functional $A_2\xi = 2\xi(0)+\xi(2)$ based on observations of the sequence $\{\xi(j)\}$ at points $\mathbb{Z}\setminus\{0,2\}$. The system of equations that determine the Fourier coefficients of the least favourable spectral density in the class D_0^- is of the form*

$$r_0 p_0 = 2;$$
$$r_2 p_0 = 1.$$

It follows from these equations that $r_0 = P, r_2 = r_{-2} = \frac{1}{2}P$. *The least favourable spectral density is of the form*

$$f_0 = 1/\left|x + ye^{2i\lambda}\right|,$$

where

$$x = \pm\sqrt{\frac{P}{2}}, y = \pm\sqrt{\frac{P}{2}}.$$

The minimax spectral characteristic can by calculated be the formula

$$h(f_0) = -\sqrt{\frac{P}{2}}e^{-2i\lambda}.$$

2.4. Least Favourable Spectral Densities in the Class $D = D_W$

Consider the problem of minimax estimation of the functional $A_s\xi$, based on observations of the sequence $\xi(j)$ at points $j \in \mathbb{Z}\backslash S$, in the case where the spectral density of the sequence is from the set of spectral densities with moments restrictions of the function $f^{-1}(\lambda)$. Let

$$D_W = \left\{ f(\lambda) \left| \frac{1}{2\pi}\int_{-\pi}^{\pi} f^{-1}(\lambda)\cos(w\lambda)d\lambda = r_w\ , w = 0, 1, \ldots, W \right.\right\},$$

where $r_w, w = 0, 1, \ldots, W$ is a strictly positive sequence. There is an infinite number of functions in the class D_W (see M. G. Krein and A. A. Nudelman [157]) and the function

$$f^{-1}(\lambda) = \sum_{w=-W}^{W} r_{|w|}e^{iw\lambda} > 0, \quad \lambda \in [-\pi, \pi].$$

In order to find solutions of the constrained optimization problem (1.17) for the set D_W of admissible spectral densities we use the method of Lagrange multipliers and find the equation

$$\left|\sum_{l=0}^{s-1}\sum_{k=M_l}^{M_l+N_{l+1}} c(k)e^{ik\lambda}\right|^2 = \sum_{w=0}^{W} \alpha_w \cos(w\lambda) = \left|\sum_{w=0}^{W} p(w)e^{iw\lambda}\right|^2, \tag{1.22}$$

where $\alpha_w, w = 0, 1, \ldots, W$ are the Lagrange multipliers and $c(k), k = 0, \ldots, W$ are solutions of the equation

$$B_s^0 \vec{\mathbf{c}}_s = \vec{\mathbf{a}}_s.$$

Consider two cases: $W \geq M_{s-1} + N_s$ and $W < M_{s-1} + N_s$. Let, first, $W \geq M_{s-1} + N_s$. In this case the given Fourier coefficients r_w define the matrix B_s^0 and the optimization problem (1.16) is degenerated. Let $p(M_{s-1} + N_s + 1) = \ldots = p(W) = 0$ and $p(k) = 0, k \notin S$. Components $p(k), k \in S$ of the vector p_s can be found from the equation

$$B_s^0 p_s = a_s.$$

Hence, relation (1.22) is true. Thus every density $f(\lambda) \in D_W$ is the least favourable and the density of the autoregressive stochastic sequence

$$f_0(\lambda) = \frac{1}{\sum\limits_{w=-W}^{W} r_{|w|} e^{iw\lambda}} = \frac{1}{\left|\sum\limits_{k=0}^{W} \gamma_k e^{ik\lambda}\right|} \tag{1.23}$$

is the least favourable, as well.

Let $W < M_{s-1} + N_s$. Then the matrix B_s is determined by the known $r_w, w \in S \cap \{0, \ldots, W\}$ and the unknown Fourier coefficients

$$r_w, w \in S \setminus \{0, \ldots, W\}$$

of the function $f^{-1}(\lambda)$. The unknown coefficients $p(k)$, $k \in S \cap \{0, \ldots, W\}$ and $r_w, w \in S \setminus \{0, \ldots, W\}$ can be found from the equation

$$B_s p_s^0 = a_s$$

with $p_s^0 = (p(0), \ldots, p(W_1), 0, \ldots, 0)$, where W_1 is determined from the relation $\{0, \ldots, W_1\} = \{0, \ldots, W\} \cap S$.

The last equation can be presented as a system of equations

$$r_0 p_0 + r_1 p_1 + \ldots + r_{-W_1} p_{W_1} = a(0);$$
$$r_1 p_0 + r_0 p_0 + \ldots + r_{-W_1+1} p_{W_1} = a(1);$$
$$\ldots$$
$$r_{W_1} p_0 + r_{W_1-1} p_0 + \ldots + r_0 p_{W_1} = a(W_1);$$
$$\ldots$$
$$r_{M_{s-1}+N_s} p_0 + r_1 p_1 + \ldots + r_0 p_{W_1} = a(M_{s-1} + N_s).$$

From the first W_1 equations we find the unknown coefficients $p(k)$ of the vector p_s^0 and from the next equations we find the Fourier coefficients $r_w, w \in S \setminus \{0,\dots,W\}$ of $f^{-1}(\lambda)$.

If the sequence $r_w, w \in S$, that is constructed from the strictly positive sequence $r_w, w \in S \cap \{0,\dots,W\}$ and calculated coefficients $r_w, w \in S \setminus \{0,\dots,W\}$, is also strictly positive, then the least favourable spectral density $f_0(\lambda)$ is determined by the Fourier coefficients $r_w, w \in S$ of the function $f_0^{-1}(\lambda)$

$$f_0(\lambda) = \frac{1}{\sum\limits_{k=0}^{M_{s-1}+N_s} (r_k e^{ik\lambda} + r_{-k} e^{-ik\lambda})} = \frac{1}{\left| \sum\limits_{k=0}^{M_{s-1}+N_s} \gamma_k e^{ik\lambda} \right|^2}. \tag{1.24}$$

Let us summarize our results and present them in the form of a theorem.

Theorem 1.6. *The least favourable spectral density in the class D_W for the optimal linear estimation of the functional $A_s\xi$ in the case where $W \geq M_{s-1} + N_s$ is the spectral density (1.23) of autoregressive stochastic sequence of order W, determined by coefficients $r_w, w = 0,1,\dots,W$. In the case where $W < M_{s-1} + N_s$ and solutions $r_w, w \in S \setminus \{0,\dots,W\}$ of equation $B_s p_s^0 = a_s$ altogether with coefficients $r_w, w \in S \cap \{0,\dots,W\}$ form a strictly positive sequence, the least favourable spectral density in D_W is the density (1.24) of autoregressive stochastic sequence of order $M_{s-1} + N_s$. The minimax characteristic of an estimate is calculated by formula (1.13).*

2.5. Least Favourable Spectral Densities in the Class $D = D_v^u$

Consider the problem of minimax estimation of the functional $A_s\xi$, based on observations of a sequence $\xi(j)$ at points $j \in \mathbb{Z} \setminus S$, in the case where we do not have complete information on the spectral density of the sequence, while we know that the spectral density $f(\lambda)$ belongs to the class of admissible spectral densities

$$D_v^u = \left\{ f(\lambda) \,|\, 0 \leq v(\lambda) \leq f(\lambda) \leq u(\lambda), \ \frac{1}{2\pi} \int_{-\pi}^{\pi} f^{-1}(\lambda) d\lambda = P \right\},$$

where $v(\lambda), u(\lambda)$ are given bounded spectral densities.

Let the sequence $a(j), j \in S$, that determines the functional $A_s\xi$, be strictly positive. We use the condition $0 \in \partial\Delta_D(f_0)$ to find solutions of the constrained optimization problem (1.17). It follows from the condition $0 \in \partial\Delta_D(f_0)$ for $D - D_v^u$ that the Fourier coefficients of the function f_0^{-1} satisfy both equation

$$B_s^0\vec{\mathbf{c}}_s = \vec{\mathbf{a}}_s$$

and equation

$$\left|\sum_{l=0}^{s-1}\sum_{k=M_l}^{M_l+N_{l+1}}\left(\left(B_s^0\right)^{-1}a_s\right)_k e^{ik\lambda}\right|^2 = \psi_1(\lambda)+\psi_2(\lambda)+p_0^{-2},$$

where $\psi_1(\lambda) \geq 0$ and $\psi_1(\lambda) = 0$ if $f_0(\lambda) \geq v(\lambda)$; $\psi_2(\lambda) \leq 0$ and $\psi_2(\lambda) = 0$ if $f_0(\lambda) \leq u(\lambda)$, p_0 is the Lagrange multiplier.
Therefore, in the case where $v(\lambda) \leq f_0(\lambda) \leq u(\lambda)$, the function $f_0^{-1}(\lambda)$ is of the form

$$f_0^{-1}(\lambda) = \sum_{k=0}^{M_{s-1}+N_s}(r_k e^{ik\lambda} + r_{-k}e^{-ik\lambda}) = \left|\sum_{k=0}^{M_{s-1}+N_s}\gamma_k e^{ik\lambda}\right|^2,$$

where $r_k = r_{-k} = Pa(k)a^{-1}(0)$. The spectral density of an autoregressive stochastic sequence of the order $M_{s-1}+N_s$, is the least favourable in the class D_u^v, if the following inequality is true

$$v^{-1}(\lambda) \geq \sum_{k=0}^{M_{s-1}+N_s}(r_k e^{ik\lambda} + r_{-k}e^{-ik\lambda}) = \left|\sum_{k=0}^{M_{s-1}+N_s}\gamma_k e^{ik\lambda}\right|^2 \geq u^{-1}(\lambda). \tag{1.25}$$

In general case the least favourable density is of the form

$$f_0(\lambda) = \max\left\{v(\lambda), \min\left\{u(\lambda), \left|p_0\sum_{l=0}^{s-1}\sum_{k=M_l}^{M_l+N_{l+1}}\left(\left(B_s^0\right)^{-1}a_s\right)_k e^{ik\lambda}\right|^{-2}\right\}\right\}. \tag{1.26}$$

Theorem 1.7. *Let the sequence $a(j)$, $j \in S$, be strictly positive and let coefficients $r_k = r_{-k} = Pa(k)a^{-1}(0)$, $k \in S$, satisfy the inequality (1.25). The least favourable in the class D_v^u spectral density for the optimal linear estimation of the functional $A_s\xi$ is the density (1.20) of the autoregressive stochastic sequence of order $M_{s-1}+N_s$. The minimax spectral characteristic $h(f_0)$ of the* *estimate can be calculated by the formula (1.21). In the case where inequality (1.25) is not satisfied, the least favourable spectral density in D_v^u is determined by relation (1.26) and the optimization problem (1.16). The minimax spectral characteristic of the estimate is calculated by formula (1.13).*

3. The Problem of Interpolation of Stationary Sequences Observed with Noise

In this section, the problem of mean-square optimal linear estimation of the functional

$$A_s\xi = \sum_{l=0}^{s-1} \sum_{j=M_l}^{M_l+N_{l+1}} a(j)\xi(j)$$

constructed from unobserved values of a stationary stochastic sequence $\xi(j)$ is considered. Estimates are based on observations of the sequence $\xi(j)+\eta(j)$ at points $j \in \mathbb{Z}\backslash S$, where $\eta(j)$ is a stationary correlated with $\xi(j)$ stochastic sequence. The set S of missing observation is of the form

$$S = \bigcup_{l=0}^{s-1}\{M_l, \ldots, M_l+N_{l+1}\},$$

$$M_l = \sum_{k=0}^{l}(N_k+K_k), N_0 = K_0 = 0.$$

Formulas for calculating the spectral characteristic and mean square error of the optimal linear estimate of the functional are derived in the case of spectral certainty, where the spectral densities of the stationary sequences $\xi(j)$ and $\eta(j)$ are exactly known. In the case of spectral uncertainty, where the spectral densities are not exactly known while some sets of admissible spectral densities are given, we apply the minimax method of estimation. This method allows us to find estimates that minimize the maximal values of the mean-square errors of estimates for all spectral densities from a given class of admissible spectral densities. Formulas that determine the least favourable spectral densities and the minimax robust spectral characteristics of the optimal estimates of the functional are derived for some specific classes of admissible spectral densities.

3.1. Hilbert Space Projection Method of Interpolation of Stationary Sequence Observed with Noise

Let $\xi(j)$ and $\eta(j)$ be stationary correlated stochastic sequences with zero mathematical expectations, $E\xi(j)=0$, $E\eta(j)=0$, and the covariance functions admitting the representation (1.3). We will suppose that the spectral densities $f(\lambda)$, $f_{\xi\eta}(\lambda)$, $f_{\eta\xi}(\lambda)$, and $g(\lambda)$ satisfy the minimality condition

$$\int_{-\pi}^{\pi} \frac{1}{f(\lambda)+f_{\xi\eta}(\lambda)+f_{\eta\xi}(\lambda)+g(\lambda)} d\lambda < \infty. \tag{1.27}$$

This condition is necessary and sufficient to make impossible the error-free interpolation of unobserved values of the sequence (see the book by Yu. A. Rozanov [10] for more details).

The stationary sequences $\xi(j)$ and $\eta(j)$ admit the spectral representation (1.2) with the orthogonal stochastic measures $Z_{\xi}(d\lambda)$ and $Z_{\eta}(\lambda)$, which are subordinated to the spectral measures $F(d\lambda)$ and $G(d\lambda)$ of the sequences $\xi(j)$ and $\eta(j)$, such that relations (1.4) hold true.

Consider the problem of mean-square optimal linear estimation of the functional

$$A_s\xi = \sum_{l=0}^{s-1} \sum_{j=M_l}^{M_l+N_{l+1}} a(j)\xi(j)$$

based on observations of the sequence $\xi(j)+\eta(j)$ at points $j \in \mathbb{Z}\backslash S$, where

$$S = \bigcup_{l=0}^{s-1} \{M_l, M_l+1, \ldots, M_l+N_{l+1}\},$$

$$M_l = \sum_{k=0}^{l} (N_k+K_k),\, N_0 = K_0 = 0.$$

Since the stationary sequence $\xi(j)$ admits the spectral representation (1.2), the functional $A_s\xi$ can be written in the form

$$A_s\xi = \int_{-\pi}^{\pi} A_s(e^{i\lambda}) Z_{\xi}(d\lambda),$$

$$A_s(e^{i\lambda}) = \sum_{l=0}^{s-1} \sum_{j=M_l}^{M_l+N_{l+1}} a(j)e^{ij\lambda}.$$

Each linear estimate $\hat{A}_s\xi$ of the functional $A_s\xi$ is of the form

$$\hat{A}_s\xi = \int_{-\pi}^{\pi} h(e^{i\lambda})(Z_\xi(d\lambda) + Z_\eta(d\lambda)),$$

where $h(e^{i\lambda}) \in L_2^s(f+g)$ is the spectral characteristic of the estimate.
The mean-square error $\Delta(h; f)$ of the estimate $\hat{A}_s\xi$ is determined by the formula

$$\Delta(h;F) = E\left|A_s\xi - \hat{A}_s\xi\right|^2 = \frac{1}{2\pi}\int_{-\pi}^{\pi} \left|A_s(e^{i\lambda}) - h(e^{i\lambda})\right|^2 f(\lambda)d\lambda$$

$$+\frac{1}{2\pi}\int_{-\pi}^{\pi} \left|h(e^{i\lambda})\right|^2 g(\lambda)d\lambda - \frac{1}{2\pi}\int_{-\pi}^{\pi} \left(A_s(e^{i\lambda}) - h(e^{i\lambda})\right)\overline{h(e^{i\lambda})} f_{\xi\eta}(\lambda)d\lambda$$

$$-\frac{1}{2\pi}\int_{-\pi}^{\pi} \overline{(A_s(e^{i\lambda}) - h(e^{i\lambda}))} h(e^{i\lambda}) f_{\eta\xi}(\lambda)d\lambda = \tag{1.28}$$

$$= \frac{1}{2\pi}\int_{-\pi}^{\pi} \left(A_s(e^{i\lambda}) - h(e^{i\lambda}), -h(e^{i\lambda})\right) F(\lambda) \left(\overline{A_s(e^{i\lambda}) - h(e^{i\lambda})}, \overline{-h(e^{i\lambda})}\right)^{\top} d\lambda,$$

where

$$F(\lambda) = \begin{pmatrix} f(\lambda) & f_{\xi\eta}(\lambda) \\ f_{\eta\xi}(\lambda) & g(\lambda) \end{pmatrix}$$

is the spectral density matrix.

Denote by $H^s(\xi+\eta)$ the closed linear subspace of the Hilbert space $H = L_2(\Omega, \mathcal{F}, P)$, generated by random variables $\{\xi(j) + \eta(j) : j \in \mathbb{Z}\backslash S\}$.
Let $L_2(f+g)$ be the Hilbert space of complex-valued on $[-\pi, \pi)$ functions, which are square integrable with respect to the measure with density $f(\lambda) + f_{\xi\eta}(\lambda) + f_{\eta\xi}(\lambda) + g(\lambda)$. Denote by $L_2^s(f+g)$ the subspace of the space $L_2(f+g)$, generated by functions $\{e^{ij\lambda}, j \in \mathbb{Z}\backslash S\}$.

According to the method of orthogonal projections, the optimal estimate of the functional $A_s\xi$ is a projection of the element $A_s\xi$ of the Hilbert space

$H = L_2(\Omega, \mathcal{F}, P)$ onto the subspace $H^s(\xi+\eta)$. This projection is determined by two conditions:

$$1)\hat{A}_s\xi \in H^s(\xi+\eta),$$
$$2)A_s\xi - \hat{A}_s\xi \perp H^s(\xi+\eta).$$

It follows from the second condition that the spectral characteristic $h(e^{i\lambda})$ for all $j \in \mathbb{Z}\backslash S$ satisfies the relation

$$E\left[\left(A_s\xi - \hat{A}_s\xi\right)\left(\overline{\xi(j)} + \overline{\eta(j)}\right)\right] =$$
$$= \frac{1}{2\pi}\int_{-\pi}^{\pi}\left(A_s(e^{i\lambda}) - h(e^{i\lambda})\right)e^{-ij\lambda}f(\lambda)d\lambda - \frac{1}{2\pi}\int_{-\pi}^{\pi}h(e^{i\lambda})e^{-ij\lambda}f_{\eta\xi}(\lambda)d\lambda +$$
$$+\frac{1}{2\pi}\int_{-\pi}^{\pi}\left(A_s(e^{i\lambda}) - h(e^{i\lambda})\right)e^{-ij\lambda}f_{\xi\eta}(\lambda)d\lambda - \frac{1}{2\pi}\int_{-\pi}^{\pi}h(e^{i\lambda})e^{-ij\lambda}g(\lambda)d\lambda = 0.$$

This relation can be written in the following form

$$\int_{-\pi}^{\pi}\Big[A_s(e^{i\lambda})(f(\lambda) + f_{\xi\eta}(\lambda)) -$$
$$- h(e^{i\lambda})\left(f(\lambda) + f_{\xi\eta}(\lambda) + f_{\eta\xi}(\lambda) + g(\lambda)\right)\Big]e^{-ij\lambda}d\lambda = 0, \quad j \in \mathbb{Z}\backslash S.$$

Hence we can conclude that the Fourier coefficients of the function

$$\left[A_s(e^{i\lambda})(f(\lambda) + f_{\xi\eta}(\lambda)) - h(e^{i\lambda})(f(\lambda) + f_{\xi\eta}(\lambda) + f_{\eta\xi}(\lambda) + g(\lambda))\right]$$

are equal to zero for $j \in \mathbb{Z}\backslash S$, and this function is of the form

$$A_s(e^{i\lambda})(f(\lambda) + f_{\xi\eta}(\lambda)) - h(e^{i\lambda})(f(\lambda) + f_{\xi\eta}(\lambda) + f_{\eta\xi}(\lambda) + g(\lambda)) = C_s(e^{i\lambda}),$$

$$C_s(e^{i\lambda}) = \sum_{l=0}^{s-1}\sum_{k=M_l}^{M_l+N_{l+1}} c(k)e^{ik\lambda},$$

where $c(k), k \in S$ are unknown coefficients that we need to find.

It follows from the last relation that the spectral characteristic of the optimal estimate $\hat{A}_s\xi$ is of the form

$$h(e^{i\lambda}) = A_s(e^{i\lambda})\frac{f(\lambda)+f_{\xi\eta}(\lambda)}{f_\zeta(\lambda)} - C_s(e^{i\lambda})\frac{1}{f_\zeta(\lambda)},$$

where

$$f_\zeta(\lambda) = f(\lambda)+f_{\xi\eta}(\lambda)+f_{\eta\xi}(\lambda)+g(\lambda).$$

It follows from the first condition, $\hat{A}_s\xi \in H^s(\xi+\eta)$, which determines the spectral characteristic of the optimal estimate, that

$$\frac{1}{2\pi}\int_{-\pi}^{\pi} h(e^{i\lambda})e^{-ij\lambda}d\lambda = 0, \; j \in S,$$

namely

$$\frac{1}{2\pi}\int_{-\pi}^{\pi}\left(A_s(e^{i\lambda})\frac{f(\lambda)+f_{\xi\eta}(\lambda)}{f_\zeta(\lambda)} - C_s(e^{i\lambda})\frac{1}{f_\zeta(\lambda)}\right)e^{-ij\lambda}d\lambda = 0, \quad j \in S.$$

We use the last equality to find the unknown coefficients $c(k)$, $k \in S$. We open brackets and get the following relation

$$\sum_{l=0}^{s-1}\sum_{k=M_l}^{M_l+N_{l+1}} a(k)\frac{1}{2\pi}\int_{-\pi}^{\pi}\frac{e^{i(k-j)\lambda}(f(\lambda)+f_{\xi\eta}(\lambda))}{f_\zeta(\lambda)}d\lambda - \sum_{l=0}^{s-1}\sum_{k=M_l}^{M_l+N_{l+1}} c(k)\frac{1}{2\pi}\int_{-\pi}^{\pi}\frac{e^{i(k-j)\lambda}}{f_\zeta(\lambda)}d\lambda = 0, \quad j \in S. \tag{1.29}$$

To simplify this relation we introduce the following notations:

$$R^s_{j,k} = \frac{1}{2\pi}\int_{-\pi}^{\pi} e^{-i(k-j)\lambda}\frac{f(\lambda)+f_{\xi\eta}(\lambda)}{f(\lambda)+f_{\xi\eta}(\lambda)+f_{\eta\xi}(\lambda)+g(\lambda)}d\lambda;$$

$$B^s_{j,k} = \frac{1}{2\pi}\int_{-\pi}^{\pi} e^{-i(k-j)\lambda}\frac{1}{f(\lambda)+f_{\xi\eta}(\lambda)+f_{\eta\xi}(\lambda)+g(\lambda)}d\lambda; \tag{1.30}$$

$$Q^s_{j,k} = \frac{1}{2\pi}\int_{-\pi}^{\pi} e^{-i(k-j)\lambda}\frac{f(\lambda)g(\lambda)-f_{\xi\eta}(\lambda)f_{\eta\xi}(\lambda)}{f(\lambda)+f_{\xi\eta}(\lambda)+f_{\eta\xi}(\lambda)+g(\lambda)}d\lambda.$$

By using the introduced notations, we can write relation (1.29) as the system of equations

$$\sum_{l=0}^{s-1}\sum_{k=M_l}^{M_l+N_{l+1}} R^s_{j,k}a(k)=\sum_{l=0}^{s-1}\sum_{k=M_l}^{M_l+N_{l+1}} B^s_{j,k}c(k),\quad j\in S.$$

These equations can be represented in the form

$$\mathbf{R}_s\vec{\mathbf{a}}_s=\mathbf{B}_s\vec{\mathbf{c}}_s. \tag{1.31}$$

In this equation $\vec{\mathbf{a}}_s$ is a vector, constructed from coefficients that determine the functional $A_s\xi$, and $\vec{\mathbf{c}}_s$ is a vector, constructed from the unknown coefficients $c(k), k\in S$, and $\mathbf{B}_s, \mathbf{R}_s$ are $q\times q$ matrices, $q=N_1+N_2+\ldots+N_s+s$, with elements $\mathbf{B}_s(j,k)=B^s_{j,k}$, $\mathbf{R}_s(j,k)=R^s_{j,k}$, $j,k\in S$, of the form

$$\mathbf{B}_s=\begin{pmatrix} B_{11} & B_{12} & \ldots & B_{1s}\\ B_{21} & B_{22} & \ldots & B_{2s}\\ \vdots & \vdots & \ddots & \vdots\\ B_{s1} & B_{s2} & \ldots & B_{ss}\end{pmatrix}, \tag{1.32}$$

where B_{mn} are matrices of dimension $(N_m+1)\times(N_n+1)$

$$B_{mn}(k,j)=B^s_{j,k},\quad m,n=1,\ldots,s,$$
$$k=M_{m-1},\ldots,M_{m-1}+N_m,\quad j=M_{n-1},\ldots,M_{n-1}+N_n.$$

From equation (1.31) we derive that the unknown coefficients $c(k), k\in S$, are calculated by the formula

$$c(k)=(\mathbf{B}_s^{-1}\mathbf{R}_s\vec{\mathbf{a}}_s)_k,$$

where $(\mathbf{B}_s^{-1}\mathbf{R}_s\vec{\mathbf{a}}_s)_k$ is the k-th component of the vector $\mathbf{B}_s^{-1}\mathbf{R}_s\vec{\mathbf{a}}_s$.
Hence, the spectral characteristic $h(e^{i\lambda})$ of estimate $\hat{A}_s\xi$ is calculated by the formula

$$h(e^{i\lambda})=\frac{A_s(e^{i\lambda})\left(f(\lambda)+f_{\xi\eta}(\lambda)\right)}{f(\lambda)+f_{\xi\eta}(\lambda)+f_{\eta\xi}(\lambda)+g(\lambda)}-$$
$$-\frac{\sum_{l=0}^{s-1}\sum_{k=M_l}^{M_l+N_{l+1}}(\mathbf{B}_s^{-1}\mathbf{R}_s\vec{\mathbf{a}}_s)_k e^{ik\lambda}}{f(\lambda)+f_{\xi\eta}(\lambda)+f_{\eta\xi}(\lambda)+g(\lambda)}. \tag{1.33}$$

The mean-square error of the estimate $\hat{A}_s\xi$ is calculated by the formula

$$\begin{aligned}
&\Delta(h;F) = \Delta(h;f,g,f_{\xi\eta},f_{\eta\xi}) = E\left|A_s\xi - \hat{A}_s\xi\right|^2 = \\
&= \frac{1}{2\pi}\int_{-\pi}^{\pi}\left|A_s(e^{i\lambda})\frac{f_{\eta\xi}(\lambda)+g(\lambda)}{f_\zeta(\lambda)} + C_s(e^{i\lambda})\frac{1}{f_\zeta(\lambda)}\right|^2 f(\lambda)d\lambda \\
&+ \frac{1}{2\pi}\int_{-\pi}^{\pi}\left|A_s(e^{i\lambda})\frac{f(\lambda)+f_{\xi\eta}(\lambda)}{f_\zeta(\lambda)} - C_s(e^{i\lambda})\frac{1}{f_\zeta(\lambda)}\right|^2 g(\lambda)d\lambda \\
&- \frac{1}{2\pi}\int_{-\pi}^{\pi}\left(A_s(e^{i\lambda})\frac{f_{\eta\xi}(\lambda)+g(\lambda)}{f_\zeta(\lambda)} + C_s(e^{i\lambda})\frac{1}{f_\zeta(\lambda)}\right) \\
&\qquad \times \left(\overline{A_s(e^{i\lambda})}\frac{f(\lambda)+f_{\eta\xi}(\lambda)}{f_\zeta(\lambda)} - \overline{C_s(e^{i\lambda})}\frac{1}{f_\zeta(\lambda)}\right) f_{\xi\eta}(\lambda)d\lambda \\
&- \frac{1}{2\pi}\int_{-\pi}^{\pi}\left(\overline{A_s(e^{i\lambda})}\frac{f_{\xi\eta}(\lambda)+g(\lambda)}{f_\zeta(\lambda)} + \overline{C_s(e^{i\lambda})}\frac{1}{f_\zeta(\lambda)}\right) \\
&\qquad \times \left(A_s(e^{i\lambda})\frac{f(\lambda)+f_{\xi\eta}(\lambda)}{f_\zeta(\lambda)} - C_s(e^{i\lambda})\frac{1}{f_\zeta(\lambda)}\right) f_{\eta\xi}(\lambda)d\lambda \\
&= \frac{1}{2\pi}\int_{-\pi}^{\pi}\left|A_s(e^{i\lambda})\right|^2 \frac{f(\lambda)g(\lambda)-f_{\xi\eta}(\lambda)f_{\eta\xi}(\lambda)}{f_\zeta(\lambda)}d\lambda + \\
&+ \frac{1}{2\pi}\int_{-\pi}^{\pi}\left|C_s(e^{i\lambda})\right|^2 \frac{1}{f_\zeta(\lambda)}d\lambda \\
&= \frac{1}{2\pi}\int_{-\pi}^{\pi}\left|\sum_{l=0}^{s-1}\sum_{k=M_l}^{M_l+N_{l+1}} a(j)e^{ij\lambda}\right|^2 \frac{f(\lambda)g(\lambda)-f_{\xi\eta}(\lambda)f_{\eta\xi}(\lambda)}{f_\zeta(\lambda)}d\lambda + \\
&+ \frac{1}{2\pi}\int_{-\pi}^{\pi}\left|\sum_{l=0}^{s-1}\sum_{k=M_l}^{M_l+N_{l+1}} (\mathbf{B}_s^{-1}\mathbf{R}_s\vec{\mathbf{a}}_s)_k e^{ik\lambda}\right|^2 \frac{1}{f_\zeta(\lambda)}d\lambda.
\end{aligned}$$

Making use of the introduced notations (1.30), we can represent the mean-square error in the form

$$\Delta(h;F) = \langle \mathbf{R}_s\vec{\mathbf{a}}_s, \mathbf{B}_s^{-1}\mathbf{R}_s\vec{\mathbf{a}}_s\rangle + \langle \mathbf{Q}_s\vec{\mathbf{a}}_s, \vec{\mathbf{a}}_s\rangle, \tag{1.34}$$

where $\langle\cdot,\cdot\rangle$ is the inner product in the space C^q, and $\mathbf{Q}_s$ is the $q\times q$ matrix with elements $\mathbf{Q}_s(j,k) = Q^s_{j,k}$, $j,k\in S$, of the form (1.32).
Summing up our reasoning, we can conclude that the following theorem is true.

Theorem 1.8. *Let $\xi(j)$ and $\eta(j)$ be stationary correlated stochastic sequences with the spectral density matrix*

$$F(\lambda) = \begin{pmatrix} f(\lambda) & f_{\xi\eta}(\lambda) \\ f_{\eta\xi}(\lambda) & g(\lambda) \end{pmatrix}$$

such that the minimality condition (1.27) is true. The spectral characteristic $h(e^{i\lambda})$ and the value of the mean-square error $\Delta(f,g)$ of the optimal linear estimate of the functional $A_s\xi$ based on observations of the sequence $\xi(j)+\eta(j)$, $j\in\mathbb{Z}\backslash S$, can be calculated by formulas (1.33), (1.34).

Estimates Based on Observations of Uncorrelated Sequences

Consider the case, where the stationary stochastic sequences $\xi(j)$ and $\eta(j)$ are not correlated. In this case the spectral densities $f_{\xi\eta}(\lambda) = f_{\eta\xi}(\lambda) = 0$ and the following corollary is true

Corollary 1.1. *Let $\xi(j)$ and $\eta(j)$ be uncorrelated stationary stochastic sequences with the spectral densities $f(\lambda)$ and $g(\lambda)$ satisfying the minimality condition*

$$\int_{-\pi}^{\pi} \frac{1}{f(\lambda)+g(\lambda)} d\lambda < \infty. \tag{1.35}$$

The spectral characteristic $h(e^{i\lambda})$ and the mean-square error $\Delta(f,g)$ of the optimal linear estimate of the functional $A_s\xi$ based on observations of the sequence $\xi(j)+\eta(j)$, $j\in\mathbb{Z}\backslash S$, can be calculated by the formulas

$$h(e^{i\lambda}) = A_s(e^{i\lambda})\frac{f(\lambda)}{f(\lambda)+g(\lambda)} - \frac{\sum_{l=0}^{s-1}\sum_{k=M_l}^{M_l+N_{l+1}} (\mathbf{B}_s^{-1}\mathbf{R}_s\vec{\mathbf{a}}_s)_k e^{ik\lambda}}{f(\lambda)+g(\lambda)}, \tag{1.36}$$

$$\Delta(h;f,g) = E\left|A_s\xi - \hat{A}_s\xi\right|^2 =$$

$$= \frac{1}{2\pi}\int_{-\pi}^{\pi} \frac{\left|A_s(e^{i\lambda})g(\lambda) + \sum_{l=0}^{s-1}\sum_{k=M_l}^{M_l+N_{l+1}} (\mathbf{B}_s^{-1}\mathbf{R}_s\vec{\mathbf{a}}_s)_k e^{ik\lambda}\right|^2}{(f(\lambda)+g(\lambda))^2} f(\lambda)d\lambda +$$

$$+ \frac{1}{2\pi}\int_{-\pi}^{\pi} \frac{\left|A_s(e^{i\lambda})f(\lambda) - \sum_{l=0}^{s-1}\sum_{k=M_l}^{M_l+N_{l+1}} (\mathbf{B}_s^{-1}\mathbf{R}_s\vec{\mathbf{a}}_s)_k e^{ik\lambda}\right|^2}{(f(\lambda)+g(\lambda))^2} g(\lambda)d\lambda = \qquad (1.37)$$

$$= \langle \mathbf{R}_s\vec{\mathbf{a}}_s, \mathbf{B}_s^{-1}\mathbf{R}_s\vec{\mathbf{a}}_s\rangle + \langle \mathbf{Q}_s\vec{\mathbf{a}}_s, \vec{\mathbf{a}}_s\rangle, \qquad (1.38)$$

where $\mathbf{R}_s$, $\mathbf{B}_s$, $\mathbf{Q}_s$ *are matrices of the form* (1.32) *determined by elements of the following form*

$$R^s_{j,k} = \frac{1}{2\pi}\int_{-\pi}^{\pi} e^{-i(j-k)\lambda} \frac{f(\lambda)}{f(\lambda)+g(\lambda)} d\lambda,$$

$$B^s_{j,k} = \frac{1}{2\pi}\int_{-\pi}^{\pi} e^{-i(j-k)\lambda} \frac{1}{f(\lambda)+g(\lambda)} d\lambda,$$

$$Q^s_{j,k} = \frac{1}{2\pi}\int_{-\pi}^{\pi} e^{-i(j-k)\lambda} \frac{f(\lambda)g(\lambda)}{f(\lambda)+g(\lambda)} d\lambda.$$

Estimates Based on Observations without Noise

It follows from the last corollary that in the case of observations without noise we have the next corollary is true.

Corollary 1.2. *Let* $\xi(j)$ *be a stationary stochastic sequence with the spectral density* $f(\lambda)$ *such that the function* $f^{-1}(\lambda)$ *is integrable. The spectral characteristic* $h(e^{i\lambda})$ *and the mean-square error* $\Delta(f)$ *of the optimal linear estimate of the functional* $A_s\xi$ *based on observations* $\xi(j)$, $j \in \mathbb{Z}\backslash S$, $S = \bigcup_{l=0}^{s-1}\{M_l, \ldots, M_l + N_{l+1}\}$, *can be calculated by the formulas*

$$h(e^{i\lambda}) = A_s(e^{i\lambda}) - C_s(e^{i\lambda})f^{-1}(\lambda), \qquad (1.39)$$

$$\Delta(h;f) = \frac{1}{2\pi}\int_{-\pi}^{\pi}\left|C_s(e^{i\lambda})\right|^2 f^{-1}(\lambda)d\lambda = \langle \mathbf{B}_s^{-1}\vec{\mathbf{a}}_s, \vec{\mathbf{a}}_s\rangle, \qquad (1.40)$$

$$C_s(e^{i\lambda}) = \sum_{l=0}^{s-1}\sum_{j=M_l}^{M_l+N_{l+1}} \left(\mathbf{B}_s^{-1}\vec{\mathbf{a}}_s\right)_j e^{ij\lambda},$$

where $\mathbf{B}_s$ *is the matrix of dimension* $q\times q$ *of the form* (1.32) *and constructed from the Fourier coefficients of the function* $f^{-1}(\lambda)$,

$$\mathbf{B}_s(k,j) = \frac{1}{2\pi}\int_{-\pi}^{\pi} f^{-1}(\lambda)e^{-i(k-j)\lambda}d\lambda, \quad k,j\in S.$$

Example 1.4. *Consider the problem of finding the mean-square optimal linear estimate of the functional*

$$A_2\xi = \xi(0)+\xi(3)$$

based on observations of the stationary sequence $\xi(j)+\eta(j)$ *at points* $j\in \mathbb{Z}\backslash\{0,3\}$.
Consider the problem in the case where two uncorrelated stationary stochastic sequences $\xi(j)$ *and* $\eta(j)$ *have the spectral densities*

$$f(\lambda) = \left|1-ae^{-i\lambda}\right|^2, \qquad |a|<1,$$

$$g(\lambda) = \left|1-be^{-i\lambda}\right|^2, \qquad |b|<1.$$

In this case the spectral density of the sequence $\xi(j)+\eta(j)$ *is of the form*

$$f(\lambda)+g(\lambda) = \left|1-ae^{-i\lambda}\right|^2 + \left|1-be^{-i\lambda}\right|^2 = \left|x-ye^{-i\lambda}\right|^2,$$

$$x = \frac{1}{2}\left(\pm\sqrt{(1+a)^2+(1+b)^2}\pm\sqrt{(1-a)^2+(1-b)^2}\right),$$

$$y = \frac{a+b}{x}.$$

Since $|a|<1$, $|b|<1$, *we have* $\left|\frac{y}{x}\right|<1$.

We use the decomposition of the function $\frac{1}{1-t}$ *into a power series in order to factorize functions* $(f(\lambda)+g(\lambda))^{-1}$, $f(\lambda)(f(\lambda)+g(\lambda))^{-1}$ *and* $f(\lambda)g(\lambda)(f(\lambda)+g(\lambda))^{-1}$. *We get the following factorizations*

$$\frac{1}{f(\lambda)+g(\lambda)} = \frac{1}{\left|x-ye^{-i\lambda}\right|^2} = \left|\sum_{k=0}^{\infty}\frac{y^k}{x^{k+1}}e^{-ik\lambda}\right|^2,$$

$$\frac{f(\lambda)}{f(\lambda)+g(\lambda)} = \frac{\left|1-ae^{-i\lambda}\right|^2}{\left|x-ye^{-i\lambda}\right|^2} = \left|\sum_{k=0}^{\infty}\frac{y^k}{x^{k+1}}e^{-ik\lambda} - \sum_{k=0}^{\infty}\frac{ay^k}{x^{k+1}}e^{-i(k+1)\lambda}\right|^2 =$$

$$= \left|\frac{1}{x} + \sum_{k=0}^{\infty}\left(\frac{y^{k+1}}{x^{k+2}} - \frac{ay^k}{x^{k+1}}\right)e^{-i(k+1)\lambda}\right|^2,$$

$$\frac{f(\lambda)g(\lambda)}{f(\lambda)+g(\lambda)} = \frac{\left|1-ae^{-i\lambda}\right|^2\cdot\left|1-be^{-i\lambda}\right|^2}{\left|x-ye^{-i\lambda}\right|^2} = \frac{\left|1-(a+b)e^{-i\lambda}+abe^{-i2\lambda}\right|^2}{\left|x-ye^{-i\lambda}\right|^2} =$$

$$= \left|\sum_{k=0}^{\infty}\frac{y^k}{x^{k+1}}e^{-ik\lambda} - \sum_{k=0}^{\infty}\frac{(a+b)y^k}{x^{k+1}}e^{-i(k+1)\lambda} + \sum_{k=0}^{\infty}\frac{aby^k}{x^{k+1}}e^{-i(k+2)\lambda}\right|^2 =$$

$$= \left|\frac{1}{x} + \frac{y-ax-bx}{x^2}e^{-i\lambda} + \sum_{k=0}^{\infty}\left(\frac{y^{k+2}}{x^{k+3}} - \frac{(a+b)y^{k+1}}{x^{k+2}} + \frac{aby^k}{x^{k+1}}\right)e^{-i(k+2)\lambda}\right|^2.$$

In order to find the spectral characteristic $h(e^{i\lambda})$ *of the optimal estimate* $\hat{A}_2\xi$ *we use formula (1.36). In this particular case, the spectral characteristic* $h(e^{i\lambda})$ *is determined by the formula*

$$h(e^{i\lambda}) = (1+e^{i3\lambda})\frac{f(\lambda)}{f(\lambda)+g(\lambda)} - \frac{(\mathbf{B}_2^{-1}\mathbf{R}_2\vec{\mathbf{a}}_2)_0 + (\mathbf{B}_2^{-1}\mathbf{R}_2\vec{\mathbf{a}}_2)_3\cdot e^{i3\lambda}}{f(\lambda)+g(\lambda)},$$

where matrices $\mathbf{B}_2$, $\mathbf{R}_2$, $\mathbf{Q}_2$ *are of the form*

$$\mathbf{B}_2 = \begin{pmatrix} b_0 & b_{-3} \\ b_3 & b_0 \end{pmatrix}, \quad \mathbf{R}_2 = \begin{pmatrix} r_0 & r_{-3} \\ r_3 & r_0 \end{pmatrix}, \quad \mathbf{Q}_2 = \begin{pmatrix} q_0 & q_{-3} \\ q_3 & q_0 \end{pmatrix},$$

and b_j, r_j, q_j, $j\in\{-3,0,3\}$ *are the Fourier coefficients of the functions*

$$\frac{1}{f(\lambda)+g(\lambda)}, \quad \frac{f(\lambda)}{f(\lambda)+g(\lambda)}, \quad \frac{f(\lambda)g(\lambda)}{f(\lambda)+g(\lambda)}.$$

The vector $\mathbf{a}_2$*, constructed from coefficients that determine the functional* $A_2\xi$*, is of the form*

$$\mathbf{a}_2 = (a(0), a(3)) = (1, 1).$$

The unknown coefficients in the formula (1.36) are calculated by formulas

$$(\mathbf{B}_2^{-1}\mathbf{R}_2\vec{\mathbf{a}}_2)_0 = \frac{b_0 r_0 - b_{-3} r_3 + b_0 r_{-3} - b_{-3} r_0}{b_0^2 - b_{-3} b_3},$$

$$(\mathbf{B}_2^{-1}\mathbf{R}_2\vec{\mathbf{a}}_2)_3 = \frac{-b_3 r_0 + b_0 r_3 - b_3 r_{-3} + b_0 r_0}{b_0^2 - b_{-3} b_3},$$

$$b_0 = \sum_{k=0}^{\infty} \left| \frac{y^k}{x^{k+1}} \right|^2,$$

$$b_3 = \overline{b_{-3}} = \sum_{k=0}^{\infty} \left(\frac{y^k}{x^{k+1}} \right) \overline{\left(\frac{y^{k+3}}{x^{k+4}} \right)},$$

$$r_0 = \frac{1}{|x|^2} + \sum_{k=0}^{\infty} \left| \frac{y^{k+1}}{x^{k+2}} - \frac{a y^k}{x^{k+1}} \right|^2,$$

$$r_3 = \overline{r_{-3}} = \frac{1}{x} \overline{\left(\frac{y^3}{x^4} - \frac{a y^2}{x^3} \right)} + \sum_{k=0}^{\infty} \left(\frac{y^{k+1}}{x^{k+2}} - \frac{a y^k}{x^{k+1}} \right) \overline{\left(\frac{y^{k+4}}{x^{k+5}} - \frac{a y^{k+3}}{x^{k+4}} \right)}.$$

The mean-square error of the estimate $\hat{A}_2\xi$ *is determined by formula (1.38). In this case*

$$\Delta(h; f) = E \left| A_2\xi - \hat{A}_2\xi \right|^2 =$$
$$= \langle \mathbf{R}_2\vec{\mathbf{a}}_2, \mathbf{B}_2^{-1}\mathbf{R}_2\vec{\mathbf{a}}_2 \rangle + \langle \mathbf{Q}_2\vec{\mathbf{a}}_2, \vec{\mathbf{a}}_2 \rangle =$$

$$= \frac{1}{b_0^2 - b_{-3} b_3} \big(2 b_0 r_0^2 - b_{-3} r_0 r_3 + 2 b_0 r_0 r_{-3} - b_{-3} r_0^2 - b_{-3} r_{-3} r_3 + b_0 r_{-3}^2$$
$$- b_{-3} r_0 r_{-3} - b_3 r_0 r_3 + b_0 r_3^2 - b_3 r_3 r_{-3} + 2 b_0 r_0 r_3 - b_3 r_0^2 - b_3 r_0 r_{-3} \big)$$
$$+ 2 q_0 + q_3 + q_{-3},$$

$$q_0 = \left| \frac{1}{x} \right|^2 + \left| \frac{y - a x - b x}{x^2} \right|^2 + \sum_{k=0}^{\infty} \left| \frac{y^{k+2}}{x^{k+3}} - \frac{(a+b) y^{k+1}}{x^{k+2}} + \frac{a b y^k}{x^{k+1}} \right|^2,$$

$$q_3 = \overline{q_{-3}} = \frac{1}{x}\overline{\left(\frac{y^3}{x^4} - \frac{(a+b)y^2}{x^3} + \frac{aby}{x^2}\right)} +$$
$$+ \frac{y-ax-bx}{x^2}\overline{\left(\frac{y^4}{x^5} - \frac{(a+b)y^3}{x^4} + \frac{aby^2}{x^3}\right)} +$$
$$+ \sum_{k=0}^{\infty}\left(\frac{y^{k+2}}{x^{k+3}} - \frac{(a+b)y^{k+1}}{x^{k+2}} + \frac{aby^k}{x^{k+1}}\right)\overline{\left(\frac{y^{k+5}}{x^{k+6}} - \frac{(a+b)y^{k+4}}{x^{k+5}} + \frac{aby^{k+3}}{x^{k+4}}\right)}.$$

The considered example illustrates the application of the derived formulas for a concrete form of spectral densities and a particular form of the functional.

3.2. Minimax Robust Method of Interpolation of Stationary Sequences Observed with Noise

The derived in the previous section formulas for calculating values of the mean square error and the spectral characteristic of the optimal estimate $\hat{A}_s\xi$ of the functional $A_s\xi$, can be applied only in the case, where the spectral densities $f(\lambda)$, $f_{\xi\eta}(\lambda)$, $f_{\eta\xi}(\lambda)$, $g(\lambda)$ are exactly known. In the case of spectral uncertainty, where information about the exact form of densities is not complete, while it is known that the densities belong to a certain class of admissible spectral densities, the minimax robust method of estimation of functionals may be applied. According to this method, we find an estimate which minimizes the maximal values of mean square errors for all spectral densities from a given class D of admissible spectral densities simultaneously.

For a description of the minimax method of estimation, we propose the following definitions.

Definition 1.4. *For a given class of admissible spectral density matrices $D = \{F(\lambda)\}$, a spectral density matrix $F^0(\lambda) \in D$ is called the least favourable spectral density matrix in the class D for the optimal linear interpolation of the functional $A_s\xi$, if the following relation is true*

$$\Delta\left(F^0\right) = \Delta\left(h\left(F^0\right); F^0\right) = \max_{F \in D} \Delta\left(h\left(F\right); F\right).$$

Definition 1.5. *For a given class of spectral density matrices $D = \{F(\lambda)\}$, the spectral characteristic $h^0(e^{i\lambda})$ of the optimal linear estimate of the functional $A_s\xi$ is called minimax robust if the following conditions hold true*

$$h^0(e^{i\lambda}) \in H_D = \bigcap_{F \in D} L_2^s(f+g),$$

$$\min_{h \in H_D} \max_{F \in D} \Delta(h; F) = \max_{F \in D} \Delta\left(h^0; F\right).$$

From the given definitions and derived formulas we obtain the following statement.

Lemma 1.2. *Spectral densities $f^0(\lambda), f^0_{\xi\eta}(\lambda), f^0_{\eta\xi}(\lambda), g^0(\lambda)$, that satisfy the minimality condition (1.27), are components of the least favourable spectral density matrix $F^0(\lambda) \in D$ in the class D for the optimal linear estimation of the functional $A_s\xi$, if the Fourier coefficients* (1.30) *of the functions*

$$\frac{1}{f^0(\lambda)+f^0_{\xi\eta}(\lambda)+f^0_{\eta\xi}(\lambda)+g^0(\lambda)}, \quad \frac{f^0(\lambda)+f^0_{\xi\eta}(\lambda)}{f^0(\lambda)+f^0_{\xi\eta}(\lambda)+f^0_{\eta\xi}(\lambda)+g^0(\lambda)},$$

$$\frac{f^0(\lambda)g^0(\lambda)-f^0_{\xi\eta}(\lambda)f^0_{\eta\xi}(\lambda)}{f^0(\lambda)+f^0_{\xi\eta}(\lambda)+f^0_{\eta\xi}(\lambda)+g^0(\lambda)}$$

define matrices $\mathbf{B}_s^0, \mathbf{R}_s^0, \mathbf{Q}_s^0$, that determine a solution of the optimization problem

$$\begin{aligned}
&\max_{F \in D} \langle \mathbf{R}_s\vec{\mathbf{a}}_s, \mathbf{B}_s^{-1}\mathbf{R}_s\vec{\mathbf{a}}_s\rangle + \langle \mathbf{Q}_s\vec{\mathbf{a}}_s, \vec{\mathbf{a}}_s\rangle \\
&= \langle \mathbf{R}_s^0\vec{\mathbf{a}}_s, (\mathbf{B}_s^0)^{-1}\mathbf{R}_s^0\vec{\mathbf{a}}_s\rangle + \langle \mathbf{Q}_s^0\vec{\mathbf{a}}_s, \vec{\mathbf{a}}_s\rangle.
\end{aligned} \tag{1.41}$$

The minimax spectral characteristic $h^0 = h(F^0)$ can be calculated by the formula (1.33), if $h(F^0) \in H_D$.

In the case of observations of the uncorrelated sequences, we have the following definitions and lemma.

Definition 1.6. *For a given class of spectral densities $D = D_f \times D_g$, spectral densities $f_0(\lambda) \in D_f$, $g_0(\lambda) \in D_g$ are called the least favourable in the class*

D for the optimal linear interpolation of the functional $A_s\xi$, if the following relation is true

$$\Delta(f_0,g_0)=\Delta(h(f_0,g_0);f_0,g_0)=\max_{(f,g)\in D_f\times D_g}\Delta(h(f,g);f,g).$$

Definition 1.7. *For a given class of spectral densities $D=D_f\times D_g$, the spectral characteristic $h^0(e^{i\lambda})$ of the optimal estimate of functional $A_s\xi$ is called minimax robust if the following relations hold true*

$$h^0(e^{i\lambda})\in H_D=\bigcap_{(f,g)\in D}L_2^s(f+g),$$

$$\min_{h\in H_D}\max_{(f,g)\in D}\Delta(h;f,g)=\sup_{(f,g)\in D}\Delta\left(h^0;f,g\right).$$

Lemma 1.3. *Spectral densities $f_0(\lambda)$, $g_0(\lambda)$ satisfying the minimality condition (1.35), are the least favourable in a class $D=D_f\times D_g$ for the optimal linear interpolation of the functional $A_s\xi$, if the Fourier coefficients of the functions*

$$(f_0(\lambda)+g_0(\lambda))^{-1},\quad f_0(\lambda)(f_0(\lambda)+g_0(\lambda))^{-1},\quad f_0(\lambda)g_0(\lambda)(f_0(\lambda)+g_0(\lambda))^{-1}$$

define matrices $\mathbf{B}_s^0,\mathbf{R}_s^0,\mathbf{Q}_s^0$, which determine a solution of the optimization problem

$$\max_{(f,g)\in D}\langle\mathbf{R}_s\vec{\mathbf{a}}_s,\mathbf{B}_s^{-1}\mathbf{R}_s\vec{\mathbf{a}}_s\rangle+\langle\mathbf{Q}_s\vec{\mathbf{a}}_s,\vec{\mathbf{a}}_s\rangle=$$
$$\langle\mathbf{R}_s^0\vec{\mathbf{a}}_s,(\mathbf{B}_s^0)^{-1}\mathbf{R}_s^0\vec{\mathbf{a}}_s\rangle+\langle\mathbf{Q}_s^0\vec{\mathbf{a}}_s,\vec{\mathbf{a}}_s\rangle. \tag{1.42}$$

The minimax spectral characteristic $h^0=h(f_0,g_0)$ is calculated by the formula (1.36), if $h(f_0,g_0)\in H_D$.

The lemmas stated above follows from the introduced definitions of the least favourable spectral densities and the minimax spectral characteristic, and results obtained in the previous section. We give the following lemma for the case of observations without noise.

Lemma 1.4. *Let a spectral density $f_0(\lambda)\in D_f$ be such that the function $f_0^{-1}(\lambda)$ is integrable. The spectral density $f_0(\lambda)\in D_f$ is the least favourable in the class D_f for the optimal linear interpolation of the functional $A_s\xi$ based on* *observations of the sequence* $\xi(j)$ *at points* $j \in \mathbb{Z}_-\backslash S$, *if the Fourier coefficients of the function* $f_0^{-1}(\lambda)$ *define the matrix* $\mathbf{B}^0$ *which determines a solution of the optimization problem*

$$\max_{f \in D_f} \langle \mathbf{B}^{-1} \vec{\mathbf{a}}_s, \vec{\mathbf{a}}_s \rangle = \langle (\mathbf{B}^0)^{-1} \vec{\mathbf{a}}_s, \vec{\mathbf{a}}_s \rangle. \tag{1.43}$$

The minimax spectral characteristic $h^0 = h(f_0)$ *is calculated by formula (1.39) if* $h(f_0) \in H_{D_f}$.

For more detailed analysis of properties of the least favourable spectral densities and the minimax-robust spectral characteristics we observe that the least favourable spectral densities f_0, g_0 and the minimax spectral characteristic $h^0 = h(f_0, g_0)$ form a saddle point of the function $\Delta(h; f, g)$ on the set $H_D \times D$. The saddle point inequalities

$$\Delta(h; f_0, g_0) \geq \Delta\left(h^0; f_0, g_0\right) \geq \Delta\left(h^0; f, g\right)$$

$$\forall h \in H_D, \quad \forall f \in D_f, \quad \forall g \in D_g,$$

hold true if $h^0 = h(f_0, g_0)$ and $h(f_0, g_0) \in H_D$, where (f_0, g_0) is a solution of the constrained optimization problem

$$\sup_{(f,g) \in D_f \times D_g} \Delta(h(f_0, g_0); f, g) = \Delta(h(f_0, g_0); f_0, g_0), \tag{1.44}$$

$$\Delta(h(f_0, g_0); f, g) = \frac{1}{2\pi} \int_{-\pi}^{\pi} \frac{\left|A_s(e^{i\lambda})g(\lambda) + C_s^0(e^{i\lambda})\right|^2}{(f(\lambda) + g(\lambda))^2} f(\lambda) d\lambda$$
$$+ \frac{1}{2\pi} \int_{-\pi}^{\pi} \frac{\left|A_s(e^{i\lambda})f(\lambda) - C_s^0(e^{i\lambda})\right|^2}{(f(\lambda) + g(\lambda))^2} g(\lambda) d\lambda,$$

$$C_s^0(e^{i\lambda}) = \sum_{l=0}^{s-1} \sum_{j=M_l}^{M_l+N_{l+1}} ((\mathbf{B}_s^0)^{-1} \mathbf{R}_s^0 \vec{\mathbf{a}}_s)_j e^{ij\lambda}.$$

The constrained optimization problem (1.44) is equivalent to the unconstrained optimization problem (see B. N. Pshenichnyj [156])

$$\Delta_D(f, g) = -\Delta(h(f_0, g_0); f, g) + \delta((f, g) \left| D_f \times D_g \right.) \to \inf, \tag{1.45}$$

where $\delta((f,g)\,|D_f \times D_g\,)$ is the indicator function of the set $D = D_f \times D_g$, namely $\delta((f,g)\,|D_f \times D_g\,) = 0$ if $(f,g) \in D_f \times D_g$ and $\delta((f,g)\,|D_f \times D_g\,) = +\infty$ if $(f,g) \notin D_f \times D_g$. A solution of the problem (1.45) is characterized by the condition $0 \in \partial\Delta_D(f_0,g_0)$. This condition is necessary and sufficient for the pair (f^0,g^0) to belong to the set of minima of the convex functional $\Delta(h(f^0,g^0);f,g)$. By $\partial\Delta_D(f_0,g_0)$ we denote the subdifferential of the functional $\Delta_D(f,g)$ at point (f_0,g_0), namely, the set of all linear bounded functionals Λ on $L_1 \times L_1$ that satisfy the inequality (see R. T. Rockafellar [158])

$$\Delta_D(f,g) - \Delta_D(f^0,g^0) \geq \Lambda((f,g)-(f_0,g_0)) \quad \forall (f,g) \in D_f \times D_g.$$

The form of the functional $\Delta(h(f_0,g_0);f,g)$ is convenient for finding derivatives and differentials of the functional in the $L_1 \times L_1$ space. Thus, the complexity of solving the optimization problem (1.45) is determined by the complexity of calculating the subdifferentials of the indicator functions $\delta((f,g)|D_f \times D_g)$ of sets $D_f \times D_g$.

Lemma 1.5. *Let (f_0,g_0) be a solution of the optimization problem (1.45). Spectral densities $f_0(\lambda)$, $g_0(\lambda)$ are the least favourable in the class $D = D_f \times D_g$, and $h^0 = h(f_0,g_0)$ is the minimax spectral characteristic of the optimal linear estimate of the functional $A_s\xi$, if $h(f_0,g_0) \in H_D$.*

3.3. Least Favourable Spectral Densities in the Class $D_f^0 \times D_g^0$

Consider the problem of minimax estimation of the functional $A_s\xi$, based on observations of the sequence $\xi(j)+\eta(j)$ at points $j \in \mathbb{Z}\backslash S$, under the condition that the stationary sequences $\xi(j)$ and $\eta(j)$ are not correlated, and the spectral densities $f(\lambda)$ and $g(\lambda)$ belong to the class of admissible spectral densities $D = D_f^0 \times D_g^0$, where

$$D_f^0 = \left\{ f(\lambda) \left| \frac{1}{2\pi} \int_{-\pi}^{\pi} f(\lambda)d\lambda \leq P_1 \right. \right\},$$

$$D_g^0 = \left\{ g(\lambda) \left| \frac{1}{2\pi} \int_{-\pi}^{\pi} g(\lambda)d\lambda \leq P_2 \right. \right\}.$$

Let the spectral densities $f_0(\lambda) \in D_f^0$, $g_0(\lambda) \in D_g^0$, and let the functions, defined by the formulas

$$h_f(f_0, g_0) = \frac{\left|A_s(e^{i\lambda})g_0(\lambda) + C_s^0(e^{i\lambda})\right|^2}{(f_0(\lambda)+g_0(\lambda))^2}, \tag{1.46}$$

$$h_g(f_0, g_0) = \frac{\left|A_s(e^{i\lambda})f_0(\lambda) - C_s^0(e^{i\lambda})\right|^2}{(f_0(\lambda)+g_0(\lambda))^2}, \tag{1.47}$$

be bounded, In this case

$$\Delta(h(f_0,g_0);f,g) = \frac{1}{2\pi}\int_{-\pi}^{\pi} h_f(f_0,g_0)f(\lambda)d\lambda + \frac{1}{2\pi}\int_{-\pi}^{\pi} h_g(f_0,g_0)g(\lambda)d\lambda$$

is a continuous linear functional in the space $L_1 \times L_1$, and thus we can apply the method of Lagrange multipliers to solve the constrained optimization problem (1.45). We get the following equation, which determines the least spectral densities $f^0 \in D_f^0$, $g^0 \in D_g^0$,

$$-\frac{1}{2\pi}\int_{-\pi}^{\pi} h_f(f_0,g_0)\rho(f(\lambda))d\lambda - \frac{1}{2\pi}\int_{-\pi}^{\pi} h_g(f_0,g_0)\rho(g(\lambda))d\lambda$$

$$+\alpha_1\frac{1}{2\pi}\int_{-\pi}^{\pi}\rho(f(\lambda))d\lambda + \alpha_2\frac{1}{2\pi}\int_{-\pi}^{\pi}\rho(g(\lambda))d\lambda = 0.$$

Here $\rho(f(\lambda))$ and $\rho(g(\lambda))$ are variations of the functions $f(\lambda)$ and $g(\lambda)$, and $\alpha_1 \geq 0$ and $\alpha_2 \geq 0$ are constants. By analysing this equation, we get that the least favourable densities $f_0(\lambda) \in D_f^0$, $g_0(\lambda) \in D_g^0$ satisfy equations

$$\left|A_s(e^{i\lambda})g_0(\lambda) + C_s^0(e^{i\lambda})\right| = \alpha_1(f_0(\lambda)+g_0(\lambda)), \tag{1.48}$$

$$\left|A_s(e^{i\lambda})f_0(\lambda) - C_s^0(e^{i\lambda})\right| = \alpha_2(f_0(\lambda)+g_0(\lambda)). \tag{1.49}$$

Note, that $\alpha_1 \neq 0$, if

$$\frac{1}{2\pi}\int_{-\pi}^{\pi} f_0(\lambda)d\lambda = P_1,$$

and $\alpha_2 \neq 0$, if

$$\frac{1}{2\pi}\int_{-\pi}^{\pi} g_0(\lambda)d\lambda = P_2.$$

Hence, the following theorems hold true.

Theorem 1.9. *Let the spectral densities $f_0(\lambda) \in D_f^0$, $g_0(\lambda) \in D_g^0$ satisfy the minimality condition (1.35) and let the functions $h_f(f_0,g_0)$, $h_g(f_0,g_0)$, calculated by formulas (1.46), (1.47), be bounded. Spectral densities $f_0(\lambda), g_0(\lambda)$ are the least favourable spectral densities in the class $D = D_f^0 \times D_g^0$ for the optimal linear interpolation of the functional $A_s\xi$ if they satisfy the system of equations (1.48), (1.49), and determine a solution of the optimization problem (1.42). The function, calculated by formula (1.36), is the minimax spectral characteristic of the optimal linear estimate of the functional $A_s\xi$.*

Theorem 1.10. *Suppose that the spectral density $f(\lambda)$ is known, and the unknown spectral density $g_0(\lambda) \in D_g^0$. Assume that the function $(f(\lambda)+g_0(\lambda))^{-1}$ is integrable, and the function $h_g(f,g_0)$, defined by formula (1.47), is bounded. The spectral density $g_0(\lambda)$ is the least favourable in the class D_g^0 for the optimal linear interpolation of the functional $A_s\xi$ if it is of the form*

$$g_0(\lambda) = \max\left\{0, \alpha_2^{-1}\left|A_s(e^{i\lambda})f_0(\lambda) - C_s^0(e^{i\lambda})\right| - f(\lambda)\right\}$$

and the pair $(f(\lambda), g_0(\lambda))$ determines a solution of the optimization problem (1.42). The function, calculated by formula (1.36), is the minimax spectral characteristic of the optimal linear estimate of the functional $A_s\xi$.

Theorem 1.11. *Let the spectral density $f_0(\lambda) \in D_f^0$, let the function $f_0^{-1}(\lambda)$ be integrable, and let the function $h(f_0)$, determined by formula (1.39), be bounded. The least favourable spectral density in the class D_f^0 for the optimal linear interpolation of the functional $A_s\xi$, based on observations of the sequence $\xi(j)$ at points $j \in \mathbb{Z}\backslash S$, is the density $f_0(\lambda)$, if it satisfies the relation*

$$f_0(\lambda) = \alpha_1\left|C_s^0(e^{i\lambda})\right|$$

and determines a solution of the optimization problem (1.43). The function $h^0 = h(f_0)$, calculated by formula (1.39), is the minimax spectral characteristic of the optimal linear estimate of the functional $A_s\xi$.

3.4. Least Favourable Spectral Densities in the Class $D_v^u \times D_\varepsilon$

Consider the problem of minimax estimation of the functional $A_s\xi$ under the condition that stationary sequences $\xi(j)$ and $\eta(j)$ are not correlated, and spectral densities $f(\lambda)$ and $g(\lambda)$ belong to the class of admissible spectral densities $D = D_v^u \times D_\varepsilon$,

$$D_v^u = \left\{ f(\lambda) \left| v(\lambda) \leq f(\lambda) \leq u(\lambda), \frac{1}{2\pi}\int_{-\pi}^{\pi} f(\lambda)d\lambda \leq P_1 \right. \right\},$$

$$D_\varepsilon = \left\{ g(\lambda) \left| g(\lambda) = (1-\varepsilon)g_1(\lambda) + \varepsilon w(\lambda), \frac{1}{2\pi}\int_{-\pi}^{\pi} g(\lambda)d\lambda \leq P_2 \right. \right\},$$

where the spectral densities $u(\lambda), v(\lambda), g_1(\lambda)$ are known and fixed, and spectral densities $u(\lambda)$ and $v(\lambda)$ are bounded.

Suppose that the spectral densities $f^0(\lambda) \in D_v^u$, $g^0(\lambda) \in D_\varepsilon$ determine the bounded functions $h_f(f_0, g_0)$, $h_g(f_0, g_0)$ by formulas (1.46), (1.47). From the condition $0 \in \partial\Delta_{D_{f,g}}(f_0, g_0)$, we derive equations which the least favourable densities should satisfy

$$\left| A_s(e^{i\lambda})g_0(\lambda) + C_s^0(e^{i\lambda}) \right| = (f_0(\lambda) + g_0(\lambda))(\gamma_1(\lambda) + \gamma_2(\lambda) + \alpha_1^{-1}), \quad (1.50)$$

$$\left| A_s(e^{i\lambda})f_0(\lambda) - C_s^0(e^{i\lambda}) \right| = (f_0(\lambda) + g_0(\lambda))(\varphi(\lambda) + \alpha_2^{-1}). \quad (1.51)$$

Here $\gamma_1 \leq 0$ and $\gamma_1 = 0$, when $f_0(\lambda) \geq v(\lambda)$; $\gamma_2 \geq 0$ and $\gamma_2 = 0$, when $f_0(\lambda) \leq u(\lambda)$, $\varphi(\lambda) \leq 0$ and $\varphi(\lambda) = 0$, when $g_0(\lambda) \geq (1-\varepsilon)g_1(\lambda)$.

The following theorems hold true.

Theorem 1.12. *Suppose that the spectral densities $f_0(\lambda) \in D_v^u$, $g_0(\lambda) \in D_\varepsilon$ satisfy the minimality condition (1.35), and the functions, defined by formulas (1.46), (1.47), are bounded. Spectral densities $f_0(\lambda), g_0(\lambda)$ are the least favourable spectral densities in the class $D_v^u \times D_\varepsilon$ for the optimal linear interpolation of the functional $A_s\xi$ if they satisfy the system of equations (1.50), (1.51), and determine a solution of the optimization problem (1.42). The function $h^0 = h(f_0)$, calculated by formula (1.36), is the minimax spectral characteristic of the optimal linear estimate of the functional $A_s\xi$.*

Theorem 1.13. *Assume that the spectral density $f(\lambda)$ is known while the unknown spectral density $g_0(\lambda) \in D_\varepsilon$. Let the function $f(\lambda)+g_0(\lambda)$ satisfy the minimality condition (1.35), and let the function $h_g(f,g_0)$, calculated by formula(1.47), be bounded. The spectral density $g_0(\lambda)$ is the least favourable in the class D_ε for the optimal linear interpolation of the functional $A_s\xi$, if it is of the form*

$$g_0(\lambda) = \max\left\{(1-\varepsilon)g_1(\lambda), \alpha_2\left|A_s(e^{i\lambda})f_0(\lambda) - C_s^0(e^{i\lambda})\right| - f(\lambda)\right\}$$

and the pair $(f(\lambda), g_0(\lambda))$ determine a solution of the optimization problem (1.42). The function, calculated by formula (1.36), is the minimax spectral characteristic of the optimal linear estimate of the functional $A_s\xi$.

Theorem 1.14. *Let the spectral density $f_0(\lambda) \in D_v^u$, let the function $f_0^{-1}(\lambda)$ be integrable, and let the function $h(f_0)$, defined by formula (1.39), be bounded. The spectral density $f_0(\lambda)$ is the least favourable in the class D_v^u for the optimal linear interpolation of the functional $A_s\xi$, based on observation of the sequence $\xi(j)$, $j \in \mathbb{Z}\backslash S$ if it is of the form*

$$f_0(\lambda) = \max\left\{v(\lambda), \min\left\{u(\lambda), \alpha_1\left|C_s^0(e^{i\lambda})\right|\right\}\right\}$$

and determines a solution of the optimization problem (1.43). The minimax spectral characteristic $h^0 = h(f_0)$ of the optimal linear estimate $\hat{A}_s\xi$ is calculated by formula (1.39).

3.5. Least Favourable Spectral Densities in the Class $D_{2\varepsilon_1} \times D_{1\varepsilon_2}$

Consider the problem of minimax estimation of the functional $A_s\xi$ under the condition that the stationary sequences $\xi(j)$ and $\eta(j)$ are not correlated, and the spectral densities $f(\lambda)$ and $g(\lambda)$ belong to the class of admissible spectral densities $D_{\varepsilon_1}^2 \times D_{\varepsilon_2}^1$,

$$D_{\varepsilon_1}^2 = \left\{ f(\lambda) \left| \frac{1}{2\pi}\int_{-\pi}^{\pi} |f(\lambda) - f_1(\lambda)|^2 \, d\lambda \le \varepsilon_1 \right.\right\},$$

$$D_{\varepsilon_2}^1 = \left\{ g(\lambda) \left| \frac{1}{2\pi}\int_{-\pi}^{\pi} |g(\lambda) - g_1(\lambda)| \, d\lambda \le \varepsilon_2 \right.\right\}.$$

where the spectral densities $f_1(\lambda)$ and $g_1(\lambda)$ are fixed and bounded.

Let the spectral densities $f_0(\lambda) \in D^2_{\varepsilon_1}$, $g_0(\lambda) \in D^1_{\varepsilon_2}$ determine the bounded functions $h_f(f_0, g_0)$, $h_g(f_0, g_0)$ by formulas (1.46), (1.47). It follows from the condition $0 \in \partial\Delta_{D_{f,g}}(f^0, g^0)$, where $D = D^2_{\varepsilon_1} \times D^1_{\varepsilon_2}$, that equations which determine the least favourable densities are as follows

$$\left|A_s(e^{i\lambda})g_0(\lambda) + C^0_s(e^{i\lambda})\right|^2 = (f_0(\lambda) + g_0(\lambda))^2(f_0(\lambda) - f_1(\lambda))\alpha_1, \tag{1.52}$$

$$\left|A_s(e^{i\lambda})f_0(\lambda) - C^0_s(e^{i\lambda})\right|^2 = (f_0(\lambda) + g_0(\lambda))\Psi(\lambda)\alpha_2. \tag{1.53}$$

Here $|\Psi(\lambda)| \leq 1$ and $\Psi(\lambda) = \text{sign}(g_0(\lambda) - g_1(\lambda))$, when $g_0(\lambda) \neq g_1(\lambda)$, α_1, α_2 are constants.

Equations (1.52), (1.53) together with the optimization problem (1.42) and the normalization conditions

$$\frac{1}{2\pi}\int_{-\pi}^{\pi} |f(\lambda) - f_1(\lambda)|^2\, d\lambda = \varepsilon_1, \tag{1.54}$$

$$\frac{1}{2\pi}\int_{-\pi}^{\pi} |g(\lambda) - g_1(\lambda)|\, d\lambda = \varepsilon_2, \tag{1.55}$$

determine the least favourable spectral densities in the class $D = D^2_{\varepsilon_1} \times D^1_{\varepsilon_2}$.

The following theorems hold true.

Theorem 1.15. *Assume that the spectral densities $f_0(\lambda) \in D^2_{\varepsilon_1}$, $g_0(\lambda) \in D^2_{\varepsilon_2}$ satisfy the minimality condition (1.35) and the functions, calculated by formulas (1.46), (1.47), are bounded. Spectral densities $f_0(\lambda)$, $g_0(\lambda)$ are the least favourable in the class $D^2_{\varepsilon_1} \times D^1_{\varepsilon_2}$ for the optimal linear interpolation of the functional $A_s\xi$ if they satisfy equations (1.52)–(1.55) and determine a solution of the optimization problem (1.42). The function $h(f_0, g_0)$, calculated by formula (1.36), is the minimax spectral characteristic of the optimal linear estimate of the functional $A_s\xi$.*

Theorem 1.16. *Suppose that the spectral density $f(\lambda)$ is known, and the unknown spectral density $g_0(\lambda) \in D^1_{\varepsilon_2}$. Let the function $(f(\lambda) + g_0(\lambda))^{-1}$ be integrable, and let the function $h_g(f, g_0)$, defined by formula (1.47), be bounded.*

The spectral density $g_0(\lambda)$ is the least favourable in the class $D^1_{\varepsilon_2}$ for the optimal interpolation of the functional $A_s\xi$, if it is of the form

$$g_0(\lambda) = \max\left\{g_1(\lambda), \alpha_2\left|A_s(e^{i\lambda})f_0(\lambda) - C^0_s(e^{i\lambda})\right| - f(\lambda)\right\}$$

and the pair $(f(\lambda), g_0(\lambda))$ determines a solution of the optimization problem (1.42). The minimax spectral characteristic of the optimal linear estimate $\hat{A}_s\xi$ is calculated by the formula (1.36).

Theorem 1.17. *Consider the problem of optimal interpolation of the functional $A_s\xi$ in the case of observations without noise. Suppose that the spectral density $f_0(\lambda) \in D^2_{\varepsilon_1}$, and the function $f_0^{-1}(\lambda)$ is integrable. Suppose that the function $h(f_0)$, calculated by formula(1.39), is bounded. The spectral density $f_0(\lambda)$ is the least favourable in the class $D^2_{\varepsilon_1}$ for the optimal interpolation of the functional $A_s\xi$ based on observations of the sequence $\xi(j)$, $j \in \mathbb{Z}\backslash S$, if it satisfies the relation*

$$\left|C^0_s(e^{i\lambda})\right|^2 = (f_0(\lambda))^2(f_0(\lambda) - f_1(\lambda))\alpha_1$$

and determines a solution of the optimization problem (1.43). The minimax spectral characteristic $h^0 = h(f_0)$ of the optimal linear estimate of the functional $A_s\xi$ is calculated by formula(1.39).

4. The Problem of Extrapolation of Stationary Sequences with Missing Observations

In this section, we propose results of the investigation of the problem of mean-square optimal linear estimation of the functional

$$A\xi = \sum_{j\in S} a(j)\xi(j) + \sum_{j=0}^{\infty} a(j)\xi(j)$$

constructed from unobserved values of a stationary stochastic sequence $\xi(j)$. Estimates are based on observations of the stationary sequence $\xi(j) + \eta(j)$ at points

$$j \in \mathbb{Z}_-\backslash S = \{\ldots, -2, -1\}\backslash S,$$

$$S = \bigcup_{l=1}^{s} \{-M_l - N_l, -M_l - N_l + 1, \ldots, -M_l\},$$

$$M_l = \sum_{k=0}^{l} (N_k + K_k), N_0 = 0, K_0 = 0,$$

where $\eta(j)$ is a stationary correlated with $\xi(j)$ stochastic sequence. Formulas for calculating the spectral characteristic and mean square error of the optimal linear estimate of the functional are derived in the case of spectral certainty, where the spectral densities of the stationary sequences $\xi(j)$ and $\eta(j)$ are exactly known. In the case of spectral uncertainty, where the spectral densities are not exactly known while some sets of admissible spectral densities are given, we apply the minimax method of estimation. This method allows us to find estimates that minimize the maximal values of the mean-square errors of estimates for all spectral densities from a given class of admissible spectral densities. Formulas that determine the least favourable spectral densities and the minimax-robust spectral characteristics of the optimal estimates of the functional are proposed for some specific classes of admissible spectral densities.

4.1. Hilbert Space Projection Method of Extrapolation

Consider stationary correlated stochastic sequences $\xi(j)$ and $\eta(j)$ with zero mathematical expectations, $E\xi(j) = 0$, $E\eta(j) = 0$, and the covariance functions $R_\xi(k)$, $R_{\xi\eta}(k)$, $R_{\eta\xi}(k)$, $R_\eta(k)$ admitting the spectral representation (1.3).

We will suppose that the spectral densities $f(\lambda)$, $f_{\xi\eta}(\lambda)$, $f_{\eta\xi}(\lambda)$, and $g(\lambda)$ satisfy the minimality condition (1.27).

The stationary sequences $\xi(j)$ and $\eta(j)$ admit the spectral representation (1.2) with the orthogonal stochastic measures $Z_\xi(d\lambda)$ and $Z_\eta(d\lambda)$ satisfying relations (1.4).

Consider the problem of mean-square optimal linear estimation of the functional

$$A\xi = \sum_{j\in S} a(j)\xi(j) + \sum_{j=0}^{\infty} a(j)\xi(j)$$

based on observations of the sequence $\xi(j) + \eta(j)$ at points $j \in \mathbb{Z}_- \backslash S$, where

$$S = \bigcup_{l=1}^{s} \{-M_l - N_l, \ldots, -M_l\}.$$

Making use of the spectral representation (1.2) of the stationary sequence $\xi(j)$, we can represent the functional $A\xi$ in the form

$$A\xi = \int_{-\pi}^{\pi} A(e^{i\lambda}) Z_\xi(d\lambda),$$

where

$$A(e^{i\lambda}) = \sum_{j\in S} a(j)e^{ij\lambda} + \sum_{j=0}^{\infty} a(j)e^{ij\lambda}.$$

For convenience, denote by $T = S \cup \{0, 1, \ldots\}$ the set of all points used in determining the functional $A\xi$. Assume that coefficients $\{a(j), j \in T\}$, that determine the functional $A\xi$, satisfy the following conditions

$$\sum_{k\in T} |a(k)| < \infty, \quad \sum_{k\in T} (k+1)\,|a(k)|^2 < \infty. \tag{1.56}$$

It follows from the first condition that the functional $A\xi$ has a finite second moment, since

$$E|A\xi|^2 \le R_\xi(0) \left(\sum_{k\in T} |a(k)| \right)^2.$$

The second condition ensures compactness of operators in the space ℓ_2 that will be introduced later (see the book by N. I. Akhiezer and I. M. Glazman [159]).

Let $\hat{A}\xi$ denote the optimal linear estimate of the functional $A\xi$ based on observations of the sequence $\xi(j) + \eta(j)$ at points $j \in \mathbb{Z}_- \setminus S$. We will search it in the form

$$\hat{A}\xi = \int_{-\pi}^{\pi} h(e^{i\lambda})(Z_\xi(d\lambda) + Z_\eta(d\lambda)), \tag{1.57}$$

where $h(e^{i\lambda})$ is the spectral characteristic of the estimate $\hat{A}\xi$.

Let

$$\Delta(h; f, g) = E\left|A\xi - \hat{A}\xi\right|^2$$

denote the mean-square error of the estimate $\hat{A}\xi$.

We first suppose that the spectral densities of the stationary sequences $\xi(j)$ and $\eta(j)$ are exactly known. In this case we can apply the method of orthogonal projections in the Hilbert space to find the estimate $\hat{A}\xi$.

Denote by $H^s(\xi+\eta)$ the closed linear subspace, generated by values $\{\xi(j)+\eta(j) : j \in \mathbb{Z}_-\backslash S\}$ in the Hilbert space $H = L_2(\Omega, \mathcal{F}, P)$. Denote by $L_2^s(f+g)$ the subspace, generated by functions $\{e^{ij\lambda}, j \in \mathbb{Z}_-\backslash S\}$ in the space $L_2(f+g)$. According to the method of orthogonal projections, the optimal estimate of the functional $A\xi$ is a projection of the element $A\xi$ of the space $H = L_2(\Omega, \mathcal{F}, P)$ onto the subspace $H^s(\xi+\eta)$. It is determined by the conditions:

$$1)\hat{A}\xi \in H^s(\xi+\eta),$$
$$2)A\xi - \hat{A}\xi \perp H^s(\xi+\eta).$$

It follows from the second condition, that the spectral characteristic $h(e^{i\lambda})$ of the optimal linear estimate $\hat{A}\xi$ is of the form

$$h(e^{i\lambda}) = A(e^{i\lambda})\frac{f(\lambda)+f_{\xi\eta}(\lambda)}{f_\zeta(\lambda)} - C(e^{i\lambda})\frac{1}{f_\zeta(\lambda)}, \tag{1.58}$$

$$C(e^{i\lambda}) = \sum_{j\in T} c(j)e^{ij\lambda},$$

where

$$f_\zeta(\lambda) = f(\lambda)+f_{\xi\eta}(\lambda)+f_{\eta\xi}(\lambda)+g(\lambda),$$

and $c(j), j \in T$, are unknown coefficients that we need to find.

From the first condition, $\hat{A}\xi \in H^s(\xi+\eta)$, that determines the optimal estimate of the functional $A\xi$, we obtain the relation

$$\frac{1}{2\pi}\int_{-\pi}^{\pi} h(e^{i\lambda})e^{-ij\lambda}d\lambda = 0,\ j \in T,$$

namely

$$\frac{1}{2\pi}\int_{-\pi}^{\pi}\left(A(e^{i\lambda})\frac{f(\lambda)+f_{\xi\eta}(\lambda)}{f_\zeta(\lambda)} - C(e^{i\lambda})\frac{1}{f_\zeta(\lambda)}\right)e^{-ij\lambda}d\lambda = 0, \quad j \in T.$$

We open the brackets and write the last equality in the form

$$\sum_{k\in T} a(k)\frac{1}{2\pi}\int_{-\pi}^{\pi}\frac{e^{i(k-j)\lambda}(f(\lambda)+f_{\xi\eta}(\lambda))}{f_{\zeta}(\lambda)}d\lambda- \\ -\sum_{l\in T} c(l)\frac{1}{2\pi}\int_{-\pi}^{\pi}\frac{e^{i(l-j)\lambda}}{f_{\zeta}(\lambda)}d\lambda=0, \quad j\in T. \tag{1.59}$$

To write these equations in more appropriate form we introduce notations

$$\begin{aligned} b_{k-j} &= \frac{1}{2\pi}\int_{-\pi}^{\pi} e^{-i(k-j)\lambda}\frac{1}{f(\lambda)+f_{\xi\eta}(\lambda)+f_{\eta\xi}(\lambda)+g(\lambda)}d\lambda, \\ r_{k-j} &= \frac{1}{2\pi}\int_{-\pi}^{\pi} e^{-i(k-j)\lambda}\frac{f(\lambda)+f_{\xi\eta}(\lambda)}{f(\lambda)+f_{\xi\eta}(\lambda)+f_{\eta\xi}(\lambda)+g(\lambda)}d\lambda, \\ q_{k-j} &= \frac{1}{2\pi}\int_{-\pi}^{\pi} e^{-i(k-j)\lambda}\frac{f(\lambda)g(\lambda)-f_{\xi\eta}(\lambda)f_{\eta\xi}(\lambda)}{f(\lambda)+f_{\xi\eta}(\lambda)+f_{\eta\xi}(\lambda)+g(\lambda)}d\lambda. \end{aligned} \tag{1.60}$$

Denote by $\vec{\mathbf{a}}=(\vec{a}_s,\vec{a}_{s-1},\dots,\vec{a}_1,\vec{a})$ the vector, constructed from the coefficients determining the functional $A\xi$. Here

$$\vec{a}_k=(a(-M_k-N_k),a(-M_k-N_k+1),\dots,a(-M_k)), \\ k=1,2,\dots,s; \quad \vec{a}=(a(0),a(1),\dots). \tag{1.61}$$

By using the introduced notations, we represent relation (1.59) in the form of equation

$$\mathbf{R}\vec{\mathbf{a}}=\mathbf{B}\vec{\mathbf{c}}, \tag{1.62}$$

where the vector $\vec{\mathbf{c}}$ is constructed from the unknown coefficients $c(k), k\in T$, and $\mathbf{B}$, $\mathbf{R}$ are linear operators in the space ℓ_2, defined by matrices B, R correspondingly. The matrix B is of the form

$$B=\begin{pmatrix} B_{s,s} & B_{s,s-1} & \dots & B_{s,1} & B_{s,n} \\ B_{s-1,s} & B_{s-1,s-1} & \dots & B_{s-1,1} & B_{s-1,n} \\ \vdots & \vdots & \ddots & \vdots & \vdots \\ B_{1,s} & B_{1,s-1} & \dots & B_{1,1} & B_{1,n} \\ B_{n,s} & B_{n,s-1} & \dots & B_{n,1} & B_{n,n} \end{pmatrix}, \tag{1.63}$$

where elements in the last column and the last row are matrices with elements that are defined by relations

$$B_{l,n}(k,j) = b_{k-j}, \quad l = 1,2,\dots,s;$$
$$k = -M_l - N_l,\dots,-M_l; \quad j = 0,1,2,\dots,$$
$$B_{n,m}(k,j) = b_{k-j}, \quad m = 1,2,\dots,s;$$
$$k = 0,1,2,\dots; \quad j = -M_m - N_m,\dots,-M_m,$$
$$B_{n,n}(k,j) = b_{k-j}, \quad k,j = 0,1,2,\dots,$$

and all other elements are matrices with elements of the form

$$B_{l,m}(j,k) = b_{k-j}, \quad l,m = 1,2,\dots,s;$$
$$k = -M_l - N_l,\dots,-M_l; j = -M_m - N_m,\dots,-M_m.$$

The matrix R is determined in the same way.
In addition, we denote by $\mathbf{Q}$ a linear operator in the space ℓ_2, determined by the matrix Q with elements $Q(k,j) = q_{k-j}$, $k,j \in T$, of the form similar to (1.63).
It follows from formula (1.62) that we can calculate the unknown coefficients $c(k)$, by formula

$$c(k) = (\mathbf{B}^{-1}\mathbf{R}\vec{\mathbf{a}})_k, \quad k \in T,$$

where $(\mathbf{B}^{-1}\mathbf{R}\vec{\mathbf{a}})_k$ is the k-th element of the vector $\mathbf{B}^{-1}\mathbf{R}\vec{\mathbf{a}}$.
Hence, the spectral characteristic $h(e^{i\lambda})$ of the optimal estimate $\hat{A}\xi$ is calculated by the formula

$$h(e^{i\lambda}) = \frac{A(e^{i\lambda})\left(f(\lambda) + f_{\xi\eta}(\lambda)\right) - \sum\limits_{k\in T}(\mathbf{B}^{-1}\mathbf{R}\vec{\mathbf{a}})_k e^{ik\lambda}}{f(\lambda) + f_{\xi\eta}(\lambda) + f_{\eta\xi}(\lambda) + g(\lambda)}. \tag{1.64}$$

The mean-square error of the estimate $\hat{A}\xi$ is calculated by the formula

$$\Delta(h;F) = \frac{1}{2\pi}\int_{-\pi}^{\pi}\left|\sum_{k\in T} a(j)e^{ij\lambda}\right|^2 \frac{f(\lambda)g(\lambda) - f_{\xi\eta}(\lambda)f_{\eta\xi}(\lambda)}{f_\zeta(\lambda)} d\lambda +$$
$$+ \frac{1}{2\pi}\int_{-\pi}^{\pi}\left|\sum_{k\in T}(\mathbf{B}^{-1}\mathbf{R}\vec{\mathbf{a}})_k e^{ik\lambda}\right|^2 \frac{1}{f_\zeta(\lambda)} d\lambda.$$

By using the introduced notations and operators **B**, **R**, **Q**, we can write the formula for calculating the mean-square error in the form

$$\Delta(h;F) = \langle \mathbf{R}\vec{\mathbf{a}}, \mathbf{B}^{-1}\mathbf{R}\vec{\mathbf{a}}\rangle + \langle \mathbf{Q}\vec{\mathbf{a}}, \vec{\mathbf{a}}\rangle, \tag{1.65}$$

where $\langle a,c\rangle = \sum_k a_k\overline{c}_k$ is the inner product in the space ℓ_2.

Summarising our reasoning, we can conclude that the following theorem is true.

Theorem 1.18. *Let $\{\xi(j), j \in \mathbb{Z}\}$ and $\{\eta(j), j \in \mathbb{Z}\}$ be stationary correlated sequences with the spectral density matrix*

$$F(\lambda) = \begin{pmatrix} f(\lambda) & f_{\xi\eta}(\lambda) \\ f_{\eta\xi}(\lambda) & g(\lambda) \end{pmatrix}$$

such that the minimality condition (1.27) is satisfied. The spectral characteristic $h(e^{i\lambda})$ and the mean-square error $\Delta(h; f, g)$ of the optimal linear estimate of the functional $A\xi$, based on observations of the sequence $\xi(j)+\eta(j)$, $j \in \mathbb{Z}_-\backslash S$, can be calculated by formulas (1.64), (1.65).

Estimates Based on Observations of Uncorrelated Sequences

Consider the problem of the optimal linear estimation of the functional $A\xi$ based on observations $\xi(j)+\eta(j)$, $j \in \mathbb{Z}_-\backslash S$ of the uncorrelated sequences $\{\xi(j), j \in \mathbb{Z}\}$ and $\{\eta(j), j \in \mathbb{Z}\}$. In this case, the spectral densities $f_{\xi\eta}(\lambda) = 0$, $f_{\eta\xi}(\lambda) = 0$, and the following corollary is true.

Corollary 1.3. *Assume that stationary sequences $\{\xi(j), j \in \mathbb{Z}\}$ and $\{\eta(j), j \in \mathbb{Z}\}$ are uncorrelated, and their spectral densities $f(\lambda)$ and $g(\lambda)$ satisfy the minimality condition* (1.35). *The spectral characteristic $h(e^{i\lambda})$ and the mean-square error $\Delta(f, g)$ of the optimal linear estimate of the functional $A\xi$, based on observations of the sequence $\xi(j)+\eta(j)$ at points $j \in \mathbb{Z}_-\backslash S$, can be calculated by formulas*

$$h(e^{i\lambda}) = A(e^{i\lambda})\frac{f(\lambda)}{f(\lambda)+g(\lambda)} - \frac{\sum_{k\in T}(\mathbf{B}^{-1}\mathbf{R}\vec{\mathbf{a}})_k e^{ik\lambda}}{f(\lambda)+g(\lambda)}, \tag{1.66}$$

$$\Delta(h;f,g) = \frac{1}{2\pi}\int_{-\pi}^{\pi} \frac{\left|A(e^{i\lambda})g(\lambda) + \sum_{k\in T}(\mathbf{B}^{-1}\mathbf{R}\vec{\mathbf{a}})_k e^{ik\lambda}\right|^2}{(f(\lambda)+g(\lambda))^2} f(\lambda)d\lambda +$$

$$+\frac{1}{2\pi}\int_{-\pi}^{\pi} \frac{\left|A(e^{i\lambda})f(\lambda) - \sum_{k\in T}(\mathbf{B}^{-1}\mathbf{R}\vec{\mathbf{a}})_k e^{ik\lambda}\right|^2}{(f(\lambda)+g(\lambda))^2} g(\lambda)d\lambda =$$

$$+\langle \mathbf{R}\vec{\mathbf{a}}, \mathbf{B}^{-1}\mathbf{R}\vec{\mathbf{a}}\rangle + \langle \mathbf{Q}\vec{\mathbf{a}}, \vec{\mathbf{a}}\rangle, \tag{1.67}$$

where $\mathbf{B},\mathbf{R},\mathbf{Q}$ *are linear operators in the space* ℓ_2 *defined by matrices with elements* $b_{k-j}, r_{k-j}, q_{k-j}$, $k, j \in T$, *that are Fourier coefficients of the functions*

$$(f(\lambda)+g(\lambda))^{-1},\ f(\lambda)(f(\lambda)+g(\lambda))^{-1},\ f(\lambda)g(\lambda)(f(\lambda)+g(\lambda))^{-1},$$

respectively:

$$b_{k-j} = \frac{1}{2\pi}\int_{-\pi}^{\pi} e^{-i(k-j)\lambda}\frac{1}{f(\lambda)+g(\lambda)}d\lambda,$$

$$r_{k-j} = \frac{1}{2\pi}\int_{-\pi}^{\pi} e^{-i(k-j)\lambda}\frac{f(\lambda)}{f(\lambda)+g(\lambda)}d\lambda,$$

$$q_{k-j} = \frac{1}{2\pi}\int_{-\pi}^{\pi} e^{-i(k-j)\lambda}\frac{f(\lambda)g(\lambda)}{f(\lambda)+g(\lambda)}d\lambda.$$

Estimates Based on Observations without Noise

Next, consider the problem in the case where the stationary sequence $\xi(j)$ is observed without noise. In this case $g(\lambda)=0$. The spectral characteristic of the optimal linear estimate $\hat{A}\xi$ is of the form

$$h(e^{i\lambda}) = A(e^{i\lambda}) - C(e^{i\lambda})f^{-1}(\lambda),$$

$$C(e^{i\lambda}) = \sum_{j\in T} c(j)e^{ij\lambda},$$

and the system of equations (1.62) can be written as follows

$$\vec{\mathbf{a}} = \mathbf{B}\vec{\mathbf{c}}, \tag{1.68}$$

where the linear operator $\mathbf{B}$ in the space ℓ_2 is defined by the matrix with elements that are Fourier coefficients of the function $f^{-1}(\lambda)$.
Hence, the unknown coefficients $c(j)$, are calculated by formula

$$c(j) = \left(\mathbf{B}^{-1}\vec{\mathbf{a}}\right)_j, \quad j \in T,$$

where $\left(\mathbf{B}^{-1}\vec{\mathbf{a}}\right)_j$ is the j-th element of the vector $\mathbf{B}^{-1}\vec{\mathbf{a}}$. The formula for calculating the spectral characteristic of the estimate $\hat{A}\xi$ is the following

$$h(e^{i\lambda}) = \left(\sum_{k\in T} a(j)e^{ij\lambda}\right) - \left(\sum_{j\in T} \left(\mathbf{B}^{-1}\vec{\mathbf{a}}\right)_j e^{ij\lambda}\right) f^{-1}(\lambda). \tag{1.69}$$

The mean-square error of the estimate of the functional is calculated by the formula

$$\Delta(h; f) = \langle \mathbf{B}^{-1}\vec{\mathbf{a}}, \vec{\mathbf{a}} \rangle. \tag{1.70}$$

We formulate the obtained result as the following statement.

Corollary 1.4. *Let $\xi(j)$ be a stationary stochastic sequence with the spectral density $f(\lambda)$ that satisfy the minimality condition*

$$\int_{-\pi}^{\pi} f^{-1}(\lambda)d\lambda < \infty. \tag{1.71}$$

The spectral characteristic $h(e^{i\lambda})$ and the mean-square error $\Delta(h, f)$ of the optimal linear estimate $\hat{A}\xi$ of the functional $A\xi$, based on observations of the sequence $\xi(j)$ at points $j \in \mathbb{Z}_- \setminus S$, $S = \bigcup_{l=1}^{s} \{-M_l - N_l, \dots, -M_l\}$, can be calculated by formulas (1.69), (1.70).

Estimation of the Functional $A_N\xi$

Suppose that the stationary sequences $\xi(j)$ and $\eta(j)$ are uncorrelated. Consider the problem of mean-square optimal linear estimation of the functional

$$A_N\xi = \sum_{j\in S} a(j)\xi(j) + \sum_{j=0}^{N} a(j)\xi(j)$$

from unobserved values of the stationary sequence $\xi(j)$. Estimates are based on observations of the sequence $\xi(j)+\eta(j)$ at points $j \in \mathbb{Z}_-\backslash S$, where $S = \bigcup_{l=1}^{s}\{-M_l - N_l, -M_l - N_l + 1, \ldots, -M_l\}$. In order to find the spectral characteristic $h_N(e^{i\lambda})$ and the mean-square error $\Delta(h_N; f)$ of the estimate

$$\hat{A}_N\xi = \int_{-\pi}^{\pi} h_N(e^{i\lambda})(Z_\xi(d\lambda) + Z_\eta(d\lambda))$$

we determine the vector

$$\vec{\mathbf{a}}_N = (\vec{a}_s, \vec{a}_{s-1}, \ldots, \vec{a}_1, \vec{a}_N),$$

with elements $\vec{a}_k$, $k = 1, \ldots, s$, which are defined by formula (1.61), and the last element is of the form $\vec{a}_N = (a(0), a(1), \ldots, a(N), 0, 0, \ldots)$.

Let $\mathbf{R}_N$ be a linear operator in the space ℓ_2, determined by the matrix $\mathbf{R}_N(k,j) = \mathbf{R}(k,j)$, $j \leq N$, $\mathbf{R}_N(k,j) = 0$, $j > N$.

The spectral characteristic of the optimal linear estimate $\hat{A}_N\xi$ can be calculated by the formula

$$h_N(e^{i\lambda}) = A_N(e^{i\lambda})\frac{f(\lambda)}{f(\lambda)+g(\lambda)} - \frac{\sum_{k\in T}(\mathbf{B}^{-1}\mathbf{R}_N\vec{\mathbf{a}}_N)_k e^{ik\lambda}}{f(\lambda)+g(\lambda)}, \tag{1.72}$$

$$A_N(e^{i\lambda}) = \sum_{j\in S} a(j)e^{ij\lambda} + \sum_{j=0}^{N} a(j)e^{ij\lambda}.$$

The mean-square error of the estimate $\hat{A}_N\xi$ can be calculated by the formula

$$\begin{aligned}\Delta(h_N;f,g) &= E\left|A_N\xi-\hat{A}_N\xi\right|^2\\ &= \frac{1}{2\pi}\int_{-\pi}^{\pi}\frac{\left|A_N(e^{i\lambda})g(\lambda)+\sum_{k\in T}(\mathbf{B}^{-1}\mathbf{R}_N\vec{\mathbf{a}}_N)_k e^{ik\lambda}\right|^2}{(f(\lambda)+g(\lambda))^2}f(\lambda)d\lambda\\ &+\frac{1}{2\pi}\int_{-\pi}^{\pi}\frac{\left|A_N(e^{i\lambda})f(\lambda)-\sum_{k\in T}(\mathbf{B}^{-1}\mathbf{R}_N\vec{\mathbf{a}}_N)_k e^{ik\lambda}\right|^2}{(f(\lambda)+g(\lambda))^2}g(\lambda)d\lambda\\ &= \langle\mathbf{R}_N\vec{\mathbf{a}}_N,\mathbf{B}^{-1}\mathbf{R}_N\vec{\mathbf{a}}_N\rangle+\langle\mathbf{Q}_N\vec{\mathbf{a}}_N,\vec{\mathbf{a}}_N\rangle,\end{aligned} \tag{1.73}$$

where $\mathbf{Q}_N$ is a linear operator in the space ℓ_2 such that $\mathbf{Q}_N(k,j)=\mathbf{Q}(k,j), k,j\leq N$, $\mathbf{Q}_N(k,j)=0$, $k,j>N$.

The linear operators **B**, **R**, **Q** are defined in Corollary 1.3.

Corollary 1.5. *Let $\xi(j)$ and $\eta(j)$ be uncorrelated stationary stochastic sequences with the spectral densities $f(\lambda)$ and $g(\lambda)$, that satisfy the minimality condition (1.35). The spectral characteristic $h_N(e^{i\lambda})$ and the mean-square error $\Delta(h_N;f,g)$ of the optimal linear estimate of the functional $A_N\xi$, based on observations of the sequence $\xi(j)+\eta(j)$, $j\in\mathbb{Z}_-\backslash S$, can be calculated by formulas (1.72), (1.73).*

For the case of observations without noise, we have the following corollary.

Corollary 1.6. *Let $\xi(j)$ be a stationary stochastic sequence with the spectral density $f(\lambda)$ satisfying the minimality condition (1.71). The spectral characteristic $h_N(e^{i\lambda})$ and the mean-square error $\Delta(h_N,f)$ of the optimal linear estimate $\hat{A}_N\xi$ of the functional $A_N\xi$, based on observations of the sequence $\xi(j)$ at points $j\in\mathbb{Z}_-\backslash S$, can be calculated by formulas*

$$\begin{aligned}h_N(e^{i\lambda}) &= \left(\sum_{j\in S}a(j)e^{ij\lambda}+\sum_{j=0}^{N}a(j)e^{ij\lambda}\right)\\ &-\left(\sum_{j\in T}\left(\mathbf{B}^{-1}\vec{\mathbf{a}}_N\right)_j e^{ij\lambda}\right)f^{-1}(\lambda),\end{aligned} \tag{1.74}$$

$$\Delta(h_N; f) = \langle \mathbf{B}^{-1}\vec{\mathbf{a}}_N, \vec{\mathbf{a}}_N \rangle, \tag{1.75}$$

where the linear operator $\mathbf{B}$ *is defined in Corollary 1.4.*

In the case where one needs to find the optimal linear estimate of the functional

$$A\xi = \sum_{j=0}^{\infty} a(j)\xi(j)$$

based on observations of the sequence $\xi(j)$ at points $j \in \mathbb{Z}_- \backslash S$, the vector $\vec{\mathbf{a}}$ is of the form

$$\vec{\mathbf{a}} = (0, 0, \ldots, 0, \vec{a}),$$

where first $|S| = \sum_{k=1}^{s} (N_k + 1)$ elements of this vector are equal to zero, and the last element $\vec{a} = (a(0), a(1) \ldots)$ consists of the coefficients determining the functional $A\xi$.

Example 1.5. *Consider the problem of linear extrapolation of the functional*

$$A_1\xi = a(0)\xi(0) + a(1)\xi(1)$$

constructed from unobserved values of a stationary stochastic sequence $\xi(j)$. *Estimates are based on observations of the sequence* $\xi(j)$ *at points* $j \in \mathbb{Z}_- \backslash S$, *where* $S = \{-3, -2\}$. *Suppose that the spectral density of the sequence* $\xi(j)$ *is of the form*

$$f(\lambda) = |1 - \alpha e^{-i\lambda}|^{-2}, \; |\alpha| < 1.$$

In this case the function $f^{-1}(\lambda)$ *can be written in the form*

$$f^{-1}(\lambda) = |1 - \alpha e^{-i\lambda}|^2 = b_{-1}e^{-i\lambda} + b_0 + b_1 e^{i\lambda},$$

where $b_0 = 1 + |\alpha|^2$, $b_{-1} = -\alpha$, $b_1 = -\bar{\alpha}$, $b_p = 0, |p| > 1$ *are the Fourier coefficients of the function* $f^{-1}(\lambda)$. *According to Corollary 1.6 the spectral characteristic of the optimal estimate* $\hat{A}_1\xi$ *of the functional* $A_1\xi$ *is calculated by the formula*

$$h_1(e^{i\lambda}) = \left(a(0) + a(1)e^{i\lambda}\right) -$$
$$- \left(\sum_{j \in T} \left(\mathbf{B}^{-1}\vec{\mathbf{a}}_1\right)_j e^{ij\lambda}\right) \left(b_{-1}e^{-i\lambda} + b_0 + b_1 e^{i\lambda}\right),$$

where $\vec{\mathbf{a}}_1 = (0,0,1,2,0,0,\dots)$.
To find the unknown coefficients

$$c(j) = \left(\mathbf{B}^{-1}\vec{\mathbf{a}}_1\right)_j, \quad j \in T = S \cup \{0,1,2,\dots\}$$

we use equation (1.68). The vector $\vec{\mathbf{c}}$ *in this equation is of the form* $\vec{\mathbf{c}} = (c(-3),c(-2),c(0),c(1),c(2),c(3),\dots)$. *The operator* $\mathbf{B}$ *is defined by the matrix*

$$B = \begin{pmatrix} b_0 & b_{-1} & 0 & 0 & 0 & \dots \\ b_1 & b_0 & 0 & 0 & 0 & \dots \\ 0 & 0 & b_0 & b_{-1} & 0 & \dots \\ 0 & 0 & b_1 & b_0 & b_{-1} & \dots \\ 0 & 0 & 0 & b_1 & b_0 & \dots \\ \vdots & \vdots & \vdots & \vdots & \vdots & \ddots \end{pmatrix}.$$

It is necessary to find the inverse matrix B^{-1}, *which determines the operator* $\mathbf{B}^{-1}$. *First, we write the matrix* B *in the form*

$$B = \begin{pmatrix} B_0 & 0 \\ 0 & B_1 \end{pmatrix},$$

$$B_0 = \begin{pmatrix} b_0 & b_{-1} \\ b_1 & b_0 \end{pmatrix}, \quad B_1 = \begin{pmatrix} b_0 & b_{-1} & 0 & \dots \\ b_1 & b_0 & b_{-1} & \dots \\ 0 & b_1 & b_0 & \dots \\ \vdots & \vdots & \vdots & \ddots \end{pmatrix}.$$

The matrix B^{-1} *can be represented in the form*

$$B^{-1} = \begin{pmatrix} B_0^{-1} & 0 \\ 0 & B_1^{-1} \end{pmatrix},$$

where B_0^{-1}, B_1^{-1} *are the inverse matrices to matrices* B_0, B_1 *respectively.*
The matrix B_0^{-1} *can be easily found. It is of the form*

$$B_0^{-1} = \begin{pmatrix} \frac{b_0}{b_0^2 - b_1 b_{-1}} & \frac{-b_{-1}}{b_0^2 - b_1 b_{-1}} \\ \frac{-b_1}{b_0^2 - b_1 b_{-1}} & \frac{b_0}{b_0^2 - b_1 b_{-1}} \end{pmatrix}.$$

In order to find the matrix $(B_1)^{-1}$ *we use the following method. The matrix* B_1 *consists of the Fourier coefficients of the function* $f^{-1}(\lambda)$

$$B_1(k,j) = b_{k-j}, \quad k,j = 0,1,2,\dots.$$

The density $f^{-1}(\lambda) = |1-\alpha e^{-i\lambda}|^2$ *admits the factorization*

$$\frac{1}{f(\lambda)} = \sum_{p=-\infty}^{\infty} b_p e^{ip\lambda} = \left|\sum_{j=0}^{\infty} \psi_j e^{-ij\lambda}\right|^2 = \left|\sum_{j=0}^{\infty} \theta_j e^{-ij\lambda}\right|^{-2},$$
$$b_0 = 1+|\alpha|^2, b_{-1} = -\alpha, b_1 = -\bar{\alpha}, b_p = 0, |p| > 1,$$
$$\psi_0 = 1, \psi_1 = -\alpha, \psi_j = 0, j > 1,$$
$$\theta_j = \alpha^j, j \geq 0.$$

It follows from the last expression that we can represent the Fourier coefficients of the function $f^{-1}(\lambda)$ *in the form*

$$b_p = \sum_{k=0}^{\infty} \psi_k \overline{\psi}_{k+p}, p \geq 0,$$

and $b_{-p} = \overline{b_p}$*,* $p \geq 0$*. Therefore*

$$b_{i-j} = \sum_{l=\max(i,j)}^{\infty} \psi_{l-i} \overline{\psi}_{l-j}.$$

Denote by Ψ *and* Θ *the linear in the space* ℓ_2 *operators determined by matrices with elements*

$$\Psi_{i,j} = \psi_{i-j},\ \Theta_{i,j} = \theta_{i-j}, \quad 0 \leq j \leq i,$$
$$\Psi_{i,j} = 0,\ \Theta_{i,j} = 0, \quad 0 \leq i < j.$$

Elements of the matrix B_1 *can be represented in the form* $B_1(i,j) = (\Psi' \overline{\Psi})_{i,j}$*. Since*

$$\Psi\Theta = \Theta\Psi = I,$$

elements of the matrix B_1^{-1} *can be calculated by the formula*

$$B_1^{-1}(i,j) = (\overline{\Theta}\Theta')_{i,j} = \sum_{l=0}^{\min(i,j)} \overline{\theta}_{i-l}\theta_{j-l}.$$

The matrix $(B_1)^{-1}$ is of the form

$$B_1^{-1} = \begin{pmatrix} 1 & \alpha & \alpha^2 & \alpha^3 & \dots \\ \overline{\alpha} & \overline{\alpha}\alpha+1 & \overline{\alpha}\alpha^2+\alpha & \overline{\alpha}\alpha^3+\alpha^2 & \dots \\ \overline{\alpha^2} & \overline{\alpha^2}\alpha+\overline{\alpha} & \overline{\alpha^2}\alpha^2+\overline{\alpha}\alpha+1 & \overline{\alpha^2}\alpha^3+\overline{\alpha}\alpha^2+\alpha & \dots \\ \dots & & & & \end{pmatrix}.$$

From equation (1.68) we can find the unknown coefficients $c(j), j \in T$,

$$c(-3) = 0, \quad c(-2) = 0,$$
$$c(0) = a(0) + a(1)\alpha,$$
$$c(1) = a(0)\overline{\alpha} + a(1)(\overline{\alpha}\alpha + 1),$$
$$c(2) = a(0)\overline{\alpha^2} + a(1)(\overline{\alpha^2}\alpha + \overline{\alpha}),$$
$$\dots$$

$$c(i) = a(0)\overline{\alpha^i} + a(1)(\overline{\alpha^i}\alpha + \overline{\alpha^{i-1}}), \quad i > 2.$$

Hence, the spectral characteristic is calculated by the formula

$$\begin{aligned} h_1(e^{i\lambda}) = & \left(a(0) + a(1)e^{i\lambda}\right) - \\ & - (c(-3)e^{-i3\lambda} + c(-2)e^{-i2\lambda} + c(0) + c(1)e^{i\lambda} + c(2)e^{i2\lambda} + \\ & + \sum_{j>2} c(j)e^{ij\lambda}) \left(b_{-1}e^{-i\lambda} + b_0 + b_1 e^{i\lambda}\right) = \\ & = -c(0)b_{-1}e^{-i\lambda} - c(1)b_1 e^{i2\lambda} - c(2)b_0 e^{i2\lambda} - c(2)b_1 e^{i3\lambda} - \\ & - \sum_{j>2} c(j)e^{ij\lambda}) \left(b_{-1}e^{-i\lambda} + b_0 + b_1 e^{i\lambda}\right). \end{aligned}$$

Since coefficients $b_1 c(j-1) + b_0 c(j) + b_{-1} c(j+1)$ at $e^{ij\lambda}$, $j \geq 2$, are equal to zero, the spectral characteristic of the estimate $\hat{A}_1\xi$ is of the form

$$h_1(e^{i\lambda}) = -c(0)b_{-1}e^{-i\lambda} = (a(0) + a(1)\alpha)\alpha e^{-i\lambda}.$$

The mean-square error of the estimate of the functional $A_1\xi$ is calculated by the formula

$$\Delta(h_1; f) = < B^{-1}\vec{\mathbf{a}}_1, \vec{\mathbf{a}}_1 >,$$

In this case it is of the form

$$\Delta(h_1;f) = (a(0))^2 + a(0)a(1)\alpha + a(0)a(1)\overline{\alpha} + (a(1))^2|\alpha|^2 + (a(1))^2.$$

Note, that for the given spectral density the mean-square error of the optimal estimate of the functional $A_1\xi$, based on observations of a stationary sequence $\xi(j), j<0$, without missing values, is the same as in the case of missing observations.

Example 1.6. *Let us find the mean-square error of extrapolation of the functional $A\xi = \sum_{k=0}^{\infty} a(k)\xi(k)$ based on observations of the sequence $\xi(j)$ at points $j \in Z_-$, when there are no missing observations (case 1). And then find the mean-square error of extrapolation of the functional $A\xi$ based on observations of the sequence $\xi(j)$ at points $j \in Z_- \backslash S$, when there are some missing observations (case 2).*

Thus, we can see the impact of the missing observations on the mean-square error. For easy computation, consider the case, where only one value is missed, $S = \{-n\}$.

Suppose that function $f^{-1}(\lambda)$ admits the factorization

$$\frac{1}{f(\lambda)} = \sum_{p=-\infty}^{\infty} b_p e^{ip\lambda} = \left|\sum_{j=0}^{\infty} \psi_j e^{-ij\lambda}\right|^2 = \left|\sum_{j=0}^{\infty} \theta_j e^{-ij\lambda}\right|^{-2}.$$

First, we find the mean-square error of extrapolation of the functional $A\xi$ in the first case, where there are no missing observations. In this case we use the formula (see papers by M. P. Moklyachuk [55, 56] for more details)

$$\Delta_1(h;f) = \langle \mathrm{B}_1^{-1}\vec{a}, \vec{a}\rangle = \langle\, \mathrm{A}\vec{\theta}, \mathrm{A}\vec{\theta}\rangle,$$

where $\mathbf{B}_1$ is a linear in the space ℓ_2 operator, defined by the matrix B_1 with elements that are the Fourier coefficients of the function $f^{-1}(\lambda)$

$$B_1 = \begin{pmatrix} b_0 & b_{-1} & b_{-2} & \dots \\ b_1 & b_0 & b_{-1} & \dots \\ b_2 & b_1 & b_0 & \dots \\ \dots & & & \end{pmatrix},$$

A is a linear in the space ℓ_2 operator, defined by relation

$$(\mathbf{A}\vec{\theta})_k = \sum_{l=0}^{\infty} a(k+l)\theta_l,$$

where $\vec{a} = (a(0), a(1), a(2), \ldots)$ and $\vec{\theta} = (\theta(0), \theta(1), \theta(2), \ldots)$. Note that under condition (1.56) the operator A *is compact.*
Making use of the introduced linear operators Ψ and Θ in the space ℓ_2, we can calculate elements of the matrix B_1 by the formula

$$B_1(i,j) = (\Psi'\overline{\Psi})_{i,j} = \sum_{l=\max(i,j)}^{\infty} \psi_{l-i}\overline{\psi}_{l-j},$$

and elements of the matrix B_1^{-1}, which determines the inverse operator B_1^{-1}, we can calculate by the formula

$$B_1^{-1}(i,j) = (\overline{\Theta}\Theta')_{i,j} = \sum_{l=0}^{\min(i,j)} \overline{\theta}_{i-l}\theta_{j-l} = \omega_{i,j}.$$

Hence, the mean-square error of the optimal estimate of the functional $A\xi$ based on observations of the sequence without missing observations can be calculated by the formula

$$\Delta_1(h;f) = \sum_{k=0}^{\infty}\left(\sum_{j=0}^{\infty} \omega_{k,j} a(j)\right) a(k) = \langle\, \mathrm{B}_1^{-1}\vec{a}, \vec{a}\rangle = \langle \Theta'\vec{a}, \Theta'\vec{a}\rangle.$$

Next, we find the mean-square error in the case of one missing observation by the formula

$$\Delta_2(\tilde{h};f) = \langle \mathrm{B}_2^{-1}\vec{\mathbf{a}}, \vec{\mathbf{a}}\rangle,$$

where B_2 is a linear in the space ℓ_2 operator, defined by the matrix B_2 with elements that are the Fourier coefficients of the function $f^{-1}(\lambda)$

$$B_2 = \begin{pmatrix} b_0 & b_{-n} & b_{-n-1} & b_{-n-2} & \ldots \\ b_n & b_0 & b_{-1} & b_{-2} & \ldots \\ b_{n+1} & b_1 & b_0 & b_{-1} & \ldots \\ \ldots & & & & \end{pmatrix},$$

and where $\vec{\mathbf{a}} = (0, a(0), a(1), a(2), \dots)$.
In order to find the inverse matrix B_2^{-1}, *we use the Frobenius formula (see, for example, the book by F. R. Gantmacher [160] for more details). We represent matrix* B_2 *in the form*

$$B_2 = \begin{pmatrix} B_{00} & B_{01} \\ B_{10} & B_{11} \end{pmatrix},$$

$$B_{00} = \begin{pmatrix} b_0 \end{pmatrix}, \quad B_{01} = \begin{pmatrix} b_{-n} & b_{-n-1} & b_{-n-2} & \dots \end{pmatrix} = \vec{\overline{b}}_n^{\,\prime},$$

$$B_{10} = \begin{pmatrix} b_n \\ b_{n+1} \\ b_{n+2} \\ \dots \end{pmatrix} = \vec{b}_n, \quad B_{11} = \begin{pmatrix} b_0 & b_{-1} & b_{-2} & \dots \\ b_1 & b_0 & b_{-1} & \dots \\ b_2 & b_1 & b_0 & \dots \\ \dots & & & \end{pmatrix}.$$

Then the inverse matrix B_2^{-1} *is of the form*

$$B_2^{-1} = \begin{pmatrix} V^{-1} & -V^{-1}B_{01}B_{11}^{-1} \\ -B_{11}^{-1}B_{10}V^{-1} & B_{11}^{-1} + B_{11}^{-1}B_{10}V^{-1}B_{01}B_{11}^{-1} \end{pmatrix},$$

where

$$V = B_{00} - B_{01}B_{11}^{-1}B_{10}.$$

The matrix B_{11} *is the same as matrix* B_1 *from this example. Using the relation*

$$\Psi'\vec{\psi}_n = \begin{pmatrix} b_n \\ b_{n+1} \\ b_{n+2} \\ \dots \end{pmatrix} = \vec{b}_n, \quad \vec{\psi}_n = \begin{pmatrix} \psi_n \\ \psi_{n+1} \\ \dots \end{pmatrix},$$

we represent the value of V *in terms of operators* Ψ *and* Θ

$$V = b_0 - (\vec{\overline{b}}_n)'\overline{\Theta}\Theta'\vec{b}_n = b_0 - \langle \Theta'\Psi'\vec{\psi}_n, \Theta'\Psi'\vec{\psi}_n \rangle =$$
$$= b_0 - \langle \vec{\psi}_n, \vec{\psi}_n \rangle = \sum_{k=0}^{\infty} |\psi_k|^2 - \sum_{k=n}^{\infty} |\psi_k|^2 = \sum_{k=0}^{n-1} |\psi_k|^2.$$

Denote by

$$\tilde{\omega}_{k,j} = (B_{11}^{-1} + B_{11}^{-1}B_{10}V^{-1}B_{01}B_{11}^{-1})_{k,j}.$$

Then for $k, j > 0$, we have

$$\tilde{\omega}_{k,j} = \omega_{k,j} + \sum_{i=0}^{\infty} \omega_{k,i} b_{i,-n} \tau_{-n,-n} \sum_{l=0}^{\infty} b_{-n,l} \omega_{l,j} = \omega_{k,j} + \gamma_{k,j}. \tag{1.76}$$

Using the introduced notations, we can represent the element $B_{11}^{-1} B_{10} V^{-1} B_{01} B_{11}^{-1}$ and, as a result, $\gamma_{k,j}$ in the following way

$$B_{11}^{-1} B_{10} V^{-1} B_{01} B_{11}^{-1} = V^{-1} \left(\overline{\Theta}\Theta' \vec{b}_n\right) \left(\overline{\Theta}\Theta' \vec{b}_n\right)^{\top} =$$

$$V^{-1} \left(\overline{\Theta}\Theta' \Psi' \vec{\psi}_n\right) \left(\overline{\Theta}\Theta' \Psi' \vec{\psi}_n\right)^{\top} = V^{-1} \left(\overline{\Theta}\vec{\psi}_n\right) \left(\overline{\Theta}\vec{\psi}_n\right)^{\top},$$

$$\gamma_{k,j} = \left(V^{-1} \left(\overline{\Theta}\vec{\psi}_n\right) \left(\overline{\Theta}\vec{\psi}_n\right)^{\top}\right)_{k,j}.$$

Now we calculate the mean-square error of extrapolation of the functional $A\xi$, based on observations with missing data.

$$\begin{aligned}
\Delta_2(\tilde{h}; f) &= \sum_{k \in T} \left(\sum_{j \in T} \tilde{\omega}_{k,j} a(j) \right) a(k) = \\
&= \sum_{k \in S} \left(\sum_{j \in T} \tilde{\omega}_{k,j} a(j) \right) a(k) + \sum_{k=0}^{\infty} \left(\sum_{j \in T} \tilde{\omega}_{k,j} a(j) \right) a(k) = \\
&= \sum_{k=0}^{\infty} \left(\sum_{j \in S} \tilde{\omega}_{k,j} a(j) + \sum_{j=0}^{\infty} \tilde{\omega}_{k,j} a(j) \right) a(k) \\
&= \sum_{k=0}^{\infty} \left(\sum_{j=0}^{\infty} \tilde{\omega}_{k,j} a(j) \right) a(k), \quad T = \{-n\} \cup \{0, 1, \ldots\}.
\end{aligned}$$

It follows from (1.76) and the last relation that

$$\begin{aligned}
\Delta_2(\tilde{h}; f) &= \Delta_1(h; f) + \sum_{k=0}^{\infty} \left(\sum_{j=0}^{\infty} \gamma_{k,j} a(j) \right) a(k) = \\
&= \langle \Theta' \vec{a}, \Theta' \vec{a} \rangle + V^{-1} \left| \langle \vec{\psi}_n, \Theta' \vec{a} \rangle \right|^2 = \\
&= \|\Theta' \vec{a}\|^2 + V^{-1} \|(\overline{\vec{\psi}}_n)' \Theta' \vec{a}\|^2.
\end{aligned}$$

Thus, we can estimate the influence of the missed observation on the value of the mean-square error.

In particular case of estimation of unobserved value $\xi(0)$, *based on observations* $\xi(k), k=-1,-2,\ldots$ *of a stationary sequence with one missing observation* $\xi(-n)$, *the vector* $\vec{a}=(1,0,0,\ldots)$, *and the mean-square error is calculated by the formula*

$$\Delta_2(\tilde{h};f)=|\theta_0|^2+V^{-1}|\theta_0|^2|\psi_n|^2=$$
$$=|\theta_0|^2\left(1+\frac{|\psi_n|^2}{\sum\limits_{k=0}^{n-1}|\psi_k|^2}\right).$$

A similar result was obtained using the representation of a stationary sequence $\xi(k)$ *in the autoregressive form, that is*

$$E|\xi(0)-\hat{\xi}(0)|^2=\sigma^2\left(1+\frac{\psi_n^2}{\sum\limits_{i=0}^{n-1}\psi_i^2}\right),$$
$$\sigma^2=\exp\left[\int_{-\pi}^{\pi}\ln f(\lambda)d\lambda\right],$$

where ψ_k *are coefficients of the autoregressive representation of* $\xi(k)$ *(see papers by P. Bondon [128], R. Cheng and M. Pourahmadi [131] for more details).*

4.2. Minimax Robust Method of Extrapolation

The obtained in the previous section formulas can be applied to calculate the mean square error and the spectral characteristic of the optimal estimate of the functional $A\xi$) only in the case of spectral certainty, where spectral densities $f(\lambda), f_{\xi\eta}(\lambda), f_{\eta\xi}(\lambda), g(\lambda)$, that determine the spectral density matrix $F(\lambda)$, are exactly known. In the case of spectral uncertainty, where spectral densities are not exactly known, while some sets of admissible spectral densities are given, application of the minimax method of estimation is reasonable. This method allows us to find estimates that minimize the maximal values of the mean-square errors of estimates for all spectral densities from a given class of admissible spectral densities.

For a description of the minimax method of estimation, we propose the following definitions.

Definition 1.8. *For a given class of spectral density matrices $D = \{F(\lambda)\}$ a matrix $F^0(\lambda) \in D$ is called the least favourable in the class D for the optimal linear extrapolation of the functional $A\xi$, if the following relation is true*

$$\Delta\left(F^0\right) = \Delta\left(h\left(F^0\right); F^0\right) = \max_{F \in D} \Delta\left(h\left(F\right); F\right).$$

Definition 1.9. *For a given class of spectral density matrices $D = \{F(\lambda)\}$ the spectral characteristic $h^0(e^{i\lambda})$ of the optimal linear estimate of the functional $A\xi$ is called minimax robust if the following conditions hold true*

$$h^0(e^{i\lambda}) \in H_D = \bigcap_{F \in D} L_2^s(f+g),$$

$$\min_{h \in H_D} \max_{F \in D} \Delta\left(h; F\right) = \max_{F \in D} \Delta\left(h^0; F\right).$$

From the given definitions and derived relations we obtain the following statement.

Lemma 1.6. *Spectral densities $f^0(\lambda), f^0_{\xi\eta}(\lambda), f^0_{\eta\xi}(\lambda), g^0(\lambda)$, that satisfy the minimality condition (1.27), determine the least favourable spectral density matrix $F^0(\lambda) \in D$ in the class D for the optimal linear extrapolation of the functional $A\xi$, if the Fourier coefficients* (1.60) *of functions*

$$\frac{1}{f^0(\lambda) + f^0_{\xi\eta}(\lambda) + f^0_{\eta\xi}(\lambda) + g^0(\lambda)}, \quad \frac{f^0(\lambda) + f^0_{\xi\eta}(\lambda)}{f^0(\lambda) + f^0_{\xi\eta}(\lambda) + f^0_{\eta\xi}(\lambda) + g^0(\lambda)},$$

$$\frac{f^0(\lambda)g^0(\lambda) - f^0_{\xi\eta}(\lambda)f^0_{\eta\xi}(\lambda)}{f^0(\lambda) + f^0_{\xi\eta}(\lambda) + f^0_{\eta\xi}(\lambda) + g^0(\lambda)}$$

define operators $\mathbf{B}^0, \mathbf{R}^0, \mathbf{Q}^0$, respectively, that determine a solution of the constrain optimization problem

$$\max_{F \in D} \left(\langle \mathbf{R}\vec{\mathbf{a}}, \mathbf{B}^{-1}\mathbf{R}\vec{\mathbf{a}}\rangle + \langle \mathbf{Q}\vec{\mathbf{a}}, \vec{\mathbf{a}}\rangle\right) = \langle \mathbf{R}^0\vec{\mathbf{a}}, (\mathbf{B}^0)^{-1}\mathbf{R}^0\vec{\mathbf{a}}\rangle + \langle \mathbf{Q}^0\vec{\mathbf{a}}, \vec{\mathbf{a}}\rangle. \qquad (1.77)$$

The minimax spectral characteristic $h^0 = h(F^0)$ is calculated by formula (1.64), if $h(F^0) \in H_D$.

In the case of uncorrelated stationary sequences, the following definitions are applied.

Definition 1.10. *For a given set of spectral densities $D = D_f \times D_g$ the spectral densities $f_0(\lambda) \in D_f$ and $g_0(\lambda) \in D_g$ are called the least favourable for the optimal extrapolation of the functional $A\xi$, if the following relation is true*

$$\Delta(f_0, g_0) = \Delta(h(f_0, g_0); f_0, g_0) = \max_{(f,g) \in D_f \times D_g} \Delta(h(f,g); f, g).$$

Definition 1.11. *For a given set of spectral densities $D = D_f \times D_g$ the spectral characteristic $h^0(e^{i\lambda})$ of the optimal linear estimate of the functional A_ξ is called minimax robust if the following conditions hold true*

$$h^0(e^{i\lambda}) \in H_D = \bigcap_{(f,g) \in D} L_2^s(f+g),$$

$$\min_{h \in H_D} \max_{(f,g) \in D} \Delta(h; f, g) = \max_{(f,g) \in D} \Delta\left(h^0; f, g\right).$$

Using the obtained in the previous section relations and the introduced definitions, we can ensure that the following statement is true.

Lemma 1.7. *Spectral densities $f_0(\lambda) \in D_f$, $g_0(\lambda) \in D_g$, that satisfy the minimality condition (1.35), are the least favourable in the class $D = D_f \times D_g$ for the optimal extrapolation of the functional $A\xi$, if the Fourier coefficients of functions*

$$(f_0(\lambda) + g_0(\lambda))^{-1},\ f_0(\lambda)(f_0(\lambda) + g_0(\lambda))^{-1},\ f_0(\lambda)g_0(\lambda)(f_0(\lambda) + g_0(\lambda))^{-1}$$

define operators $\mathbf{B}^0, \mathbf{R}^0, \mathbf{Q}^0$, that determine a solution of the constrain optimization problem

$$\max_{(f,g) \in D_f \times D_g} \left(\langle \mathbf{R}\vec{\mathbf{a}}, \mathbf{B}^{-1}\mathbf{R}\vec{\mathbf{a}}\rangle + \langle \mathbf{Q}\vec{\mathbf{a}}, \vec{\mathbf{a}}\rangle\right) = \\ = \langle \mathbf{R}^0\vec{\mathbf{a}}, (\mathbf{B}^0)^{-1}\mathbf{R}^0\vec{\mathbf{a}}\rangle + \langle \mathbf{Q}^0\vec{\mathbf{a}}, \vec{\mathbf{a}}\rangle. \tag{1.78}$$

The minimax spectral characteristic $h^0 = h(f_0, g_0)$ is determined by formula (1.66), if $h(f_0, g_0) \in H_D$.

Corollary 1.7. *Suppose that the spectral density $f_0(\lambda) \in D_f$ is such that the function $f_0^{-1}(\lambda)$ is integrable. The spectral density $f_0(\lambda) \in D_f$ is the least favourable in the class D_f for the optimal extrapolation of the functional $A\xi$, if the Fourier coefficients of the function $f_0^{-1}(\lambda)$ define the operator $\mathbf{B}^0$, that determines a solution of the constrain optimization problem*

$$\max_{f\in D_f}\langle \mathbf{B}^{-1}\vec{\mathbf{a}}, \vec{\mathbf{a}}\rangle = \langle (\mathbf{B}^0)^{-1}\vec{\mathbf{a}}, \vec{\mathbf{a}}\rangle. \tag{1.79}$$

The minimax spectral characteristic $h^0 = h(f_0)$ can be calculated by the formula (1.69), if $h(f_0) \in H_{D_f}$.

For more detailed analysis of properties of the least favourable spectral densities and the minimax-robust spectral characteristics we observe that the least favourable spectral densities f_0, g_0 and the minimax spectral characteristic $h^0 = h(f_0, g_0)$ form a saddle point of the function $\Delta(h; f, g)$ on the set $H_D \times D$. The saddle point inequalities

$$\Delta(h; f_0, g_0) \geq \Delta(h^0; f_0, g_0) \geq \Delta\left(h^0; f, g\right) \quad \forall h \in H_D, \forall f \in D_f, \forall g \in D_g$$

are satisfied if $h^0 = h(f_0, g_0)$ and $h(f_0, g_0) \in H_D$, where (f_0, g_0) is a solution of the constrained optimization problem

$$\sup_{(f,g)\in D_f\times D_g} \Delta(h(f_0, g_0); f, g) = \Delta(h(f_0, g_0); f_0, g_0), \tag{1.80}$$

$$\Delta(h(f_0, g_0); f, g) = \frac{1}{2\pi}\int_{-\pi}^{\pi} \frac{\left|A(e^{i\lambda})g_0(\lambda) + C^0(e^{i\lambda})\right|^2}{(f_0(\lambda) + g_0(\lambda))^2} f(\lambda)d\lambda$$
$$+ \frac{1}{2\pi}\int_{-\pi}^{\pi} \frac{\left|A(e^{i\lambda})f_0(\lambda) - C^0(e^{i\lambda})\right|^2}{(f_0(\lambda) + g_0(\lambda))^2} g(\lambda)d\lambda,$$
$$C^0(e^{i\lambda}) = \sum_{j\in T}((\mathbf{B}^0)^{-1}\mathbf{R}^0\vec{\mathbf{a}})_j e^{ij\lambda}.$$

The constrained optimization problem (1.80) is equivalent to the unconstrained optimization problem (see B. N. Pshenichnyj [156])

$$\Delta_D(f, g) = -\Delta(h(f_0, g_0); f, g) + \delta((f, g)\left|D_f \times D_g\right.) \to \inf, \tag{1.81}$$

where $\delta((f,g)\,|D_f \times D_g)$ is the indicator function of the set $D = D_f \times D_g$. A solution of the problem (1.81) is characterized by the condition $0 \in \partial\Delta_D(f_0, g_0)$, where $\partial\Delta_D(f_0, g_0)$ is the subdifferential of the convex functional $\Delta_D(f, g)$ at point (f_0, g_0) (see R. T. Rockafellar [158]).

The form of the functional $\Delta(h(f_0, g_0); f, g)$ is convenient for finding derivatives and differentials of the functional $\Delta(h(f_0, g_0); f, g)$ in the $L_1 \times L_1$ space. Thus the complexity of solving the optimization problem (1.81) is determined by the complexity of calculating subdifferentials of the indicator functions $\delta((f,g)|D_f \times D_g)$ of sets $D_f \times D_g$ (see book by A. D. Ioffe and V. M. Tihomirov [161]).

Lemma 1.8. *Assume that (f_0, g_0) is a solution to the optimization problem (1.81). The spectral densities $f_0(\lambda)$, $g_0(\lambda)$ are the least favourable spectral densities in the class $D = D_f \times D_g$, and $h^0 = h(f_0, g_0)$ is the minimax spectral characteristic of the optimal linear estimate of the functional $A\xi$, if condition $h(f_0, g_0) \in H_D$ is satisfied.*

4.3. Least Favourable Spectral Densities in the Class $D_W \times D_0$

Consider the problem of minimax extrapolation of the functional $A\xi$, based on observations of the sequence $\{\xi(j) + \eta(j) : j \in \mathbb{Z}_- \backslash S\}$, under the condition that the stationary sequences $\xi(j)$ and $\eta(j)$ are not correlated and the spectral densities $f(\lambda)$ and $g(\lambda)$ belong to the class of admissible spectral densities $D = D_W \times D_0$, where

$$D_W = \left\{ f(\lambda) \left| \frac{1}{2\pi} \int_{-\pi}^{\pi} f(\lambda) \cos(w\lambda) d\lambda = r_w, w = 0, 1, \ldots, W \right. \right\},$$

$$D_0 = \left\{ g(\lambda) \left| \frac{1}{2\pi} \int_{-\pi}^{\pi} g(\lambda) d\lambda \le P_1 \right. \right\}.$$

Assume that the sequence $\{r_w, w = 0, 1, \ldots, W\}$ is strictly positive. The moment problem in this case is indeterminate, and the set D_W contains an infinite number of densities (see books by M. G. Krein and A. A. Nudelman [157], N. I. Akhiezer and M. G. Krein [162]). Suppose that the densities $f_0(\lambda) \in D_W$

and $g_0(\lambda) \in D_0$, and the functions, defined by formulas

$$h_f(f_0, g_0) = \frac{\left|A(e^{i\lambda})g_0(\lambda) + C^0(e^{i\lambda})\right|^2}{(f_0(\lambda) + g_0(\lambda))^2}, \tag{1.82}$$

$$h_g(f_0, g_0) = \frac{\left|A(e^{i\lambda})f_0(\lambda) - C^0(e^{i\lambda})\right|^2}{(f_0(\lambda) + g_0(\lambda))^2}, \tag{1.83}$$

are bounded. In this case the linear functional

$$\Delta(h(f_0, g_0); f, g) = \frac{1}{2\pi}\int_{-\pi}^{\pi} h_f(f_0, g_0) f(\lambda) d\lambda + \frac{1}{2\pi}\int_{-\pi}^{\pi} h_g(f_0, g_0) g(\lambda) d\lambda$$

is continuous and bounded in the space $L_1 \times L_1$. We can use the method of Lagrange multipliers and find equations which the least favourable densities should satisfy

$$\left|A(e^{i\lambda})g_0(\lambda) + C^0(e^{i\lambda})\right| = \left(\sum_{w=0}^{W} \psi_w cos(w\lambda)\right)(f_0(\lambda) + g_0(\lambda)), \tag{1.84}$$

$$\left|A(e^{i\lambda})f_0(\lambda) - C^0(e^{i\lambda})\right| = \alpha(f_0(\lambda) + g_0(\lambda)), \tag{1.85}$$

where constants $\psi_w \geq 0, \alpha \geq 0$.

So we ensure that the following statements are true.

Theorem 1.19. *Let the spectral densities $f_0(\lambda) \in D_W$ and $g_0(\lambda) \in D_0$ satisfy the minimality condition (1.35). Suppose that functions, determined by formulas (1.82), (1.83), are bounded. Then the densities $f_0(\lambda)$, $g_0(\lambda)$, that satisfy equations (1.84), (1.85), are the least favourable spectral densities in the class $D_W \times D_0$ for the optimal estimation of the functional $A\xi$, if they determine a solution of the optimization problem (1.78). The function $h(f_0, g_0)$, determined by formula(1.66), is the minimax spectral characteristic of the optimal estimate of the functional $A\xi$.*

Theorem 1.20. *Let the spectral densities $f_0(\lambda) \in D_0$ and $g_0(\lambda) \in D_0$ satisfy the minimality condition (1.35). Let the functions, defined by formulas (1.82), (1.83), be bounded. The functions $f_0(\lambda)$, $g_0(\lambda)$, determined from the equations*

$$\left|A(e^{i\lambda})g_0(\lambda) + C^0(e^{i\lambda})\right| = \alpha_1(f_0(\lambda) + g_0(\lambda)), \tag{1.86}$$

$$\left|A(e^{i\lambda})f_0(\lambda)-C^0(e^{i\lambda})\right| = \alpha_2(f_0(\lambda)+g_0(\lambda)), \tag{1.87}$$

where $\alpha_1 \geq 0$, $\alpha_2 \geq 0$, *are the least favourable spectral densities in the class* $D_0 \times D_0$, *if they determine a solution of the optimization problem (1.78). Note that* $\alpha_1 \neq 0$, *if*

$$\frac{1}{2\pi}\int_{-\pi}^{\pi} f_0(\lambda)d\lambda = P_1,$$

and $\alpha_2 \neq 0$, *if*

$$\frac{1}{2\pi}\int_{-\pi}^{\pi} g_0(\lambda)d\lambda = P_1.$$

The function $h(f_0, g_0)$, *calculated by formula (1.66), is the minimax spectral characteristic of the optimal estimate of the functional* $A\xi$.

Corollary 1.8. *Assume that the spectral density* $f(\lambda)$ *is known, and the unknown spectral density* $g_0(\lambda) \in D_0$. *Let the function* $f(\lambda)+g_0(\lambda)$ *satisfy the minimality condition (1.35), and let the function* $h_g(f, g_0)$, *determined by formula (1.83), be bounded. The spectral density* $g_0(\lambda)$ *is the least favourable in the class* D_0 *for the optimal linear extrapolation of the functional* $A\xi$, *if it is of the form*

$$g_0(\lambda) = \max\left\{0, \alpha^{-1}\left|A(e^{i\lambda})f(\lambda)-C^0(e^{i\lambda})\right| - f(\lambda)\right\},$$

and the pair $(f(\lambda), g_0(\lambda))$ *determines a solution of the optimization problem (1.78). The minimax spectral characteristic of the linear estimate of the functional* $A\xi$ *can be calculated by the formula(1.66).*

Corollary 1.9. *Consider the case where the spectral density* $g(\lambda)$ *is known, and the unknown spectral density* $f_0(\lambda) \in D_W$. *Suppose that the function* $f_0(\lambda)+g(\lambda)$ *satisfy the minimality condition (1.35), and the function* $h_f(f_0, g)$, *determined by formula(1.82), is bounded. The spectral density* $f_0(\lambda)$ *is the least favourable in the class* D_W *for the optimal linear extrapolation of the functional* $A\xi$, *if it is of the form*

$$f_0(\lambda) = \frac{\left|A(e^{i\lambda})g(\lambda)+C^0(e^{i\lambda})\right|}{\sum_{w=0}^{W}\psi_w cos(w\lambda)} - g(\lambda),$$

and the pair $(f_0(\lambda), g(\lambda))$ *determines a solution of the optimization problem (1.78). The function, calculated by the formula(1.66), is the minimax spectral characteristic of the optimal estimate* $\hat{A}\xi$.

Corollary 1.10. *Suppose we know the exact form of the spectral density* $g(\lambda)$, *and we know that the unknown spectral density* $f_0(\lambda) \in D_0$. *Let the function* $f_0(\lambda) + g(\lambda)$ *satisfy the minimality condition (1.35), and let the function* $h_f(f_0, g)$, *determined by formula (1.82), be bounded. The spectral density* $f_0(\lambda)$ *is the least favourable in the class* D_0 *for the optimal linear extrapolation of the functional* $A\xi$, *if it is of the form*

$$f_0(\lambda) = \max\left\{0, \alpha^{-1}\left|A(e^{i\lambda})g(\lambda) + C^0(e^{i\lambda})\right| - g(\lambda)\right\},$$

and the pair $(f_0(\lambda), g(\lambda))$ *determines a solution of the optimization problem (1.78). The minimax spectral characteristic of the linear estimate of the functional* $A\xi$ *can be calculated by the formula(1.66).*

Corollary 1.11. *Let the spectral density* $f(\lambda)$ *be known, and let the unknown spectral density* $g_0(\lambda) \in D_W$. *Assume that the function* $f(\lambda) + g_0(\lambda)$ *satisfy the minimality condition (1.35), and the function* $h_g(f, g_0)$, *determined by formula (1.83), is bounded. The spectral density* $g_0(\lambda)$ *is the least favourable in the class* D_W *for the optimal linear extrapolation of the functional* $A\xi$, *if it is of the form*

$$g_0(\lambda) = \frac{\left|A(e^{i\lambda})f(\lambda) - C^0(e^{i\lambda})\right|}{\sum_{w=0}^{W} \psi_w cos(w\lambda)} - f(\lambda),$$

and the pair $(f(\lambda), g_0(\lambda))$ *determines a solution to optimization problem (1.78). The function, calculated by formula(1.66), is the minimax spectral characteristic of the optimal estimate of the functional* $A\xi$.

Let us consider the case of extrapolation of the functional $A\xi$ based on observations of the sequence without noise. In this case $g(\lambda) = 0$. Hence, we get the following corollaries.

Corollary 1.12. *Let a spectral density* $f_0(\lambda) \in D_0$, *and let the function* $f_0^{-1}(\lambda)$ *satisfy the minimality condition (1.71). The spectral density* $f_0(\lambda)$ *is the least* *favourable in the class D_0 for the optimal linear extrapolation of the functional $A\xi$, if it is of the form*

$$f_0(\lambda) = \alpha^{-1}\left|C^0(e^{i\lambda})\right|,$$

and the function $f_0(\lambda)$ determines a solution of the optimization problem (1.79). The minimax spectral characteristic of the optimal linear estimate of the functional $A\xi$ is determined by the formula (1.69).

Corollary 1.13. *Let a spectral density $f_0(\lambda) \in D_W$, and let the function $f_0^{-1}(\lambda)$ satisfy the minimality condition (1.71). The spectral density $f_0(\lambda)$ is the least favourable in the class D_W for the optimal linear extrapolation of the functional $A\xi$, if it is of the form*

$$f_0(\lambda) = \frac{\left|C^0(e^{i\lambda})\right|}{\sum\limits_{w=0}^{W} \psi_w cos(w\lambda)},$$

and the function $f_0(\lambda)$ determines a solution of the optimization problem (1.79). The function, determined by formula (1.69), is the minimax spectral characteristic of the optimal estimate of the functional $A\xi$.

4.4. Least Favourable Spectral Densities in the Class $D_v^u \times D_\varepsilon$

Consider the problem of minimax extrapolation of the functional $A\xi$, based on observations of the sequence $\{\xi(j)+\eta(j) : j \in \mathbb{Z}_- \backslash S\}$, under the condition that the stationary sequences $\xi(j)$ and $\eta(j)$ are not correlated, and the spectral densities $f(\lambda)$ and $g(\lambda)$ belong to the class of admissible spectral densities $D = D_v^u \times D_\varepsilon$,

$$D_v^u = \left\{ f(\lambda) \left| v(\lambda) \leq f(\lambda) \leq u(\lambda), \frac{1}{2\pi}\int_{-\pi}^{\pi} f(\lambda)d\lambda \leq P_1 \right.\right\},$$

$$D_\varepsilon = \left\{ g(\lambda) \left| g(\lambda) = (1-\varepsilon)g_1(\lambda) + \varepsilon w(\lambda), \frac{1}{2\pi}\int_{-\pi}^{\pi} g(\lambda)d\lambda \leq P_2 \right.\right\},$$

where the spectral densities $u(\lambda), v(\lambda), g_1(\lambda)$ are known and fixed, and the densities $u(\lambda)$ and $v(\lambda)$ are bounded.

Let the spectral densities $f_0(\lambda) \in D_v^u$, $g_0(\lambda) \in D_\varepsilon$ determine the bounded functions $h_f(f_0, g_0)$, $h_g(f_0, g_0)$ by formulas (1.82), (1.83).
From the condition $0 \in \partial\Delta_D(f_0, g_0)$, where $D = D_v^u \times D_\varepsilon$, we derive equations that the least favourable spectral densities should satisfy

$$\left|A(e^{i\lambda})g_0(\lambda) + C^0(e^{i\lambda})\right| = (f_0(\lambda) + g_0(\lambda))(\gamma_1(\lambda) + \gamma_2(\lambda) + \alpha_1^{-1}), \qquad (1.88)$$

$$\left|A(e^{i\lambda})f_0(\lambda) - C^0(e^{i\lambda})\right| = (f_0(\lambda) + g_0(\lambda))(\varphi(\lambda) + \alpha_2^{-1}). \qquad (1.89)$$

Here $\gamma_1 \leq 0$ and $\gamma_1 = 0$, when $f_0(\lambda) \geq v(\lambda)$; $\gamma_2 \geq 0$ and $\gamma_2 = 0$, when $f_0(\lambda) \leq u(\lambda)$; $\varphi(\lambda) \leq 0$ and $\varphi(\lambda) = 0$, when $g_0(\lambda) \geq (1-\varepsilon)g_1(\lambda)$.

Thus, the following theorem is true.

Theorem 1.21. *Suppose that spectral densities $f_0(\lambda) \in D_v^u$, $g_0(\lambda) \in D_\varepsilon$ satisfy the minimality condition (1.35). Let the functions, determined by formulas (1.82), (1.83), be bounded. The functions $f_0(\lambda)$, $g_0(\lambda)$, determined by formulas (1.88), (1.89), are the least favourable spectral densities in the class $D_v^u \times D_\varepsilon$ for the optimal linear extrapolation of the functional $A\xi$, if they determine a solution of the optimization problem (1.78). The function $h(f_0, g_0)$, calculated by formula (1.66), is the minimax spectral characteristic of the optimal estimate of the functional $A\xi$.*

Corollary 1.14. *Suppose that we know the spectral density $f(\lambda)$, and the unknown spectral density $g_0(\lambda) \in D_\varepsilon$. Let the function $f(\lambda) + g_0(\lambda)$ satisfy the minimality condition (1.35), and let the function $h_g(f, g_0)$, determined by (1.83), be bounded. The spectral density $g_0(\lambda)$ is the least favourable in the class D_ε for the optimal linear extrapolation of the functional $A\xi$, if it is of the form*

$$g_0(\lambda) = \max\{(1-\varepsilon)g_1(\lambda), f_1(\lambda)\},$$
$$f_1(\lambda) = \alpha_2\left|A(e^{i\lambda})f(\lambda) - C^0(e^{i\lambda})\right| - f(\lambda),$$

and the pair $(f(\lambda), g_0(\lambda))$ determines a solution of the optimization problem (1.78). The function $h(f, g_0)$, calculated by formula (1.66), is the minimax spectral characteristic of the optimal estimate of the functional $A\xi$.

Corollary 1.15. *Let a spectral density $g(\lambda)$ be known, and let the unknown spectral density $f_0(\lambda) \in D_v^u$. Assume that the function $f_0(\lambda) + g(\lambda)$ satisfy the* *minimality condition (1.35), and the function $h_f(f_0, g)$, determined by (1.82), is bounded. The spectral density $f_0(\lambda)$ is the least favourable density in the class D_v^u for the optimal linear extrapolation of the functional $A\xi$, if it is of the form*

$$f_0(\lambda) = \min\left\{\max\left\{\alpha_1 \left|A(e^{i\lambda})g(\lambda) + C^0(e^{i\lambda})\right| - g(\lambda), v(\lambda)\right\}, u(\lambda)\right\},$$

and the pair $(f_0(\lambda), g(\lambda))$ determines a solution of the optimization problem (1.78). The minimax spectral characteristic $h(f_0, g)$ is calculated by formula (1.66).

4.5. Least Favourable Spectral Densities in the Class $D_{\varepsilon_1}^1 \times D_{\varepsilon_2}$

Consider the problem of minimax extrapolation of the functional $A\xi$, based on observations of the sequence $\{\xi(j) + \eta(j) : j \in \mathbb{Z}_- \setminus S\}$, under the condition that stationary sequences $\xi(j)$ and $\eta(j)$ are not correlated, and the spectral densities $f(\lambda)$ and $g(\lambda)$ belong to the class of admissible spectral densities $D_{\varepsilon_1}^1 \times D_{\varepsilon_2}$,

$$D_{\varepsilon_1}^1 = \left\{ f(\lambda) \left| \frac{1}{2\pi} \int_{-\pi}^{\pi} |f(\lambda) - f_1(\lambda)| d\lambda \le \varepsilon_1 \right.\right\},$$

$$D_{\varepsilon_2} = \left\{ g(\lambda) \left| g(\lambda) = (1-\varepsilon_2) g_1(\lambda) + \varepsilon_2 w(\lambda), \frac{1}{2\pi} \int_{-\pi}^{\pi} g(\lambda) d\lambda \le P_1 \right.\right\},$$

where the spectral densities $f_1(\lambda), g_1(\lambda)$ are known and fixed.
Densities from the class $D_{\varepsilon_1}^1$ form the "ε–neighbourhood" model in the space L_1 of a given bounded spectral density $f_1(\lambda)$.
Densities from the class D_ε form the model of "ε–contamination" of a given bounded spectral density $g_1(\lambda)$.

Assume that the spectral densities $f^0(\lambda) \in D_{\varepsilon_1}^1$, $g^0(\lambda) \in D_{\varepsilon_2}$ determine the bounded functions $h_f(f_0, g_0)$, $h_g(f_0, g_0)$ by formulas (1.82), (1.83). From the condition $0 \in \partial\Delta_D(f_0, g_0)$, where $D_{\varepsilon_1}^1 \times D_{\varepsilon_2}$, we derive the following equations that the least favourable spectral densities should satisfy

$$\left|A(e^{i\lambda})g_0(\lambda) + C^0(e^{i\lambda})\right| = (f_0(\lambda) + g_0(\lambda))\Psi(\lambda)\alpha_1, \qquad (1.90)$$

$$\left|A(e^{i\lambda})f_0(\lambda) - C^0(e^{i\lambda})\right| = (f_0(\lambda) + g_0(\lambda))(\varphi(\lambda) + \alpha_2^{-1}). \tag{1.91}$$

Here $|\Psi(\lambda)| \leq 1$ and $\Psi(\lambda) = \text{sign}(f_0(\lambda) - f_1(\lambda))$, when $f_0(\lambda) \neq f_1(\lambda)$, α_1, α_2 are constants, $\varphi(\lambda) \leq 0$, and $\varphi(\lambda) = 0$, when $g_0(\lambda) \geq (1-\varepsilon_2)g_1(\lambda)$.

Equations (1.90), (1.91) together with the optimization problem (1.78) and the normalization condition

$$\frac{1}{2\pi}\int_{-\pi}^{\pi} |f(\lambda) - f_1(\lambda)|\,d\lambda = \varepsilon_1 \tag{1.92}$$

determine the least favourable densities in the class $D_{\varepsilon_1}^1 \times D_{\varepsilon_2}$.

The following theorem is true.

Theorem 1.22. *Assume that the spectral densities $f_0(\lambda) \in D_{\varepsilon_1}^1$ and $g_0(\lambda) \in D_{\varepsilon_2}$ satisfy the minimality condition (1.35), and the functions $h_f(f_0, g_0)$, $h_g(f_0, g_0)$, calculated by formulas (1.82), (1.83), are bounded. The spectral densities $f_0(\lambda)$, $g_0(\lambda)$ are the least favourable in the class $D_{\varepsilon_1}^1 \times D_{\varepsilon_2}$ for the optimal linear extrapolation of the functional $A\xi$, if they satisfy equations (1.90)–(1.92), and determine a solution of the optimization problem (1.78). The function $h(f_0, g_0)$, calculated by formula (1.66), is the minimax spectral characteristic of the optimal estimate of the functional $A\xi$.*

Corollary 1.16. *Let the spectral density $g(\lambda)$ be known, and let the unknown spectral density $f_0(\lambda) \in D_{\varepsilon_1}^1$. Suppose that the function $f_0(\lambda) + g(\lambda)$ satisfy the minimality condition (1.35), and the function $h_f(f_0, g)$, calculated by formula (1.82), is bounded. The spectral density $f_0(\lambda)$ is the least favourable in the class $D_{\varepsilon_1}^1$ for the optimal linear extrapolation of the functional $A\xi$, if it is of the form*

$$f_0(\lambda) = \max\left\{f_1(\lambda), \alpha_1\left|A(e^{i\lambda})g(\lambda) + C^0(e^{i\lambda})\right| - g(\lambda)\right\},$$

and the pair $(f_0(\lambda), g(\lambda))$ determines a solution of the optimization problem (1.78). The minimax spectral characteristic $h(f_0, g)$ is calculated by formula (1.66).

Corollary 1.17. *Suppose that we known the density $f(\lambda)$, and the unknown spectral density $g_0(\lambda) \in D_{\varepsilon_2}$. Let the function $f(\lambda) + g_0(\lambda)$ satisfy the minimality condition (1.35), and the function $h_g(f, g_0)$, determined by formula (1.83), be* *bounded. The spectral density $g_0(\lambda)$ is the least favourable in the class D_{ε_2} for the optimal linear extrapolation of the functional $A\xi$, if it is of the form*

$$g_0(\lambda) = \max\left\{(1-\varepsilon_2)g_1(\lambda), \alpha_2\left|A(e^{i\lambda})f(\lambda) - C^0(e^{i\lambda})\right| - f(\lambda)\right\},$$

and the pair $(f(\lambda), g_0(\lambda))$ determines a solution of the optimization problem (1.78). The function, calculated by formula(1.66), is the minimax spectral characteristic of the optimal estimate of the functional $A\xi$.

5. The Problem of Filtering of Stationary Sequences with Missing Observations

In this section, we deal with the problem of mean-square optimal linear estimation of the functional

$$A\xi = \sum_{j\in Z^S} a(j)\xi(-j)$$

constructed from unobserved values of a stationary stochastic sequence $\{\xi(j), j\in\mathbb{Z}\}$. Estimates are based on observations of the sequence $\xi(j)+\eta(j)$ at points $j\in\mathbb{Z}_-\backslash S$, where $\{\eta(j), j\in\mathbb{Z}\}$ is stationary correlated with $\xi(j)$ stochastic sequence,

$$S = \bigcup_{l=1}^{s}\{-(M_l+N_l),\ldots,-M_l\}, \quad Z^S = \{1,2,\ldots\}\backslash S^+,$$

$$S^+ = \bigcup_{l=1}^{s}\{M_l,\ldots,M_l+N_l\}, M_0 = 0, N_0 = 0.$$

The problem is investigated in the case of spectral certainty and in the case of spectral uncertainty. Formulas for calculating the spectral characteristic and mean square error of the optimal linear estimate of the functional are derived in the case of spectral certainty, where spectral densities of the stationary sequences $\xi(j)$ and $\eta(j)$ are exactly known. In the case of spectral uncertainty, where the spectral densities are not exactly known while some sets of admissible spectral densities are given the minimax robust method of estimation is applied. Formulas that determine the least favourable spectral densities and the minimax-robust spectral characteristics of the optimal estimates of the functional are derived for some specific classes of admissible spectral densities.

5.1. Hilbert Space Projection Method of Filtering

Let $\xi(j)$ and $\eta(j)$ be stationary correlated stochastic sequences with zero mean values, $E\xi(j)=0$, $E\eta(j)=0$, and covariance functions $R_{\xi}(k)$, $R_{\xi\eta}(k)$, $R_{\eta\xi}(k)$, $R_{\eta}(k)$, that admit the spectral representation (1.3). Let the spectral densities $f(\lambda)$, $f_{\xi\eta}(\lambda)$, $f_{\eta\xi}(\lambda)$, $g(\lambda)$ of the sequences satisfy the minimality condition (1.27).

The stationary sequences $\xi(j)$ and $\eta(j)$ admit the spectral representation (1.2) with the orthogonal stochastic measures $Z_{\xi}(d\lambda)$ and $Z_{\eta}(d\lambda)$, that correspond to the spectral measures $F(d\lambda)$ and $G(d\lambda)$ of the sequences $\xi(j)$ and $\eta(j)$, and satisfy relations (1.4).

Consider the problem of mean-square optimal linear estimation of the functional

$$A\xi = \sum_{j\in Z^S} a(j)\xi(-j)$$

based on observations of the sequence $\xi(j)+\eta(j)$ at points $j\in\mathbb{Z}_-\backslash S$, where

$$S=\bigcup_{l=1}^{s}\{-(M_l+N_l),\ldots,-M_l\},$$

$$Z^S=\{1,2,\ldots\}\backslash S^+,\quad S^+=\bigcup_{l=1}^{s}\{M_l,\ldots,M_l+N_l\}.$$

Assume that coefficients $\{a(j), j\in Z^S\}$, that determine the functional $A\xi$, satisfy conditions

$$\sum_{k\in Z^S}|a(k)|<\infty,\quad \sum_{k\in Z^S}(k+1)\,|a(k)|^2<\infty. \tag{1.93}$$

According to the first condition the functional $A\xi$ has a finite second moment. The second condition ensures that operators in the space ℓ_2, which will be introduced later, are compact.

Using the spectral representation of the stationary sequence $\xi(j)$ the functional $A\xi$ can be written in the form

$$A\xi=\int_{-\pi}^{\pi}A(e^{i\lambda})Z_{\xi}(d\lambda),$$

$$A(e^{i\lambda}) = \sum_{j \in Z^S} a(j) e^{-ij\lambda}.$$

Denote by $H^s(\xi+\eta)$ the closed linear subspace, generated by values $\{\xi(j)+\eta(j) : j \in \mathbb{Z}_- \backslash S\}$ of the stationary sequence $\xi(j)+\eta(j)$ in the Hilbert space $H = L_2(\Omega, \mathcal{F}, P)$.

Denote by $L_2^s(f+g)$ the subspace, generated by functions $\{e^{ij\lambda}, j \in \mathbb{Z}_- \backslash S\}$ in the space $L_2(f+g)$.

Suppose that $\hat{A}\xi$ is a linear estimate of the functional $A\xi$ based on the known observations of the sequence $\xi(j)+\eta(j)$. It is of the form (1.57).

According the method of orthogonal projections in the Hilbert space, the optimal estimate of the functional $A\xi$ is a projection of the element $A\xi$ on the subspace $H^s(\xi+\eta)$. This projection is determined by the following two conditions:

$$1)\hat{A}\xi \in H^s(\xi+\eta),$$
$$2)A\xi - \hat{A}\xi \perp H^s(\xi+\eta).$$

From the second condition we get that the spectral characteristic $h(e^{i\lambda})$ of the estimate $\hat{A}\xi$ is of the form

$$h(e^{i\lambda}) = \frac{A(e^{i\lambda})\left(f(\lambda)+f_{\xi\eta}(\lambda)\right) - C(e^{i\lambda})}{f_\zeta(\lambda)},$$

$$C(e^{i\lambda}) = \sum_{j \in S} c(j) e^{ij\lambda} + \sum_{j=0}^{\infty} c(j) e^{ij\lambda},$$

where

$$f_\zeta(\lambda) = f(\lambda) + f_{\xi\eta}(\lambda) + f_{\eta\xi}(\lambda) + g(\lambda), \quad T = S \cup \{0,1,\ldots\},$$

and $\{c(j), j \in T\}$ are unknown coefficients. Our aim is to find relations that determine these coefficients.

It follows from the first condition, $\hat{A}\xi \in H^s(\xi+\eta)$, that the following equation holds true

$$\frac{1}{2\pi}\int_{-\pi}^{\pi} h(e^{i\lambda}) e^{-ik\lambda} d\lambda = 0, k \in T,$$

namely

$$\frac{1}{2\pi}\int_{-\pi}^{\pi}\left(A(e^{i\lambda})\frac{f(\lambda)+f_{\xi\eta}(\lambda)}{f_{\zeta}(\lambda)}-C(e^{i\lambda})\frac{1}{f_{\zeta}(\lambda)}\right)e^{-ij\lambda}d\lambda=0,\quad j\in T.$$

The last relation allow us to find equations for determining the unknown coefficients $c(j), j\in T$. We open the brackets and write the last equalities in the form

$$\begin{aligned}&\sum_{j\in Z^S}a(j)\frac{1}{2\pi}\int_{-\pi}^{\pi}\frac{e^{-i(k+j)\lambda}(f(\lambda)+f_{\xi\eta}(\lambda))}{f_{\zeta}(\lambda)}d\lambda-\\&-\sum_{j\in S}c(j)\frac{1}{2\pi}\int_{-\pi}^{\pi}\frac{e^{-i(k-j)\lambda}}{f(\lambda)+g(\lambda)}d\lambda+\\&+\sum_{j=0}^{\infty}c(j)\frac{1}{2\pi}\int_{-\pi}^{\pi}\frac{e^{-i(k-j)\lambda}}{f_{\zeta}(\lambda)}d\lambda=0,\quad k\in T.\end{aligned}\tag{1.94}$$

To simplify these equations we introduce notations

$$\begin{aligned}b_{k-j}&=\frac{1}{2\pi}\int_{-\pi}^{\pi}e^{-i(k-j)\lambda}\frac{1}{f(\lambda)+f_{\xi\eta}(\lambda)+f_{\eta\xi}(\lambda)+g(\lambda)}d\lambda,\\r_{k-j}&=\frac{1}{2\pi}\int_{-\pi}^{\pi}e^{-i(k+j)\lambda}\frac{f(\lambda)+f_{\xi\eta}(\lambda)}{f(\lambda)+f_{\xi\eta}(\lambda)+f_{\eta\xi}(\lambda)+g(\lambda)}d\lambda,\\q_{k-j}&=\frac{1}{2\pi}\int_{-\pi}^{\pi}e^{-i(k-j)\lambda}\frac{f(\lambda)g(\lambda)-f_{\xi\eta}(\lambda)f_{\eta\xi}(\lambda)}{f(\lambda)+f_{\xi\eta}(\lambda)+f_{\eta\xi}(\lambda)+g(\lambda)}d\lambda.\end{aligned}\tag{1.95}$$

Now we can write equalities (1.94) as a system of equations

$$\sum_{j\in Z^S}r_{k,j}a(j)=\sum_{j\in S}b_{k,j}c(j)+\sum_{j=0}^{\infty}b_{k,j}c(j),\qquad k\in T.$$

Let us introduce the following notations $a(j)=0, j\in S$, $a(0)=0$ and $a(j)=0, j\in S^+$. Making use of these notations, we can write the system of equations in as follows

$$\sum_{j\in T} r_{k,j}a(j) = \sum_{j\in S} b_{k,j}c(j) + \sum_{j=0}^{\infty} b_{k,j}c(j), \qquad k\in T.$$

These equations can be presented in the form

$$\mathbf{R}\vec{\mathbf{a}} = \mathbf{B}\vec{\mathbf{c}}, \tag{1.96}$$

where the vector $\vec{\mathbf{c}}$ consists of unknown coefficients $c(j), j\in T$, and the vector $\vec{\mathbf{a}}$ is of the same dimension as vector $\vec{\mathbf{c}}$. It is of the form

$$\vec{\mathbf{a}} = (\vec{0}_0, \vec{a}_1, \vec{0}_1, \vec{a}_2, \vec{0}_2, \ldots \vec{a}_i, \vec{0}_i, \ldots, \vec{a}_s, \vec{0}_s, \vec{a}_{s+1}),$$

where $\vec{0}_0$ is a vector that consists of $|S|+1$ zeros, where $|S| = \sum_{k=1}^{s}(N_k+1)$ is a number of missing values, and vectors $\vec{0}_i$, $i=1,2,\ldots,s$, have N_i+1 zeros. The vectors

$$\begin{aligned} \vec{a}_1 &= (a(1),\ldots,a(M_1-1)), \\ \vec{a}_i &= (a(M_{i-1}+N_{i-1}+1),\ldots,a(M_i-1)), \quad i=2,\ldots,s, \\ \vec{a}_{s+1} &= (a(M_s+N_s+1),a(M_s+N_s+2),\ldots), \end{aligned}$$

are constructed from the coefficients that determine the functional $A\xi$.
By $\mathbf{B}$, $\mathbf{R}$ we denote linear operators in the space ℓ_2, which are defined by matrices B, R with elements (1.95), $B(k,j) = b_{k,j}, k, j\in T$, $R(k,j) = r_{k,j}$, $k,j\in T$. These matrices are of the form

$$B = \begin{pmatrix} B_{s,s} & B_{s,s-1} & \ldots & B_{s,1} & B_{s,n} \\ B_{s-1,s} & B_{s-1,s-1} & \ldots & B_{s-1,1} & B_{s-1,n} \\ \vdots & \vdots & \ddots & \vdots & \vdots \\ B_{1,s} & B_{1,s-1} & \ldots & B_{1,1} & B_{1,n} \\ B_{n,s} & B_{n,s-1} & \ldots & B_{n,1} & B_{n,n} \end{pmatrix}, \tag{1.97}$$

where elements in the last column and the last row are matrices with the following elements

$$B_{l,n}(k,j) = b_{k,j}, \quad l = 1,2,\dots,s,$$
$$k = -M_l - N_l, \dots, -M_l, \quad j = 0,1,2,\dots,$$
$$B_{n,m}(k,j) = b_{k,j}, \quad m = 1,2,\dots,s,$$
$$k = 0,1,2,\dots, \quad j = -M_m - N_m, \dots, -M_m,$$
$$B_{n,n}(k,j) = b_{k,j}, \quad k,j = 0,1,2,\dots,$$

and all other elements of the matrix B are matrices with elements

$$B_{l,m}(j,k) = b_{k,j}, \quad l,m = 1,2,\dots,s$$
$$k = -M_l - N_l, \dots, -M_l, \quad j = -M_m - N_m, \dots, -M_m.$$

The unknown coefficients $c(k), k \in T$, determined by (1.96), can be calculated by the formula $c(k) = (\mathbf{B}^{-1}\mathbf{R}\vec{\mathbf{a}})_k$, where $(\mathbf{B}^{-1}\mathbf{R}\vec{\mathbf{a}}_s)_k$ is the k-th element of the vector $\mathbf{B}^{-1}\mathbf{R}\vec{\mathbf{a}}$. Consequently, the spectral characteristic $h(e^{i\lambda})$ of the optimal estimate $\hat{A}\xi$ is calculated by the formula

$$h(e^{i\lambda}) = \frac{A(e^{i\lambda})\left(f(\lambda) + f_{\xi\eta}(\lambda)\right)}{f(\lambda) + f_{\xi\eta}(\lambda) + f_{\eta\xi}(\lambda) + g(\lambda)} - \\ - \frac{\sum\limits_{k \in T} (\mathbf{B}^{-1}\mathbf{R}\vec{\mathbf{a}})_k e^{ik\lambda}}{f(\lambda) + f_{\xi\eta}(\lambda) + f_{\eta\xi}(\lambda) + g(\lambda)}. \tag{1.98}$$

Let $\mathbf{Q}$ be a linear operator in the space ℓ_2, defined by the matrix Q of the form (1.97) with element (1.95), $Q(k,j) = q_{k,j}$, $k,j \in T$.

Similarly to (1.28) the mean-square error of the optimal estimate $\hat{A}\xi$ in the case of filtering can be calculated by the formula

$$\Delta(h;F) = \frac{1}{2\pi}\int_{-\pi}^{\pi} \left| \sum_{j \in Z^S} a(j) e^{ij\lambda} \right|^2 \frac{f(\lambda)g(\lambda) - f_{\xi\eta}(\lambda) f_{\eta\xi}(\lambda)}{f_\zeta(\lambda)} d\lambda + \\ + \frac{1}{2\pi}\int_{-\pi}^{\pi} \left| \sum_{k \in T} (\mathbf{B}^{-1}\mathbf{R}\vec{\mathbf{a}})_k e^{ik\lambda} \right|^2 \frac{1}{f_\zeta(\lambda)} d\lambda.$$

Making use of the introduced operators $\mathbf{B}$, $\mathbf{R}$ and $\mathbf{Q}$, the last formula can be written in the form

$$\Delta(h;F) = \langle \mathbf{R}\vec{\mathbf{a}}, \mathbf{B}^{-1}\mathbf{R}\vec{\mathbf{a}} \rangle + \langle \mathbf{Q}\vec{\mathbf{a}}, \vec{\mathbf{a}} \rangle. \tag{1.99}$$

Summarizing our reasoning, we can conclude that the following theorem is true.

Theorem 1.23. *Let $\xi(j)$ and $\eta(j)$ be stationary correlated stochastic sequences with the matrix of spectral densities*

$$F(\lambda) = \begin{pmatrix} f(\lambda) & f_{\xi\eta}(\lambda) \\ f_{\eta\xi}(\lambda) & g(\lambda) \end{pmatrix}$$

such that the minimal condition (1.27) is satisfied. The spectral characteristic $h(e^{i\lambda})$ and the value of the mean-square error $\Delta(f,g)$ of the optimal linear estimate of the functional $A\xi$, based on observations of the sequence $\xi(j)+\eta(j)$, $j \in \mathbb{Z}_-\setminus S$, can be calculated by formulas (1.98), (1.99).

Estimates Based on Observations of Uncorrelated Sequences

Consider filtering problem in the case where the stationary sequences $\xi(j)$ and $\eta(j)$ are not correlated. In this case $f_{\xi\eta}(\lambda) = f_{\eta\xi}(\lambda) = 0$ and the following corollary is true.

Corollary 1.18. *Let $\xi(j)$, $\eta(j)$ be uncorrelated stationary stochastic sequences with the spectral densities $f(\lambda)$ and $g(\lambda)$, that satisfy the minimality condition (1.35). The spectral characteristic $h(e^{i\lambda})$ and the mean-square error $\Delta(f,g)$ of the optimal linear estimate of the functional $A\xi$ based on observations of the sequence $\xi(j)+\eta(j)$, $j \in \mathbb{Z}_-\setminus S$, can be calculated by formulas*

$$h(e^{i\lambda}) = \frac{A(e^{i\lambda})f(\lambda)}{f(\lambda)+g(\lambda)} - \frac{\sum\limits_{k\in T}(\mathbf{B}^{-1}\mathbf{R}\vec{\mathbf{a}})_k e^{ik\lambda}}{f(\lambda)+g(\lambda)}, \tag{1.100}$$

$$\begin{aligned} \Delta(h;f,g) &= \frac{1}{2\pi}\int_{-\pi}^{\pi} \frac{\left|A(e^{i\lambda})g(\lambda) + \sum\limits_{k\in T}(\mathbf{B}^{-1}\mathbf{R}\vec{\mathbf{a}})_k e^{ik\lambda}\right|^2}{(f(\lambda)+g(\lambda))^2} f(\lambda)d\lambda \\ &+ \frac{1}{2\pi}\int_{-\pi}^{\pi} \frac{\left|A(e^{i\lambda})f(\lambda) - \sum\limits_{k\in T}(\mathbf{B}^{-1}\mathbf{R}\vec{\mathbf{a}})_k e^{ik\lambda}\right|^2}{(f(\lambda)+g(\lambda))^2} g(\lambda)d\lambda \\ &= \langle \mathbf{R}\vec{\mathbf{a}}, \mathbf{B}^{-1}\mathbf{R}\vec{\mathbf{a}} \rangle + \langle \mathbf{Q}\vec{\mathbf{a}}, \vec{\mathbf{a}} \rangle, \end{aligned} \tag{1.101}$$

where **B**, **R**, **Q** *are linear operators in the space* ℓ_2, *defined by matrices of the form* (1.97) *with elements* $b_{k,j}$, $r_{k,j}$, $q_{k,j}$, $k, j \in T$, *which are defined as follows*

$$b_{k,j} = \frac{1}{2\pi}\int_{-\pi}^{\pi} e^{-i(k-j)\lambda}\frac{1}{f(\lambda)+g(\lambda)}d\lambda,$$

$$r_{k,j} = \frac{1}{2\pi}\int_{-\pi}^{\pi} e^{-i(k+j)\lambda}\frac{f(\lambda)}{f(\lambda)+g(\lambda)}d\lambda, \tag{1.102}$$

$$q_{k,j} = \frac{1}{2\pi}\int_{-\pi}^{\pi} e^{-i(k-j)\lambda}\frac{f(\lambda)g(\lambda)}{f(\lambda)+g(\lambda)}d\lambda.$$

Estimates Based on Observations Missed on One Interval

Consider, next, the problem of mean-square optimal linear estimation of the functional

$$A\xi = \sum_{j\in Z^S} a(j)\xi(-j)$$

based on observations of the sequence $\xi(j)+\eta(j)$ at points $j \in \mathbb{Z}_-\setminus S$, where

$$S = \{-(M+N),\ldots,-M\},$$

$$S^+ = \{M,\ldots,M+N\},$$

$$Z^S = \{1,2,\ldots\}\setminus S^+.$$

In this case from Theorem 1.23 the following corollary follows.

Corollary 1.19. *Suppose that* $\xi(j)$ *and* $\eta(j)$ *are uncorrelated stationary stochastic sequences with the spectral densities* $f(\lambda)$ *and* $g(\lambda)$, *that satisfy the minimality condition (1.35). The spectral characteristic* $h(e^{i\lambda})$ *and the mean-square error* $\Delta(f,g)$ *of the optimal linear estimate of the functional* $A\xi$ *based on observations of the sequence* $\xi(j)+\eta(j)$, $j \in \mathbb{Z}_-\setminus S$ *can be calculated by formulas*

$$h(e^{i\lambda}) = A(e^{i\lambda})\frac{f(\lambda)}{f(\lambda)+g(\lambda)} - \frac{\sum_{k\in T}(\mathbf{B}^{-1}\mathbf{R}\vec{\mathbf{a}})_k e^{ik\lambda}}{f(\lambda)+g(\lambda)}, \tag{1.103}$$

$$\Delta(h; f, g) = \langle \mathbf{R}\vec{\mathbf{a}}, \mathbf{B}^{-1}\mathbf{R}\vec{\mathbf{a}} \rangle + \langle \mathbf{Q}\vec{\mathbf{a}}, \vec{\mathbf{a}} \rangle, \tag{1.104}$$

where **B**, **R**, **Q** *are linear operators in the space* ℓ_2, *defined by matrices* B, R, Q *with elements* (1.102) *of the form*

$$B = \begin{pmatrix} B_{s,s} & B_{s,n} \\ B_{n,s} & B_{n,n} \end{pmatrix},$$

and its components are matrices with elements

$$\begin{aligned}
B_{s,n}(k,j) &= b_{k,j}, \quad k = -M-N, \dots, -M, \quad j = 0,1,2,\dots, \\
B_{n,s}(k,j) &= b_{k,j}, \quad k = 0,1,2,\dots, \quad j = -M-N,\dots,-M, \\
B_{n,n}(k,j) &= b_{k,j}, \quad k,j = 0,1,2,\dots, \\
B_{s,s}(j,k) &= b_{k,j}, \quad k = -M-N,\dots,-M, \quad j = -M-N,\dots,-M.
\end{aligned}$$

Estimates Based on Observations with One Missed Observation

Consider the problem of mean-square optimal linear estimation of the functional

$$A\xi = \sum_{j \in Z^S} a(j)\xi(-j),$$

based on observations of the sequence $\xi(j) + \eta(j)$, in the case of one missing observation, namely at points $j \in \mathbb{Z}_- \backslash \{-s\}$, $Z^S = \{1,2,\dots\} \backslash \{s\}$.

In this case, the following corollary is true.

Corollary 1.20. *Let* $\xi(j)$ *and* $\eta(j)$ *be uncorrelated stationary stochastic sequences with the spectral densities* $f(\lambda)$ *and* $g(\lambda)$ *satisfying the minimality condition (1.35). The spectral characteristic* $h(e^{i\lambda})$ *and the mean-square error* $\Delta(f,g)$ *of the optimal linear estimate of the functional* $A\xi$ *based on observations of the sequence* $\xi(j) + \eta(j)$, $j \in \mathbb{Z}_- \backslash \{-s\}$ *can be calculated by formulas*

$$h(e^{i\lambda}) = A(e^{i\lambda}) \frac{f(\lambda)}{f(\lambda)+g(\lambda)} - \frac{\sum_{k \in T} (\mathbf{B}^{-1}\mathbf{R}\vec{\mathbf{a}})_k e^{ik\lambda}}{f(\lambda)+g(\lambda)}, \tag{1.105}$$

$$\Delta(h; f, g) = \langle \mathbf{R}\vec{\mathbf{a}}, \mathbf{B}^{-1}\mathbf{R}\vec{\mathbf{a}} \rangle + \langle \mathbf{Q}\vec{\mathbf{a}}, \vec{\mathbf{a}} \rangle, \tag{1.106}$$

where **B**, **R**, **Q** *are linear operators in the space* ℓ_2, *defined by matrices B, R, Q with elements* (1.102) *of the form*

$$B = \begin{pmatrix} b_{-s,-s} & B_{-s,n} \\ B_{n,-s} & B_{n,n} \end{pmatrix}.$$

Elements in the last column and the last row are matrices with the following elements

$$B_{-s,n}(k,j) = b_{k,j}, \quad k=-s, \quad j=0,1,2,\ldots,$$
$$B_{n,-s}(k,j) = b_{k,j}, \quad k=0,1,2,\ldots, \quad j=-s,$$
$$B_{n,n}(k,j) = b_{k,j}, \quad k,j=0,1,2,\ldots.$$

Estimation of the Functional $A_N\xi$

Consider the problem of mean-square optimal linear estimation of the functional

$$A_N\xi = \sum_{j\in Z^S\cap\{0,\ldots,N\}} a(j)\xi(-j),$$

based on observations of the sequence $\xi(j)+\eta(j)$ at points $j \in \mathbb{Z}_-\backslash S$, $S = \bigcup_{l=1}^{s}\{-(M_l+N_l),\ldots,-M_l\}$.

Each linear estimate of the functional $A_N\xi$ is of the form

$$\hat{A}_N\xi = \int_{-\pi}^{\pi} h_N(e^{i\lambda})(Z_\xi(d\lambda)+Z_\eta(d\lambda)).$$

Define the vector $\vec{\mathbf{a}}_N$ as follows: components of the vector with indices $T\cap(S\cup\{0,\ldots,N\})$ coincide with elements of the vector $\vec{\mathbf{a}}$ with corresponding indices, and components with indices $T\backslash(S\cup\{0,\ldots,N\})$ are equal to zero. The linear operators **B**, **R**, **Q** in the space ℓ_2 are defined in Theorem 1.23.

In this case the spectral characteristic $h_N(e^{i\lambda})$ and the mean-square error $\Delta(h_N; f, g)$ of the optimal linear estimate of the functional $A_N\xi$ are determined by formulas

$$h_N(e^{i\lambda}) = A_N(e^{i\lambda})\frac{f(\lambda)}{f(\lambda)+g(\lambda)} - \frac{\sum_{k\in T}(\mathbf{B}^{-1}\mathbf{R}\vec{\mathbf{a}}_N)_k e^{ik\lambda}}{f(\lambda)+g(\lambda)}, \tag{1.107}$$

$$\Delta(h_N; f,g) = \langle \mathbf{R}\vec{\mathbf{a}}_N, \mathbf{B}^{-1}\mathbf{R}\vec{\mathbf{a}}_N\rangle + \langle \mathbf{Q}\vec{\mathbf{a}}_N, \vec{\mathbf{a}}_N\rangle, \tag{1.108}$$

where

$$A_N(e^{i\lambda}) = \sum_{j\in Z^S\cap\{0,\dots,N\}} a(j)e^{-ij\lambda}.$$

The following corollary is true.

Corollary 1.21. *Suppose that $\xi(j)$, $\eta(j)$ are uncorrelated stationary sequences with the spectral densities $f(\lambda)$ and $g(\lambda)$, that satisfy the minimality condition (1.35). The spectral characteristic $h_N(e^{i\lambda})$ and the mean-square error $\Delta(h_N; f,g)$ of the optimal linear estimate of the functional $A_N\xi$ based on observations of the sequence $\xi(j)+\eta(j)$ at points $j\in\mathbb{Z}_-\setminus S$, can be calculated by formulas (1.107), (1.108).*

Example 1.7. *As an example, let us analyse the problem of mean-square optimal linear estimation of the functional*

$$A_2\xi = a(1)\xi(-1) + a(2)\xi(-2)$$

constructed from unobserved values $\xi(-1)$, $\xi(-2)$ of the sequence $\xi(j)$. Estimates are based on observations of the sequence $\xi(j)+\eta(j)$ at points $j\in\mathbb{Z}_-\setminus S$, where $S=\{-n,-n+1\}$.

Assume that $\xi(j)$ and $\eta(j)$ are uncorrelated stationary sequences with the spectral densities

$$f(\lambda) = |1-\alpha e^{-i\lambda}|^2, \quad g(\lambda) = |1-\beta e^{-i\lambda}|^2$$

respectively, where $|\alpha|<1$ and $|\beta|<1$.
The spectral density of the sequence $\xi(j)+\eta(j)$ is of the form

$$f(\lambda)+g(\lambda) = \left|1-ae^{-i\lambda}\right|^2 + \left|1-be^{-i\lambda}\right|^2 = \left|x-ye^{-i\lambda}\right|^2,$$

$$x = \frac{1}{2}\left(\pm\sqrt{(1+a)^2+(1+b)^2} \pm \sqrt{(1-a)^2+(1-b)^2}\right),$$

$$y = \frac{a+b}{x}.$$

Since $|a|<1$, $|b|<1$, we have $\left|\frac{y}{x}\right|<1$. Using the decomposition of the function $\frac{1}{1-t}$ into the power series, we can obtain the canonical factorizations of the functions

$$\frac{1}{f(\lambda)+g(\lambda)}, \quad \frac{f(\lambda)}{f(\lambda)+g(\lambda)}, \quad \frac{f(\lambda)g(\lambda)}{f(\lambda)+g(\lambda)}.$$

Namely

$$\frac{1}{f(\lambda)+g(\lambda)} = \frac{1}{\left|x-ye^{-i\lambda}\right|^2} = \left|\sum_{k=0}^{\infty}\frac{y^k}{x^{k+1}}e^{-ik\lambda}\right|^2,$$

$$\frac{f(\lambda)}{f(\lambda)+g(\lambda)} = \frac{\left|1-ae^{-i\lambda}\right|^2}{\left|x-ye^{-i\lambda}\right|^2} =$$

$$= \left|\sum_{k=0}^{\infty}\frac{y^k}{x^{k+1}}e^{-ik\lambda} - \sum_{k=0}^{\infty}\frac{ay^k}{x^{k+1}}e^{-i(k+1)\lambda}\right|^2 =$$

$$= \left|\frac{1}{x}+\sum_{k=0}^{\infty}\left(\frac{y^{k+1}}{x^{k+2}}-\frac{ay^k}{x^{k+1}}\right)e^{-i(k+1)\lambda}\right|^2,$$

$$\frac{f(\lambda)g(\lambda)}{f(\lambda)+g(\lambda)} = \frac{\left|1-ae^{-i\lambda}\right|^2\cdot\left|1-be^{-i\lambda}\right|^2}{\left|x-ye^{-i\lambda}\right|^2} =$$

$$= \frac{\left|1-(a+b)e^{-i\lambda}+abe^{-i2\lambda}\right|^2}{\left|x-ye^{-i\lambda}\right|^2} =$$

$$= \left|\sum_{k=0}^{\infty}\frac{y^k}{x^{k+1}}e^{-ik\lambda} - \sum_{k=0}^{\infty}\frac{(a+b)y^k}{x^{k+1}}e^{-i(k+1)\lambda} + \sum_{k=0}^{\infty}\frac{aby^k}{x^{k+1}}e^{-i(k+2)\lambda}\right|^2 =$$

$$= \left|\frac{1}{x}+\frac{y-ax-bx}{x^2}e^{-i\lambda}+\sum_{k=0}^{\infty}\left(\frac{y^{k+2}}{x^{k+3}}-\frac{(a+b)y^{k+1}}{x^{k+2}}+\frac{aby^k}{x^{k+1}}\right)e^{-i(k+2)\lambda}\right|^2.$$

The spectral characteristic of the optimal estimate $\hat{A}_2\xi$ of the functional is calculated by the formula

$$h_2(e^{i\lambda}) = (a(1)e^{-i\lambda}+a(2)e^{-2i\lambda})\frac{f(\lambda)}{f(\lambda)+g(\lambda)} - \frac{\sum_{k\in T}(\mathbf{B}^{-1}\mathbf{R}\vec{\mathbf{a}}_2)_k e^{ik\lambda}}{f(\lambda)+g(\lambda)},$$

where values $(\mathbf{B}^{-1}\mathbf{R}\vec{\mathbf{a}}_2)_k$, $k \in T = \{-n,-n+1\} \cup \{0,1,\ldots\}$ *are unknown and our aim is to find them.*

We construct the vector $\vec{\mathbf{a}}_N = (0,0,0,a(1),a(2),0,\ldots)$, *and operators* $\mathbf{B}$, $\mathbf{R}$, $\mathbf{Q}$ *defined by matrices B, R, Q with elements (1.102). These matrices are of the following form*

$$B = \begin{pmatrix} b_{-n,-n} & b_{-n,-n+1} & b_{-n,0} & b_{-n,1} & b_{-n,2} & \cdots \\ b_{-n+1,-n} & b_{-n+1,-n+1} & b_{-n+1,0} & b_{-n+1,1} & b_{-n+1,2} & \cdots \\ b_{0,-n} & b_{1,-n+1} & b_{0,0} & b_{0,1} & b_{0,2} & \cdots \\ b_{1,-n} & b_{1,-n+1} & b_{1,0} & b_{1,1} & b_{1,2} & \cdots \\ b_{2,-n} & b_{2,-n+1} & b_{2,0} & b_{2,1} & b_{2,2} & \cdots \\ \cdots \end{pmatrix}.$$

It is necessary to find the matrix B^{-1}, *that defines the operator* $\mathbf{B}^{-1}$. *We will find it in the following manner. We write the matrix B in the form*

$$B = \begin{pmatrix} K & L \\ M & N \end{pmatrix},$$

$$K = \begin{pmatrix} b_{-n,-n} & b_{-n,-n+1} \\ b_{-n+1,-n} & b_{-n+1,-n+1} \end{pmatrix},$$

$$L = \begin{pmatrix} b_{-n,0} & b_{-n,1} & b_{-n,2} & \cdots \\ b_{-n+1,0} & b_{-n+1,1} & b_{-n+1,2} & \cdots \end{pmatrix},$$

$$M = \begin{pmatrix} b_{0,-n} & b_{1,-n+1} \\ b_{1,-n} & b_{1,-n+1} \\ b_{2,-n} & b_{2,-n+1} \\ \vdots & \vdots \end{pmatrix}, \quad N = \begin{pmatrix} b_{0,0} & b_{0,1} & b_{0,2} & \cdots \\ b_{1,0} & b_{1,1} & b_{1,2} & \cdots \\ b_{2,0} & b_{2,1} & b_{2,2} & \cdots \\ \vdots & \vdots & \vdots \end{pmatrix}.$$

According to the Frobenius formula (see, for example, the book by F. R. Gantmacher [160] for additional details) the matrix B^{-1} *is of the form*

$$B^{-1} = \begin{pmatrix} V^{-1} & -V^{-1}LN^{-1} \\ -N^{-1}MV^{-1} & N^{-1}+N^{-1}MV^{-1}LN^{-1} \end{pmatrix},$$

where

$$V = K - LN^{-1}M.$$

We first find the matrix N^{-1}. The matrix N is similar to $B_{n,n}$ in Theorem 1.23. It consists of the Fourier coefficients of the function $(f(\lambda)+g(\lambda))^{-1}$.
The function

$$(f(\lambda)+g(\lambda))^{-1} = |x-ye^{-i\lambda]}|^{-2}$$

admits the factorization (z_p are the Fourier coefficients of the function $(f(\lambda)+g(\lambda))^{-1}$)

$$\frac{1}{f(\lambda)+g(\lambda)} = \frac{1}{|x-ye^{-i\lambda}|^2} =$$
$$= \sum_{p=-\infty}^{\infty} z_p e^{ip\lambda} = \left|\sum_{k=0}^{\infty} \psi_k e^{-ik\lambda}\right|^2 = \left|\sum_{j=0}^{\infty} \theta_j e^{-ij\lambda}\right|^{-2},$$
$$\psi_k = \frac{y^k}{x^{k+1}}, k \geq 0, \quad \theta_0 = x, \theta_1 = -y, \quad \theta_j = 0, j > 1.$$

Hence

$$z_p = \sum_{k=0}^{\infty} \psi_k \overline{\psi}_{k+p}, p \geq 0,$$

and $z_{-p} = \overline{z}_p$, $p \geq 0$. In the case, when $i \geq j$, we get

$$b_{i,j} = z_{i-j} = \sum_{l=i}^{\infty} \psi_{l-i} \overline{\psi}_{l-j},$$

and in the case, when $i < j$, we get

$$b_{i,j} = z_{i-j} = \overline{z}_{j-i} = \sum_{l=j}^{\infty} \psi_{l-j} \overline{\psi}_{l-i}.$$

Denote by Ψ and Θ linear operators in the space ℓ_2, defined by matrices with elements $\Psi_{i,j} = \psi_{i-j}$, $\Theta_{i,j} = \theta_{i-j}$, when $0 \leq j \leq i$, and $\Psi_{i,j} = 0$, $\Theta_{i,j} = 0$, when $0 \leq i < j$. Elements of the matrix N can be written as follows

$$N(i,j) = (\Psi' \overline{\Psi})_{i,j}.$$

It can be shown that the following relation is true

$$\Psi\Theta = \Theta\Psi = I.$$

Elements of the matrix N^{-1} are calculated by the formula

$$N^{-1}(i,j) = (\overline{\Theta}\Theta')_{i,j}.$$

Denote by

$$\gamma_{i,j} = N^{-1}(i,j) = \sum_{l=0}^{\min(i,j)} \overline{\theta}_{i-l}\theta_{j-l}. \tag{1.109}$$

The matrix N^{-1} is of the form

$$N^{-1} = \begin{pmatrix} \gamma_{0,0} & \gamma_{0,1} & \gamma_{0,2} & \cdots \\ \gamma_{1,0} & \gamma_{1,1} & \gamma_{1,2} & \cdots \\ \gamma_{2,0} & \gamma_{2,1} & \gamma_{2,2} & \cdots \\ \cdots & & & \end{pmatrix}.$$

We also use notations

$$(LN^{-1}M)_{k,j} = \beta_{k,j}, \quad (V^{-1})_{k,j} = \tau_{k,j},$$

$$(N^{-1}MV^{-1})_{k,j} = -\rho_{k,j}, \quad (V^{-1}LN^{-1})_{k,j} = -\sigma_{k,j},$$

$$(N^{-1} + N^{-1}MV^{-1}LN^{-1})_{k,j} = \omega_{k,j}.$$

As a result the matrix B^{-1} is of the form

$$B^{-1} = \begin{pmatrix} \tau_{-n,-n} & \tau_{-n,-n+1} & \sigma_{-n,0} & \sigma_{-n,1} & \sigma_{-n,2} & \cdots \\ \tau_{-n+1,-n} & \tau_{-n+1,-n+1} & \sigma_{-n+1,0} & \sigma_{-n+1,1} & \sigma_{-n+1,2} & \cdots \\ \rho_{0,-n} & \rho_{0,-n+1} & \omega_{0,0} & \omega_{0,1} & \omega_{0,2} & \cdots \\ \rho_{1,-n} & \rho_{1,-n+1} & \omega_{1,0} & \omega_{1,1} & \omega_{1,2} & \cdots \\ \rho_{2,-n} & \rho_{2,-n+1} & \omega_{2,0} & \omega_{2,1} & \omega_{2,2} & \cdots \\ \cdots & & & & & \end{pmatrix},$$

where elements of the matrix are given by the following formulas

$$\tau_{-n,-n} = (b_{-n+1,-n+1} - \beta_{-n+1,-n+1})/d,$$

$$\tau_{-n,-n+1} = -(b_{-n,-n+1} - \beta_{-n,-n+1})/d,$$

$$\tau_{-n+1,-n} = -(b_{-n+1,-n} - \beta_{-n+1,-n})/d,$$

$$\tau_{-n+1,-n+1} = (b_{-n,-n} - \beta_{-n,-n})/d,$$

$$d = (b_{-n,-n} - \beta_{-n,-n})(b_{-n+1,-n+1} - \beta_{-n+1,-n+1}) - (b_{-n,-n+1} - \beta_{-n,-n+1})(b_{-n+1,-n} - \beta_{-n+1,-n}),$$

$$\beta_{k,j} = \sum_{i=0}^{\infty}\left(\sum_{r=0}^{\infty} b_{k,r}\gamma_{r,i}\right) b_{i,j},$$

$$\rho_{k,j} = -\sum_{i=0}^{\infty}\gamma_{k,i}\sum_{r=-n}^{-n+1} b_{i,r}\tau_{r,j},$$

$$\sigma_{k,j} = -\sum_{i=-n}^{-n+1}\tau_{k,i}\sum_{r=0}^{\infty} b_{i,r}\gamma_{r,j},$$

$$\omega_{k,j} = \gamma_{k,j} + \sum_{i=0}^{\infty}\gamma_{k,i}\sum_{m=-n}^{-n+1}\sum_{r=-n}^{-n+1} b_{i,r}\tau_{r,m}\sum_{l=0}^{\infty} b_{m,l}\gamma_{l,j} = \gamma_{k,j} + \nu_{k,j}. \tag{1.110}$$

The unknown coefficients $c(k), k \in T$ are determined by the formulas

$$c(k) = \sum_{j\in T}(B^{-1}R)_{k,j}a(j) = a(1)(B^{-1}R)_{k,1} + a(2)(B^{-1}R)_{k,2}, \tag{1.111}$$

$$(B^{-1}R)_{k,j} = \sum_{i=-n}^{-n+1}\tau_{k,i}r_{i,j} + \sum_{i=0}^{\infty}\sigma_{k,i}r_{i,j}, \quad k = -n, -n+1,\ j = 1,2;$$

$$(B^{-1}R)_{k,j} = \sum_{i=-n}^{-n+1}\rho_{k,i}r_{i,j} + \sum_{i=0}^{\infty}\omega_{k,i}r_{i,j}, \quad k = 0,1,\dots,\ j = 1,2. \tag{1.112}$$

The spectral characteristic of the optimal estimate is calculated by the formula

$$h_2(e^{i\lambda}) = (a(1)e^{-i\lambda} + a(2)e^{-2i\lambda})\frac{f(\lambda)}{f(\lambda)+g(\lambda)} - \frac{\sum_{k\in T} c(k)e^{ik\lambda}}{f(\lambda)+g(\lambda)}.$$

The mean-square error of the estimate is as follows

$$\Delta(h_2; f, g) = \langle \mathbf{R}\vec{\mathbf{a}}_2, \mathbf{B}^{-1}\mathbf{R}\vec{\mathbf{a}}_2 \rangle + \langle \mathbf{Q}\vec{\mathbf{a}}_2, \vec{\mathbf{a}}_2 \rangle =$$

$$= \sum_{k \in T} (a(1)r_{k,1} + a(2)r_{k,2})\, c(k) + a(1)q_{1,1} + a(2)q_{1,2} + a(1)q_{2,1} + a(2)q_{2,2}.$$

The unknown coefficients $r_{k,j}$ can be found from the factorization of the function

$$\frac{f(\lambda)}{f(\lambda)+g(\lambda)} = \left| \sum_{k=0}^{\infty} \phi_k e^{-i(k+1)\lambda} \right|^2,$$

$$\phi_0 = \frac{1}{x}, \quad \phi_k = \frac{y^{k+1}}{x^{k+2}} - \frac{ay^k}{x^{k+1}}, \; k > 0.$$

Hence we have the following formula for calculating the coefficients

$$r_{i,j} = \sum_{l=\max(i,j)}^{\infty} \phi_{l-i}\overline{\phi}_{l-j}.$$

Coefficients $q_{k,j}$ are found from the factorization of the function

$$\frac{f(\lambda)g(\lambda)}{f(\lambda)+g(\lambda)} = \left| \sum_{k=0}^{\infty} \varphi_k e^{-i(k+2)\lambda} \right|^2,$$

$$\varphi_0 = \frac{1}{x}, \quad \varphi_1 = \frac{y - ax - bx}{x^2},$$

$$\varphi_k = \frac{y^{k+2}}{x^{k+3}} - \frac{(a+b)y^{k+1}}{x^{k+2}} + \frac{aby^k}{x^{k+1}}, \; k > 1.$$

As a result, we get the following formula that determines the coefficients

$$q_{i,j} = \sum_{l=\max(i,j)}^{\infty} \varphi_{l-i}\overline{\varphi}_{l-j}.$$

Example 1.8. *Consider the problem of mean-square optimal linear estimation of the functional*

$$A_2\xi = a(1)\xi(-1) + a(2)\xi(-2),$$

based on observations of the sequence $\xi(j)+\eta(j)$ *at all points* $j\in\mathbb{Z}_-$, *namely without missing observations, and compare values of the mean-square errors obtained in the case of missing observations and in the case of all observations.*

The mean-square error of the estimate in the case of all observations is calculated by the formula (see papers by M. P. Moklyachuk [55, 56] for more details)

$$\Delta_1(\tilde{h};f,g)=\langle\tilde{\mathbf{R}}\vec{\mathbf{a}},\tilde{\mathbf{B}}^{-1}\tilde{\mathbf{R}}\vec{\mathbf{a}}\rangle+\langle\tilde{\mathbf{Q}}\vec{\mathbf{a}},\vec{\mathbf{a}}\rangle,$$

where $\tilde{h}$ *is the spectral characteristic of the estimate,* $\vec{\mathbf{a}}=(0,a(1),a(2),0,0,\ldots)$, *and* $\tilde{\mathbf{B}},\tilde{\mathbf{R}},\tilde{\mathbf{Q}}$ *are linear operators in the space* ℓ_2, *defined as follows:*

$$(\tilde{\mathbf{B}})_{k,j}=b_{k,j},\quad k,j\geq 0,$$
$$(\tilde{\mathbf{R}})_{k,j}=r_{k,j},\quad k,j\geq 0,$$
$$(\tilde{\mathbf{Q}})_{k,j}=q_{k,j},\quad k,j\geq 0.$$

Here elements $b_{k,j},r_{k,j},q_{k,j}$ *are determined by formulas (1.102). Hence*

$$\Delta_1(\tilde{h};f,g)=\sum_{k=0}^{\infty}(a(1)r_{k,1}+a(2)r_{k,2})\,\tilde{c}(k)+$$
$$+a(1)q_{1,1}+a(2)q_{1,2}+a(1)q_{2,1}+a(2)q_{2,2},$$

where $\tilde{c}(k)=(\tilde{\mathbf{B}}^{-1}\tilde{\mathbf{R}}\vec{\mathbf{a}})_k,k\geq 0$.

Since the matrix defining the operator $\tilde{\mathbf{B}}$, *coincides with the matrix N from the preceding example, the elements of the matrix defining the inverse operator* $\tilde{\mathbf{B}}^{-1}$ *is calculated according to the formula (1.109). So,*

$$\tilde{c}(k)=\sum_{j=0}^{\infty}\sum_{i=0}^{\infty}\gamma_{k,i}r_{i,j}a(j),\quad k\geq 0.$$

Consider relation (1.112). Using formula (1.110), we can write relation (1.112) in the following form

$$(B^{-1}R)_{k,j}=\sum_{i=-n}^{-n+1}\rho_{k,i}r_{i,j}+\sum_{i=0}^{\infty}\gamma_{k,i}r_{i,j}+\sum_{i=0}^{\infty}\nu_{k,i}r_{i,j},\quad k,j\geq 0.$$

Consider, next, formula (1.111) for $k \geq 0$

$$
\begin{aligned}
c(k) &= \sum_{j\in T}(B^{-1}R)_{k,j}a(j) = \\
&= \sum_{j\in S}(B^{-1}R)_{k,j}a(j) + \sum_{j=0}^{\infty}(B^{-1}R)_{k,j}a(j) = \\
&= \sum_{j\in S}(B^{-1}R)_{k,j}a(j) + \sum_{j=0}^{\infty}\left(\sum_{i=-n}^{-n+1}\rho_{k,i}r_{i,j} + \right.\\
&\left. + \sum_{i=0}^{\infty}\gamma_{k,i}r_{i,j} + \sum_{i=0}^{\infty}\nu_{k,i}r_{i,j}\right)a(j) = \\
&= \tilde{c}(k) + \sum_{j\in S}(B^{-1}R)_{k,j}a(j) + \\
&+ \sum_{j=0}^{\infty}\left(\sum_{i=-n}^{-n+1}\rho_{k,i}r_{i,j} + \sum_{i=0}^{\infty}\nu_{k,i}r_{i,j}\right)a(j) = \\
&= \tilde{c}(k) + c_T(k).
\end{aligned}
$$

Hence,

$$
\begin{aligned}
\Delta(h_2; f, g) = \Delta_1(\tilde{h}; f, g) + \\
&+ \sum_{k=0}^{\infty}\left(a(1)r_{k,1} + a(2)r_{k,2}\right)c_T(k) + \\
&+ \sum_{k\in S}\left(a(1)r_{k,1} + a(2)r_{k,2}\right)c(k).
\end{aligned}
$$

As a result, we can estimate the effect of the missed observation on the value of the mean-square error.

5.2. Minimax Robust Method of Filtering

The derived in the previous section formulas for calculating the values of mean square error and the spectral characteristic of the optimal estimate $\hat{A}\xi$ of the functional $A\xi$, based on observations of the sequence $\xi(j) + \eta(j)$ at points $j \in \mathbb{Z}_- \backslash S$, $S = \bigcup_{l=1}^{s}\{-(M_l + N_l), \ldots, -M_l\}$, can be applied in the case where spectral densities $f(\lambda)$, $f_{\xi\eta}(\lambda)$, $f_{\eta\xi}(\lambda)$, $g(\lambda)$, are exactly known. In the case of spectral uncertainty, where information about the exact form of spectral densities is not complete, while it is known that densities belong to a certain class of spectral densities, the minimax robust method of estimation of functionals may be applied. According to this method, we find an estimate which minimizes the maximal values of the mean square errors for all spectral densities from a given class D of admissible spectral densities simultaneously.

Consider the problem in the case, where stationary sequences $\xi(j)$ and $\eta(j)$ are uncorrelated. For description of the minimax method of estimation we present the following definitions.

Definition 1.12. *For a given set of spectral densities $D = D_f \times D_g$, the spectral densities $f_0(\lambda) \in D_f$ and $g_0(\lambda) \in D_g$ are called the least favourable for the optimal filtering of the functional $A\xi$, if the following relation is satisfied*

$$\Delta(f_0, g_0) = \Delta(h(f_0, g_0); f_0, g_0) = \max_{(f,g)\in D_f \times D_g} \Delta(h(f,g); f, g).$$

Definition 1.13. *For a given set of spectral densities $D = D_f \times D_g$ the spectral characteristic $h^0(e^{i\lambda})$ of the optimal filtering of the functional $A\xi$ is called minimax robust, if the following relations hold true*

$$h^0(e^{i\lambda}) \in H_D = \bigcap_{(f,g)\in D} L_2^s(f+g),$$

$$\min_{h\in H_D} \max_{(f,g)\in D} \Delta(h; f, g) = \max_{(f,g)\in D} \Delta\left(h^0; f, g\right).$$

From the given definitions and derived formulas we obtain the following statement.

Lemma 1.9. *Spectral densities $f_0(\lambda) \in D_f$ and $g_0(\lambda) \in D_g$, that satisfy the minimality condition (1.35), are the least favourable spectral densities in the class $D = D_f \times D_g$ for the optimal linear filtering of the functional $A\xi$, if the Fourier coefficients of the functions*

$$(f_0(\lambda)+g_0(\lambda))^{-1},\ f_0(\lambda)(f_0(\lambda)+g_0(\lambda))^{-1},\ f_0(\lambda)g_0(\lambda)(f_0(\lambda)+g_0(\lambda))^{-1}$$

define operators B^0, R^0, Q^0, that determine a solution of the optimization problem

$$\max_{(f,g)\in D_f\times D_g} \left(\langle \mathbf{R}\vec{\mathbf{a}}, \mathbf{B}^{-1}\mathbf{R}\vec{\mathbf{a}}\rangle + \langle \mathbf{Q}\vec{\mathbf{a}}, \vec{\mathbf{a}}\rangle\right) = \langle \mathbf{R}^0\vec{\mathbf{a}}, (\mathbf{B}^0)^{-1}\mathbf{R}^0\vec{\mathbf{a}}\rangle + \langle \mathbf{Q}^0\vec{\mathbf{a}}, \vec{\mathbf{a}}\rangle. \tag{1.113}$$

The minimax spectral characteristic $h^0 = h(f_0, g_0)$ is calculated by the formula(1.100), if $h(f_0, g_0) \in H_D$.

For more detailed analysis of properties of the least favourable spectral densities and the minimax-robust spectral characteristics we observe that the least favourable spectral densities f_0, g_0 and the minimax spectral characteristic $h^0 = h(f_0, g_0)$ form a saddle point of the function $\Delta(h; f, g)$ on the set $H_D \times D$. The saddle point inequalities

$$\Delta(h; f_0, g_0) \geq \Delta\left(h^0; f_0, g_0\right) \geq \Delta\left(h^0; f, g\right)$$

$$\forall h \in H_D, \forall f \in D_f, \forall g \in D_g$$

hold true, if $h^0 = h(f_0, g_0)$ and $h(f_0, g_0) \in H_D$, where (f_0, g_0) is a solution of the constrained optimization problem

$$\sup_{(f,g) \in D_f \times D_g} \Delta(h(f_0, g_0); f, g) = \Delta(h(f_0, g_0); f_0, g_0), \tag{1.114}$$

$$\Delta(h(f_0, g_0); f, g) = \frac{1}{2\pi} \int_{-\pi}^{\pi} \frac{\left|A(e^{i\lambda}) g_0(\lambda) + C^0(e^{i\lambda})\right|^2}{(f_0(\lambda) + g_0(\lambda))^2} f(\lambda) d\lambda$$

$$+ \frac{1}{2\pi} \int_{-\pi}^{\pi} \frac{\left|A(e^{i\lambda}) f_0(\lambda) - C^0(e^{i\lambda})\right|^2}{(f_0(\lambda) + g_0(\lambda))^2} g(\lambda) d\lambda,$$

$$C^0(e^{i\lambda}) = \sum_{j \in T} ((\mathbf{B}^0)^{-1} \mathbf{R}^0 \vec{\mathbf{a}})_j e^{ij\lambda},$$

The constrained optimization problem (1.114) is equivalent to the unconstrained optimization problem (see B. N. Pshenichnyj [156])

$$\Delta_D(f, g) = -\Delta(h(f_0, g_0); f, g) + \delta((f, g) \left| D_f \times D_g \right.) \to \inf, \tag{1.115}$$

where $\delta((f, g) \left| D_f \times D_g \right.)$ is the indicator function of the set $D = D_f \times D_g$. A solution of the problem (1.115) is characterized by the condition $0 \in \partial \Delta_D(f_0, g_0)$, where $\partial \Delta_D(f_0)$ is the subdifferential of the convex functional $\Delta_D(f, g)$ at point (f_0, g_0) (see book by R. T. Rockafellar [158] for more details).

The form of the functional $\Delta(h(f_0, g_0); f, g)$ is convenient for application of the Lagrange method of indefinite multipliers. Thus the complexity of solving optimization problems (1.45) is determined by the complexity of calculating subdifferentials of the indicator functions $\delta((f,g)|D_f \times D_g)$ of sets $D_f \times D_g$.

The following statement is true.

Lemma 1.10. *Let (f_0, g_0) be a solution to the optimization problem (1.115). Spectral densities $f_0(\lambda)$ and $g_0(\lambda)$ are the least favourable in the class $D = D_f \times D_g$, and the spectral characteristic $h^0 = h(f_0, g_0)$ is the minimax spectral characteristic of the optimal linear estimate of the functional $A\xi$, if $h(f_0, g_0) \in H_D$.*

5.3. Least Favourable Spectral Densities in the Class $D_{\varepsilon_1} \times D_{\varepsilon_2}$

Consider the problem of minimax filtering of the functional $A\xi$, based on observations of the sequence $\xi(j)+\eta(j)$ at points $j \in \mathbb{Z}_- \setminus S$, under the condition that the stationary sequences $\xi(j)$ and $\eta(j)$ are not correlated, and the spectral densities $f(\lambda)$ and $g(\lambda)$ belong to the class of admissible spectral densities $D_{\varepsilon_1} \times D_{\varepsilon_2}$,

$$D_{\varepsilon_1} = \left\{ f(\lambda) | f(\lambda) = (1-\varepsilon_1)u_1(\lambda) + \varepsilon_1 u(\lambda), \frac{1}{2\pi}\int_{-\pi}^{\pi} f(\lambda)d\lambda = P_1 \right\},$$

$$D_{\varepsilon_2} = \left\{ g(\lambda) | g(\lambda) = (1-\varepsilon_2)v_1(\lambda) + \varepsilon_2 v(\lambda), \frac{1}{2\pi}\int_{-\pi}^{\pi} g(\lambda)d\lambda = P_2 \right\},$$

where spectral densities $u_1(\lambda)$, $v_1(\lambda)$ are known and fixed, and $u(\lambda)$, $v(\lambda)$ are unknown spectral densities.

These classes of admissible spectral densities form the "ε–contamination" models of spectral densities.

Suppose that $f_0(\lambda) \in D_f^0$, $g_0(\lambda) \in D_g^0$ and the functions, determined by formulas

$$h_f(f_0, g_0) = \frac{\left|A(e^{i\lambda})g_0(\lambda) + C^0(e^{i\lambda})\right|^2}{(f_0(\lambda)+g_0(\lambda))^2}, \tag{1.116}$$

$$h_g(f_0, g_0) = \frac{\left|A(e^{i\lambda})f_0(\lambda) - C^0(e^{i\lambda})\right|^2}{(f_0(\lambda)+g_0(\lambda))^2}. \tag{1.117}$$

are bounded. In this case,

$$\Delta(h(f_0,g_0);f,g) = \frac{1}{2\pi}\int_{-\pi}^{\pi} h_f(f_0,g_0)f(\lambda)d\lambda + \frac{1}{2\pi}\int_{-\pi}^{\pi} h_g(f_0,g_0)g(\lambda)d\lambda$$

is a linear and continuous functional in the space $L_1 \times L_1$, and we can apply the method of Lagrange multipliers and solve the constrained optimization problem (1.114) (see B. N. Pshenichnyj [156]).

From the condition $0 \in \partial\Delta_D(f_0,g_0)$ we get the following relations

$$\left|A(e^{i\lambda})g_0(\lambda) + C^0(e^{i\lambda})\right| = (f_0(\lambda)+g_0(\lambda))(\varphi_1(\lambda)+\alpha_1^{-1}), \tag{1.118}$$

$$\left|A(e^{i\lambda})f_0(\lambda) - C^0(e^{i\lambda})\right| = (f_0(\lambda)+g_0(\lambda))(\varphi_2(\lambda)+\alpha_2^{-1}), \tag{1.119}$$

where $\varphi_1(\lambda) \leq 0$ and $\varphi_1(\lambda) = 0$, when $f_0(\lambda) \geq (1-\varepsilon_1)u_1(\lambda)$, $\varphi_2(\lambda) \leq 0$ and $\varphi_2(\lambda) = 0$, when $g_0(\lambda) \geq (1-\varepsilon_2)v_1(\lambda)$.

The following statements are true.

Theorem 1.24. *Suppose that spectral densities $f_0(\lambda) \in D_{\varepsilon_1}$, $g_0(\lambda) \in D_{\varepsilon_2}$, and the minimality condition (1.35) is satisfied. Let the functions, defined by formulas (1.116), (1.117), be bounded. The densities $f_0(\lambda)$ and $g_0(\lambda)$, determined by equations (1.118), (1.119), are the least favourable spectral densities in the class $D_{\varepsilon_1} \times D_{\varepsilon_2}$ for the optimal linear filtering of the functional $A\xi$, if they determine a solution of the optimization problem (1.113). The function $h(f_0,g_0)$, calculated by formula(1.100), is the minimax spectral characteristic of the estimate of the functional $A\xi$.*

Corollary 1.22. *Assume that we know the spectral density $f(\lambda)$, and the unknown spectral density $g_0(\lambda) \in D_{\varepsilon_2}$. Let the function $f(\lambda)+g_0(\lambda)$ satisfy the minimality condition (1.35), and let the function $h_g(f,g_0)$, defined by (1.117), be bounded. The spectral density $g_0(\lambda)$ is the least favourable in the class D_{ε_2} for the optimal filtering of the functional $A\xi$, if it is of the form*

$$g_0(\lambda) = \max\left\{(1-\varepsilon_2)g_1(\lambda), \alpha_2\left|A(e^{i\lambda})f(\lambda) - C^0(e^{i\lambda})\right| - f(\lambda)\right\},$$

and the pair $f(\lambda), g_0(\lambda)$ determines a solution of the optimization problem (1.113). The function $h(f,g_0)$, determined by (1.100), is the minimax spectral characteristic of the optimal estimate $\hat{A}\xi$.

5.4. Least Favourable Spectral Densities in the Class $D_{\varepsilon}^{1}\times D_{v}^{u}$

Consider the problem of minimax filtering of the functional $A\xi$, based on observations of the sequence $\xi(j)+\eta(j)$ at points $j\in\mathbb{Z}_-\setminus S$, under the condition that the stationary sequences $\xi(j)$ and $\eta(j)$ are not correlated, and the spectral densities $f(\lambda)$ and $g(\lambda)$ belong to the class of admissible spectral densities $D=D_{\varepsilon}^{1}\times D_{v}^{u}$,

$$D_{\varepsilon}^{1}=\left\{f(\lambda)\left|\frac{1}{2\pi}\int_{-\pi}^{\pi}|f(\lambda)-f_1(\lambda)|d\lambda\leq\varepsilon\right.\right\},$$

$$D_{v}^{u}=\left\{g(\lambda)\left|v(\lambda)\leq g(\lambda)\leq u(\lambda),\frac{1}{2\pi}\int_{-\pi}^{\pi}g(\lambda)d\lambda\leq P\right.\right\},$$

where spectral densities $u(\lambda),v(\lambda),f_1(\lambda)$ are known and fixed, and densities $u(\lambda)$ and $v(\lambda)$ are bounded.

Densities from the class of admissible spectral densities D_{ε}^{1} form the "ε–neighbourhood" model in the space L_1 of a given bounded spectral density $f_1(\lambda)$.

Densities from the class of admissible spectral densities D_{v}^{u} form the "stripe" model of spectral densities.

Let the spectral densities $f^0(\lambda)\in D_{\varepsilon}^{1}$, $g^0(\lambda)\in D_{v}^{u}$ determine the bounded functions $h_f(f_0,g_0),h_g(f_0,g_0)$ by formulas (1.116), (1.117). From the condition $0\in\partial\Delta_D(f_0,g_0)$ we get the following equations that the least favourable spectral densities should satisfy

$$\left|A(e^{i\lambda})g_0(\lambda)+C^0(e^{i\lambda})\right|=(f_0(\lambda)+g_0(\lambda))\Psi(\lambda)\alpha_1, \tag{1.120}$$

$$\left|A(e^{i\lambda})f_0(\lambda)-C^0(e^{i\lambda})\right|=(f_0(\lambda)+g_0(\lambda))(\gamma_1(\lambda)+\gamma_2(\lambda)+\alpha_2^{-1}). \tag{1.121}$$

Here $|\Psi(\lambda)|\leq 1$ and $\Psi(\lambda)=\operatorname{sign}(f_0(\lambda)-f_1(\lambda))$, when $f_0(\lambda)\neq f_1(\lambda)$, α_1,α_2 are constant values, $\gamma_1\leq 0$ and $\gamma_1=0$, when $g_0(\lambda)\geq v(\lambda)$, $\gamma_2\geq 0$ and $\gamma_2=0$, when $g_0(\lambda)\leq u(\lambda)$.

Equations (1.120), (1.121) together with the optimization problem (1.113) and the normalization condition

$$\frac{1}{2\pi}\int_{-\pi}^{\pi}|f(\lambda)-f_1(\lambda)|d\lambda=\varepsilon, \tag{1.122}$$

determine the least favourable spectral densities in the class $D=D_\varepsilon^1\times D_v^u$.

The following statements are true.

Theorem 1.25. *Suppose that the spectral densities $f_0(\lambda)\in D_\varepsilon^1$, $g_0(\lambda)\in D_v^u$ satisfy the minimality condition (1.35). Let the functions, defined by formulas (1.116), (1.117), be bounded. The spectral densities $f_0(\lambda)$, $g_0(\lambda)$, determined form equations (1.120)–(1.122), are the least favourable in the class $D_\varepsilon^1\times D_v^u$ for the optimal linear filtering of the functional $A\xi$, if they determine a solution of the optimization problem (1.113). The function $h(f_0,g_0)$, calculated by formula(1.100), is the minimax spectral characteristic of the optimal estimate of the functional $A\xi$.*

Corollary 1.23. *Assume that the unknown spectral density $f_0(\lambda)\in D_\varepsilon^1$, the density $g(\lambda)$ is known and fixed, and the minimality condition (1.35) is satisfied. Let the function $h_f(f_0,g)$, defined by (1.116), be bounded. The spectral density $f_0(\lambda)$ is the least favourable in the class D_ε^1 for the optimal linear filtering of the functional $A\xi$, if it is of the form*

$$f_0(\lambda)=\max\left\{f_1(\lambda),\alpha_1\left|A(e^{i\lambda})g(\lambda)+C^0(e^{i\lambda})\right|-g(\lambda)\right\},$$

and the pair $f_0(\lambda),g(\lambda)$ defines a solution of the optimization problem (1.113). The function $h(f_0,g)$, calculated by the formula(1.100), is the minimax spectral characteristic of the optimal estimate of the functional $A\xi$.

5.5. Least Favourable Spectral Densities in the Class $D_\varepsilon^2\times D_v^u$

Consider the problem of minimax filtering of the functional $A\xi$, based on observations of the sequence $\xi(j)+\eta(j)$ at points $j\in\mathbb{Z}_-\backslash S$, under the condition that the stationary sequences $\xi(j)$ and $\eta(j)$ are not correlated, and the spectral densities $f(\lambda)$ and $g(\lambda)$ belong to the class of admissible spectral densities

$D = D_\varepsilon^2 \times D_v^u$,

$$D_\varepsilon^2 = \left\{ f(\lambda) \left| \frac{1}{2\pi} \int_{-\pi}^{\pi} |f(\lambda) - f_1(\lambda)|^2 d\lambda \leq \varepsilon \right. \right\},$$

$$D_v^u = \left\{ g(\lambda) \left| v(\lambda) \leq g(\lambda) \leq u(\lambda), \frac{1}{2\pi} \int_{-\pi}^{\pi} g(\lambda) d\lambda \leq P \right. \right\},$$

where spectral densities $u(\lambda)$, $v(\lambda)$, $f_1(\lambda)$ are known and fixed, and spectral densities $u(\lambda)$ and $v(\lambda)$ are bounded.

Densities from the class of admissible spectral densities D_ε^2 form the "ε–neighbourhood" model in the space L_2 of a given bounded spectral density $f_1(\lambda)$.

Densities from the class of admissible spectral densities D_v^u form the "stripe" model of spectral densities.

Let the spectral densities $f_0(\lambda) \in D_\varepsilon^2$, $g_0(\lambda) \in D_v^u$ be such that the functions $h_f(f_0, g_0)$, $h_g(f_0, g_0)$, defined by formulas (1.116), (1.117), are bounded.

From the condition $0 \in \partial\Delta_D(f^0, g^0)$, in the case of $D = D_\varepsilon^2 \times D_v^u$, we can get equations

$$\left| A(e^{i\lambda}) g_0(\lambda) + C^0(e^{i\lambda}) \right|^2 = (f_0(\lambda) + g_0(\lambda))^2 (f_0(\lambda) - f_1(\lambda)) \alpha_1, \qquad (1.123)$$

$$\left| A(e^{i\lambda}) f_0(\lambda) - C^0(e^{i\lambda}) \right| = (f_0(\lambda) + g_0(\lambda))(\gamma_1(\lambda) + \gamma_2(\lambda) + \alpha_2^{-1}), \qquad (1.124)$$

where α_1, α_2 are constants, $\gamma_1 \leq 0$, and $\gamma_1 = 0$, when $g_0(\lambda) \geq v(\lambda)$; $\gamma_2 \geq 0$ and $\gamma_2 = 0$, when $g_0(\lambda) \leq u(\lambda)$.

Equations (1.123), (1.124) together with the optimization problem (1.113) and the normalization condition

$$\frac{1}{2\pi} \int_{-\pi}^{\pi} |f(\lambda) - f_1(\lambda)|^2 d\lambda = \varepsilon \qquad (1.125)$$

determine the least favourable spectral densities in the class $D = D_\varepsilon^2 \times D_v^u$.

The following statements hold true.

Theorem 1.26. *Let the spectral densities $f_0(\lambda) \in D_\varepsilon^2$, $g_0(\lambda) \in D_v^u$ satisfy the minimality condition (1.35), and let the functions, calculated by formulas (1.116), (1.117), be bounded. The spectral densities $f_0(\lambda)$, $g_0(\lambda)$ are the least favourable in the class $D_\varepsilon^2 \times D_v^u$ for the optimal linear filtering of the functional $A\xi$, if they satisfy equations (1.123)–(1.125), and determine a solution of the optimization problem (1.113). The function $h(f_0, g_0)$, calculated by formula(1.100), is the minimax spectral characteristic of the optimal estimate of the functional $A\xi$.*

Corollary 1.24. *Suppose that the unknown spectral density $f_0(\lambda) \in D_\varepsilon^2$, the spectral density $g(\lambda)$ is known, and the minimality condition (1.35) is satisfied. Let the function $h_f(f_0, g)$, defined by formula (1.116), be bounded. The spectral density $f_0(\lambda)$ is the least favourable in the class D_ε^2 for the optimal filtering of the functional $A\xi$, if the following relation is true*

$$\left|A(e^{i\lambda})g(\lambda) + C^0(e^{i\lambda})\right|^2 = (f_0(\lambda) + g(\lambda))^2 (f_0(\lambda) - f_1(\lambda)),$$

and the pair $f_0(\lambda), g(\lambda)$ determines a solution of the optimization problem (1.113). The function $h(f_0, g)$, calculated by formula(1.100), is the minimax spectral characteristic of the optimal estimate of the functional $A\xi$.

Conclusion

In this chapter, we present results of the investigation of the interpolation, extrapolation, and filtering problems for linear functionals constructed from unobserved values of stationary stochastic sequences. We develop methods of finding the optimal estimates of the functionals in the case where some observations are missing. We investigate the problems under the condition of spectral certainty, where spectral densities of the involved stationary sequences are exactly known.

In the case of interpolation problem, we derived formulas for calculating values of the mean-square error and the spectral characteristic of the optimal estimate of the functional

$$A_s\xi = \sum_{l=0}^{s-1} \sum_{j=M_l}^{M_l+N_{l+1}} a(j)\xi(j)$$

constructed from unobserved values of a stationary stochastic sequence $\xi(j)$. Estimates are based on observations of the sequence $\xi(j)+\eta(j)$ at points $j \in \mathbb{Z}\backslash S$, where $\eta(j)$ is a stationary correlated with $\xi(j)$ stochastic sequence,

$$S=\bigcup_{l=0}^{s-1}\{M_l, \ldots, M_l+N_{l+1}\}, \quad M_l=\sum_{k=0}^{l}(N_k+K_k), N_0=K_0=0.$$

In the case of extrapolation problem, formulas for calculating values of the mean-square error and the spectral characteristic of the optimal estimate are derived for the functional

$$A\xi=\sum_{j\in S}a(j)\xi(j)+\sum_{j=0}^{\infty}a(j)\xi(j)$$

with depends on unobserved values of a stationary stochastic sequence $\xi(j)$. Estimates are constructed from observations of the sequence $\xi(j)+\eta(j)$ at points

$$j\in\mathbb{Z}_-\backslash S=\{\ldots,-2,-1\}\backslash S,$$

$$S=\bigcup_{l=1}^{s}\{-M_l-N_l,-M_l-N_l+1,\ldots,-M_l\}.$$

In this case, $\eta(j)$ is a stationary correlated with $\xi(j)$ stochastic sequence.

In the case of filtering problem, formulas for calculating values of the mean-square error and the spectral characteristic of the optimal estimate are established for the functional

$$A\xi=\sum_{j\in Z^S}a(j)\xi(-j)$$

from unobserved values of a stationary stochastic sequence $\xi(j)$. Estimates are constructed from observations of the sequence $\xi(j)+\eta(j)$ at points $j\in\mathbb{Z}_-\backslash S$, where $\eta(j)$ is a stationary correlated with $\xi(j)$ stationary sequence,

$$S=\bigcup_{l=1}^{s}\{-(M_l+N_l),\ldots,-M_l\},$$

$$Z^S=\{1,2,\ldots\}\backslash S^+,$$

$$S^+ = \bigcup_{l=1}^{s} \{M_l, \ldots, M_l + N_l\}.$$

Next, we investigate the problems in the case of spectral uncertainty, where the spectral densities of the involved stationary sequences are not exactly known, while some special sets of admissible spectral densities are given. In this case, we apply the minimax robust method of estimation. This method allows us to find estimates that minimize the maximal values of the mean-square errors of the optimal estimates of the functionals for all spectral densities from a given class of admissible spectral densities. We derive formulas that determine the least favourable spectral densities and the minimax spectral characteristics for some special classes of admissible spectral densities.

We came to the conclusion that the least favourable spectral densities are solutions of the optimization problem

$$\Delta_D(f,g) = -\Delta(h(f_0,g_0);f,g) + \delta((f,g)\left|D_f \times D_g\right.) \to \inf,$$

where $\delta((f,g)\left|D_f \times D_g\right.)$ is the indicator function of the set $D = D_f \times D_g$ of admissible spectral densities. A solution of this problem is characterized by the condition $0 \in \partial\Delta_D(f_0,g_0)$. This is the necessary and sufficient condition for the pair (f^0,g^0) to belong to the set of minimums of the convex functional $\Delta_D(f,g)$. By $\partial\Delta_D(f_0,g_0)$ we denote the subdifferential of the functional $\Delta_D(f,g)$ at point (f_0,g_0).

The form of the functional $\Delta(h(f_0,g_0);f,g)$ is convenient for applying of the Lagrange method of indefinite multipliers to find a solution of the optimization problem. Thus, the complexity of solving the optimization problem is determined by the complexity of calculating of the subdifferentials of the indicator functions $\delta((f,g)|D_f \times D_g)$ of the sets $D_f \times D_g$. Using the method of Lagrange multipliers and the form of subdifferentials of the indicator functions we describe relations that determine the least favourable spectral densities in some special classes of admissible spectral densities. These classes are the following. The classes D_0 and D_W of spectral densities with restrictions on moments, the classes of densities D_ε which form the "ε–contaminated" models of spectral densities, the classes of densities $D_{1\delta}$ which form the "δ–neighbourhood" models in the space L_1 of a fixed bounded spectral density, the classes of densities $D_{2\delta}$ which describe the "δ–neighbourhood" models in the space L_2 of a fixed bounded spectral density, the classes of densities D_V^U which describe the "strip" models of spectral densities.

Chapter 2

Estimation of Functionals from Stationary Stochastic Processes with Missing Observations

In this chapter, we propose results of the investigation of the problem of mean square optimal estimation of linear functionals constructed from unobserved values of a continuous time stationary stochastic process. Estimates are based on observations of the process with a noise. We develop methods of finding the optimal estimates of the functionals in the case where some observations are missing. Formulas are derived for computing values of the spectral characteristics and the mean-square errors of the optimal linear estimates of the functionals in the case of spectral certainty, where the spectral densities of the processes are exactly known. Formulas are proposed that determine the least favourable spectral densities and the minimax robust spectral characteristics of the optimal estimates of the functionals for concrete classes of spectral densities under the condition that spectral densities of the processes are not known exactly while some classes of admissible spectral densities are given.

1. Stationary Processes. Spectral Representation

In this section, we present a brief review of the main results of the spectral theory of stationary stochastic processes (see books by I. I. Gikhman and A. V. Skorokhod [153], A. M. Yaglom [8, 9] for more details).

Definition 2.1. *A complex-valued mean-square continuous stochastic process* $\xi(t), t \in \mathbb{R}$ *with finite first and second moments,* $E|\xi(t)| < \infty$, $E|\xi(t)|^2 < \infty, t \in \mathbb{R}$ *is called stationary (in the wide sense), if*

$$E\xi(t) = m, \quad E(\xi(t)\overline{\xi(s)}) = R(t-s), \quad t,s \in \mathbb{R}$$

for constant m *and complex-valued function* $R(t), t \in \mathbb{R}$.
Without loss of generality, we will consider stationary processes with $m = 0$ *throughout the work.*

Theorem 2.1. *A function* $R(t), t \in \mathbb{R}$ *is the covariance function of a mean-square continuous stationary process (in the wide sense) if there exists a finite Lebesgue-Stieltjes measure* $F = F(B)$, $B \in \mathcal{B}(\mathbb{R})$ *such that the following spectral representation is true*

$$R(t) = \int_{-\infty}^{\infty} e^{i\lambda t} F(d\lambda), \quad t \in \mathbb{R}. \tag{2.1}$$

The finite measure $F = F(B)$, $B \in \mathcal{B}(\mathbb{R})$ is called the spectral measure of the process $\xi(t)$, and the corresponding distribution function $F(\lambda) = F([-\infty, \lambda))$ is called the spectral function of the process $\xi(t)$. In the case, where the spectral measure of the process is absolutely continuous, the spectral function $F(\lambda)$ has a spectral density $f(\lambda)$, and the covariance function itself can be represented in the form

$$R(t) = \frac{1}{2\pi}\int_{-\infty}^{\infty} e^{i\lambda t} f(\lambda) d\lambda. \tag{2.2}$$

Theorem 2.2. *Let* $\xi(t)$ *be a mean-square continuous stationary stochastic process. There exists an orthogonal stochastic measure* $Z = Z(\Delta)$, $\Delta \in \mathcal{B}(\mathbb{R}))$ *such that the process* $\xi(t)$ *admits the spectral representation*

$$\xi(t) = \int_{-\infty}^{\infty} e^{i\lambda t} Z(d\lambda), \quad t \in \mathbb{R}, \tag{2.3}$$

and $E\left(Z(\Delta_1)\overline{Z(\Delta_2)}\right) = F(\Delta_1 \cap \Delta_2)$.

Suppose that $\xi(t)$ and $\eta(t)$ are uncorrelated stationary stochastic processes with the spectral functions $F(\lambda)$ and $G(\lambda)$ that have spectral densities $f(\lambda)$ and $g(\lambda)$ respectively. For the corresponding orthogonal stochastic measures $Z_\xi(\Delta_1)$, $Z_\eta(\Delta_2)$ and the spectral densities $f(\lambda)$, $g(\lambda)$ the following relations is true

$$\begin{aligned} EZ_\xi(\Delta_1)\overline{Z_\xi(\Delta_2)} = F(\Delta_1 \cap \Delta_2) = \frac{1}{2\pi}\int_{\Delta_1 \cap \Delta_2} f(\lambda)d\lambda, \\ EZ_\eta(\Delta_1)\overline{Z_\eta}(\Delta_2) = G(\Delta_1 \cap \Delta_2) = \frac{1}{2\pi}\int_{\Delta_1 \cap \Delta_2} g(\lambda)d\lambda. \end{aligned} \tag{2.4}$$

Denote by $H = L_2(\Omega, \mathcal{F}, P)$ the Hilbert space of random variables ξ with zero first moment, $E\xi = 0$, finite second moment, $E|\xi|^2 < \infty$, and the inner product $(\xi, \eta) = E\xi\overline{\eta}$, and denote by $H(\xi)$ the closed linear subspace of the Hilbert space $H = L_2(\Omega, \mathcal{F}, P)$ generated by values of a stationary stochastic process $\{\xi(t) : t \in \mathbb{R}\}$. Denote by $L_2(f)$ the Hilbert space of complex-valued functions on $(-\infty, \infty)$, which are square integrable with respect to a measure with the density $f(\lambda)$.

Theorem 2.3. *Assume that $\xi(t)$ is a stationary stochastic process which admits the spectral representation* (2.3). *For each random variable $\eta \in H(\xi)$ there exists a function $\varphi \in L_2(f)$, such that the following representation is true*

$$\eta = \int_{-\infty}^{\infty} \varphi(\lambda)Z(d\lambda).$$

Spectral properties of stationary stochastic processes described above are used in the following sections to investigate the problems of estimating the unobserved values of stationary processes.

2. The Problem of Interpolation of Stationary Processes with Missing Observations

In this section, the problem of mean-square optimal linear estimation of the functional

$$A_s\xi = \sum_{l=1}^{s} \int_{-M_l-N_l}^{-M_l} a(t)\xi(t)dt$$

constructed from unobserved values of a stationary stochastic process $\xi(t)$ is considered. Estimates are based on observations of the process $\xi(t)+\eta(t)$ at points $t \in \mathbb{R}\backslash S$,

$$S = \bigcup_{l=1}^{s}[-M_l - N_l, \dots, -M_l],$$

$$M_l = \sum_{k=0}^{l}(N_k + K_k), N_0 = 0, K_0 = 0.$$

The stationary process $\eta(t)$ is uncorrelated with the process $\xi(t)$.

The problem is investigated in the case of spectral certainty as well as in the case of spectral uncertainty. Formulas for calculating the spectral characteristic and the mean square error of the optimal linear estimate of the functional are derived in the case of spectral certainty, where the spectral densities of the stationary processes $\xi(t)$ and $\eta(t)$ are exactly known. In the case of spectral uncertainty, where the spectral densities of the stationary processes are not exactly known while some sets of admissible spectral densities are given, formulas that determine the least favourable spectral densities and the minimax-robust spectral characteristics of the optimal estimates of the functional are derived for some specific classes of admissible spectral densities.

2.1. The Classical Method of Linear Interpolation

Consider uncorrelated mean-square continuous stationary stochastic processes $\{\xi(t), t \in \mathbb{R}\}$ and $\{\eta(t), t \in \mathbb{R}\}$ with zero expectations, $E\xi(t) = 0$, $E\eta(t) = 0$, and covariance functions $R_\xi(t) = E\xi(t+s)\overline{\xi(s)}$, $R_\eta(t) = E\eta(t+s)\overline{\eta(s)}$ of the form (2.2) with the spectral densities $f(\lambda)$ and $g(\lambda)$ satisfying the minimality condition

$$\int_{-\infty}^{\infty} \frac{|\gamma(\lambda)|^2}{f(\lambda)+g(\lambda)} d\lambda < \infty. \tag{2.5}$$

Here

$$\gamma(\lambda) = \sum_{l=1}^{s} \int_{-M_l-N_l}^{-M_l} \alpha(t) e^{it\lambda} dt$$

is a nonzero function of the exponential type. This condition is sufficient to make impossible the error-free estimation of the process (see the book by Yu. A. Rozanov [10] for more details).

The processes $\xi(t)$ and $\eta(t)$ admit the spectral representation (2.3) with the orthogonal stochastic measures $Z_\xi(d\lambda)$ and $Z_\eta(d\lambda)$ such that relations (2.4) are satisfied.

Consider the problem of mean-square optimal linear estimation of the functional

$$A_s\xi = \sum_{l=1}^{s} \int_{-M_l-N_l}^{-M_l} a(t)\xi(t)dt$$

constructed from unobserved values of the stationary process $\xi(t)$. Estimates are based on observations of the process $\xi(t)+\eta(t)$ at points $t \in \mathbb{R} \backslash S$, where

$$S = \bigcup_{l=1}^{s} [-M_l - N_l, \ldots, -M_l].$$

Making use of the spectral representation (2.3) of the stationary process $\xi(t)$, we can represent the functional $A_s\xi$ in the form

$$A_s\xi = \int_{-\infty}^{\infty} A_s(e^{i\lambda}) Z_\xi(d\lambda),$$

$$A_s(e^{i\lambda}) = \sum_{l=1}^{s} \int_{-M_l-N_l}^{-M_l} a(t) e^{it\lambda} dt.$$

Denote by $\hat{A}_s\xi$ a linear estimate of the functional $A_s\xi$ based on observations of the process $\xi(t)+\eta(t)$. It is of the form

$$\hat{A}_s\xi = \int_{-\infty}^{\infty} h(e^{i\lambda})(Z_\xi(d\lambda)+Z_\eta(d\lambda)),$$

where $h(e^{i\lambda})$ is the spectral characteristic of the estimate of the functional.

The mean-square error $\Delta(h; f, g)$ of the estimate $\hat{A}_s\xi$ is calculated by the formula

$$\begin{aligned}\Delta(f,g) &= E\left|A_s\xi - \hat{A}_s\xi\right|^2 \\ &= \frac{1}{2\pi}\int_{-\infty}^{\infty}\left|A_s(e^{i\lambda}) - h(e^{i\lambda})\right|^2 f(\lambda)d\lambda + \frac{1}{2\pi}\int_{-\infty}^{\infty}\left|h(e^{i\lambda})\right|^2 g(\lambda)d\lambda.\end{aligned}$$

We first investigate the problem in the case, where spectral densities of stationary processes $\xi(t)$ and $\eta(t)$ are exactly known. In this case we use the method of orthogonal projections in the Hilbert space to find the optimal estimate $\hat{A}_s\xi$ of the functional $A_s\xi$.

Denote by $H^s(\xi+\eta)$ the closed linear subspace of the Hilbert space $H = L_2(\Omega, \mathcal{F}, P)$, generated by values $\{\xi(t)+\eta(t) : t \in \mathbb{R}\backslash S\}$. Denote by $L_2^s(f+g)$ the subspace, generated by functions $\{e^{it\lambda} : t \in \mathbb{R}\backslash S\}$ in the space $L_2(f+g)$.

According to the method of orthogonal projections, the optimal estimate of the functional $A_s\xi$ is a projection of the element $A_s\xi$ of the space $H = L_2(\Omega, \mathcal{F}, P)$ onto the subspace $H^s(\xi+\eta)$. It is determined by the following conditions:

$$\begin{aligned}&1)\, \hat{A}_s\xi \in H^s(\xi+\eta),\\ &2)\, A_s\xi - \hat{A}_s\xi \perp H^s(\xi+\eta).\end{aligned}$$

It follows from the second condition, that the spectral characteristic $h(e^{i\lambda})$ of the optimal linear estimate $\hat{A}_s\xi$ for all $t \in \mathbb{R}\backslash S$ satisfies the relation

$$E\left[\left(A_s\xi - \hat{A}_s\xi\right)\left(\overline{\xi(t)} + \overline{\eta(t)}\right)\right] =$$
$$= \frac{1}{2\pi}\int_{-\infty}^{\infty}\left(A_s(e^{i\lambda}) - h(e^{i\lambda})\right)e^{-it\lambda}f(\lambda)d\lambda$$
$$- \frac{1}{2\pi}\int_{-\infty}^{\infty} h(e^{i\lambda})e^{-it\lambda}g(\lambda)d\lambda = 0.$$

This relation can be written in the form

$$\frac{1}{2\pi}\int_{-\infty}^{\infty}\left[A_s(e^{i\lambda})f(\lambda) - h(e^{i\lambda})(f(\lambda)+g(\lambda))\right]e^{-it\lambda}d\lambda = 0,\ t \in \mathbb{R}\backslash S. \tag{2.6}$$

Denote the function in the brackets by

$$C_s(e^{i\lambda}) = A_s(e^{i\lambda})f(\lambda) - h(e^{i\lambda})(f(\lambda)+g(\lambda)), \lambda \in \mathbb{R},$$

and its Fourier transform denote by

$$\mathbf{c}(t) = \frac{1}{2\pi}\int_{-\infty}^{\infty} C_s(e^{i\lambda})e^{-it\lambda}d\lambda, \quad t \in \mathbb{R}.$$

It follows from the relation (2.6) that the function $\mathbf{c}(t)$ is nonzero only on the set S. Hence, the function $C_s(e^{i\lambda})$ is of the form

$$C_s(e^{i\lambda}) = \sum_{l=1}^{s}\int_{-M_l-N_l}^{-M_l} \mathbf{c}(t)e^{it\lambda}dt,$$

and the spectral characteristic of the estimate $\hat{A}_s\xi$ has the form

$$h(e^{i\lambda}) = A_s(e^{i\lambda})\frac{f(\lambda)}{f(\lambda)+g(\lambda)} - \frac{C_s(e^{i\lambda})}{f(\lambda)+g(\lambda)}.$$

From the first condition, $\hat{A}_s\xi \in H^s(\xi+\eta)$, which determines the optimal estimate of the functional $A_s\xi$, we get that for some function $v(t) \in L_2^s(f+g)$

$$h(e^{i\lambda}) = \frac{1}{2\pi}\int\limits_{\mathbb{R}\setminus S} v(t)e^{it\lambda}d\lambda,$$

and hence, we have

$$\int\limits_{-\infty}^{\infty}\left(A_s(e^{i\lambda})\frac{f(\lambda)}{f(\lambda)+g(\lambda)} - \frac{C_s(e^{i\lambda})}{f(\lambda)+g(\lambda)}\right)e^{-it\lambda}d\lambda = 0,\ t \in S. \tag{2.7}$$

We write the last relation in terms of linear operators in the space $L_2(S)$. To do this, we introduce the following linear operators:

$$(\mathbf{B}_s\mathbf{x})(t) = \frac{1}{2\pi}\sum_{l=1}^{s}\int\limits_{-M_l-N_l}^{-M_l}\mathbf{x}(u)\int\limits_{-\infty}^{\infty}e^{i\lambda(u-t)}\frac{1}{f(\lambda)+g(\lambda)}d\lambda du,$$

$$(\mathbf{R}_s\mathbf{x})(t) = \frac{1}{2\pi}\sum_{l=1}^{s}\int\limits_{-M_l-N_l}^{-M_l}\mathbf{x}(u)\int\limits_{-\infty}^{\infty}e^{i\lambda(u-t)}\frac{f(\lambda)}{f(\lambda)+g(\lambda)}d\lambda du,$$

$$(\mathbf{Q}_s\mathbf{x})(t) = \frac{1}{2\pi}\sum_{l=1}^{s}\int\limits_{-M_l-N_l}^{-M_l}\mathbf{x}(u)\int\limits_{-\infty}^{\infty}e^{i\lambda(u-t)}\frac{f(\lambda)g(\lambda)}{f(\lambda)+g(\lambda)}d\lambda du,$$

$$\mathbf{x}(t) \in L_2(S), \quad t \in S.$$

Now equality (2.7) can be written in the form

$$(\mathbf{R}_s\mathbf{a})(t) = (\mathbf{B}_s\mathbf{c})(t), \quad t \in S. \tag{2.8}$$

Assume that the operator $\mathbf{B}_s$ has an inverse operator. In this case the function $\mathbf{c}(t)$ can be calculated by the formula

$$\mathbf{c}(t) = (\mathbf{B}_s^{-1}\mathbf{R}_s\mathbf{a})(t), \quad t \in S.$$

Consequently, the spectral characteristic $h(e^{i\lambda})$ of the estimate $\hat{A}_s\xi$ is calculated by the formula

$$h(e^{i\lambda}) = A_s(e^{i\lambda})\frac{f(\lambda)}{f(\lambda)+g(\lambda)} - \frac{C_s(e^{i\lambda})}{f(\lambda)+g(\lambda)}, \tag{2.9}$$

where

$$C_s(e^{i\lambda}) = \sum_{l=1}^{s} \int_{-M_l-N_l}^{-M_l} (\mathbf{B}_s^{-1}\mathbf{R}_s\mathbf{a})(t)e^{it\lambda}dt, \quad t \in S.$$

The mean-square error $\hat{A}_s\xi$ is calculated by the formula

$$\begin{aligned}\Delta(h;f,g) =& E\left|A_s\xi - \hat{A}_s\xi\right|^2 = \\ =& \frac{1}{2\pi}\int_{-\infty}^{\infty} \frac{\left|A_s(e^{i\lambda})g(\lambda) + C_s(e^{i\lambda})\right|^2}{(f(\lambda)+g(\lambda))^2} f(\lambda)d\lambda + \\ &+ \frac{1}{2\pi}\int_{-\infty}^{\infty} \frac{\left|A_s(e^{i\lambda})f(\lambda) - C_s(e^{i\lambda})\right|^2}{(f(\lambda)+g(\lambda))^2} g(\lambda)d\lambda = \\ =& \langle \mathbf{R}_s\vec{\mathbf{a}}, \mathbf{B}_s^{-1}\mathbf{R}_s\vec{\mathbf{a}}\rangle + \langle \mathbf{Q}_s\vec{\mathbf{a}}, \vec{\mathbf{a}}\rangle, \end{aligned} \tag{2.10}$$

where

$$\langle A, C\rangle = \sum_{l=1}^{s} \int_{-M_l-N_l}^{-M_l} A(t)\overline{C(t)}dt$$

is the inner product in the space $L_2(S)$.

The following statement is true.

Theorem 2.4. *Let $\xi(t)$ and $\eta(t)$ be uncorrelated mean-square continuous stationary stochastic processes with the spectral densities $f(\lambda)$, $g(\lambda)$ satisfying the minimality condition (2.5). The spectral characteristic $h(e^{i\lambda})$ and the mean-square error $\Delta(h;f,g)$ of the optimal linear estimate of the functional $A_s\xi$ based on observations of the process $\xi(t)+\eta(t)$, $t \in \mathbb{R}\backslash S$ can be calculated by formulas (2.9), (2.10).*

Estimates Based on Observations without Noise

Consider the case, where the stationary process $\xi(t)$ is observed without noise. In this case the spectral characteristic of the estimate $\hat{A}_s\xi$ has the form

$$h(e^{i\lambda}) = A_s(e^{i\lambda}) - C_s(e^{i\lambda})f^{-1}(\lambda),$$

where

$$C_s(e^{i\lambda}) = \sum_{l=1}^{s} \int_{-M_l-N_l}^{-M_l} \mathbf{c}(t) e^{it\lambda} dt,$$

and the relation (2.8) is of the form

$$\mathbf{a}(t) = (\mathbf{B}_s \mathbf{c})(t), \quad t \in S.$$

If the operator $\mathbf{B}_s$ is invertible, then the unknown function $\mathbf{c}(t)$ is calculated by the formula

$$\mathbf{c}(t) = (\mathbf{B}_s^{-1} \mathbf{a})(t), \quad t \in S,$$

and the formula for the spectral characteristic of the estimate $\hat{A}_s \xi$ has the form

$$h(e^{i\lambda}) = A_s(e^{i\lambda}) - C_s(e^{i\lambda}) f^{-1}(\lambda), \tag{2.11}$$

where

$$C_s(e^{i\lambda}) = \sum_{l=1}^{s} \int_{-M_l-N_l}^{-M_l} (\mathbf{B}_s^{-1} \mathbf{a})(t) e^{it\lambda} dt, \quad t \in S.$$

The mean-square error of the estimate of the functional is of the form

$$\Delta(h; f) = \langle \mathbf{B}_s^{-1} \mathbf{a}, \mathbf{a} \rangle. \tag{2.12}$$

The obtained result we formulate as a theorem.

Theorem 2.5. *Let $\xi(t)$ be a stationary stochastic process with the spectral density $f(\lambda)$ which satisfies the minimality condition (2.5) with $g(\lambda) = 0$. The spectral characteristic $h(e^{i\lambda})$ and the mean-square error $\Delta(h, f)$ of the optimal linear estimate $\hat{A}_s \xi$ of the functional $A_s \xi$ based on observations of the process $\xi(t)$ at points $t \in \mathbb{R} \backslash S$, $S = \bigcup_{l=1}^{s} [-M_l - N_l, \ldots, -M_l]$, can be calculated by formulas (2.11), (2.12).*

2.2. Minimax Robust Method of Interpolation

Theorems 2.4 and 2.5 can be applied to calculate the spectral characteristic and the mean-square error of the optimal linear estimate $\hat{A}_s\xi$ of the functional $A_s\xi$ only in cases where the spectral densities $f(\lambda)$ and $g(\lambda)$ of the stationary processes $\xi(t)$ and $\eta(t)$ are exactly known. In the case of spectral uncertainty, where information about the exact form of densities is not complete, while it is known that densities belong to a certain class of admissible spectral densities, the minimax robust method of estimation of functionals may be applied. According to this method, we find an estimate which minimizes the maximal values of the mean square errors for all spectral densities from a given class D of admissible spectral densities simultaneously. To describe the minimax method of estimation, we propose the following definitions.

Definition 2.2. *For a given set of spectral densities $D = D_f \times D_g$ the spectral densities $f_0(\lambda) \in D_f$, $g_0(\lambda) \in D_g$ are called the least favourable for the optimal interpolation of the functional $A_s\xi$, if the following relation is true*

$$\Delta(f_0, g_0) = \Delta(h(f_0, g_0); f_0, g_0) = \max_{(f,g)\in D_f\times D_g} \Delta(h(f,g); f, g).$$

Definition 2.3. *For a given set of spectral densities $D = D_f \times D_g$ the spectral characteristic $h^0(e^{i\lambda})$ of the optimal interpolation of the functional $A_s\xi$ is called minimax robust, if the following relations are satisfied*

$$h^0(e^{i\lambda}) \in H_D = \bigcap_{(f,g)\in D} L_2^s(f+g),$$

$$\min_{h\in H_D} \max_{(f,g)\in D} \Delta(h; f, g) = \max_{(f,g)\in D} \Delta\left(h^0; f, g\right).$$

Using the obtained results and presented definitions, the following lemma can be verified.

Lemma 2.1. *The spectral densities $f_0(\lambda) \in D_f$ and $g_0(\lambda) \in D_g$, satisfying the minimality condition (2.5), are the least favourable in the class $D = D_f \times D_g$ for the optimal linear interpolation of the functional $A_s\xi$, if the Fourier transforms of functions*

$$(f_0(\lambda) + g_0(\lambda))^{-1}, \quad f_0(\lambda)(f_0(\lambda) + g_0(\lambda))^{-1},$$

$$f_0(\lambda)g_0(\lambda)(f_0(\lambda)+g_0(\lambda))^{-1}$$

define operators $\mathbf{B}_s^0, \mathbf{R}_s^0, \mathbf{Q}_s^0$, *which determine a solution of the optimization problem*

$$\max_{(f,g)\in D_f\times D_g} \left(\langle \mathbf{R}_s\vec{\mathbf{a}}, \mathbf{B}_s^{-1}\mathbf{R}_s\vec{\mathbf{a}}\rangle + \langle \mathbf{Q}_s\vec{\mathbf{a}}, \vec{\mathbf{a}}\rangle\right) = \\ = \langle \mathbf{R}_s^0\vec{\mathbf{a}}, (\mathbf{B}_s^0)^{-1}\mathbf{R}_s^0\vec{\mathbf{a}}\rangle + \langle \mathbf{Q}_s^0\vec{\mathbf{a}}, \vec{\mathbf{a}}\rangle. \tag{2.13}$$

The minimax spectral characteristic $h^0 = h(f_0, g_0)$ *is calculated by formula (2.9) if* $h(f_0, g_0) \in H_D$.

Corollary 2.1. *The spectral density* $f_0(\lambda) \in D_f$, *satisfying the minimality condition (2.5) with* $g(\lambda) = 0$, *is the least favourable in* D_f *for the optimal interpolation of the functional* $A_s\xi$ *from observations of the process* $\xi(t)$ *at points* $t \in \mathbb{R}\backslash S$, *if the Fourier transform of the function* $f_0^{-1}(\lambda)$ *defines the operator* $\mathbf{B}_s^0$, *which determine a solution of the optimization problem*

$$\max_{f\in D_f} \langle \mathbf{B}_s^{-1}\vec{\mathbf{a}}, \vec{\mathbf{a}}\rangle = \langle (\mathbf{B}_s^0)^{-1}\vec{\mathbf{a}}, \vec{\mathbf{a}}\rangle. \tag{2.14}$$

The minimax spectral characteristic $h^0 = h(f_0)$ *is calculated by formula (2.11), if* $h(f_0) \in H_{D_f}$.

For more detailed analysis of properties of the least favourable spectral densities and the minimax-robust spectral characteristics we observe that the least favourable spectral densities $f_0(\lambda)$, $g_0(\lambda)$ and the minimax spectral characteristic $h^0 = h(f_0, g_0)$ form a saddle point of the function $\Delta(h; f, g)$ on the set $H_D \times D$. The saddle point inequalities

$$\Delta\left(h^0; f, g\right) \leq \Delta\left(h^0; f_0, g_0\right) \leq \Delta\left(h; f_0, g_0\right), \; \forall h \in H_D, \forall f \in D_f, \forall g \in D_g$$

are true, if $h^0 = h(f_0, g_0)$ and $h(f_0, g_0) \in H_D$, where (f_0, g_0) is a solution of the constrained optimization problem

$$\sup_{(f,g)\in D_f\times D_g} \Delta\left(h(f_0, g_0); f, g\right) = \Delta\left(h(f_0, g_0); f_0, g_0\right), \tag{2.15}$$

$$\Delta(h(f_0,g_0);f,g)=\frac{1}{2\pi}\int\limits_{-\infty}^{\infty}\frac{\left|A_s(e^{i\lambda})g_0(\lambda)+C_s^0(e^{i\lambda})\right|^2}{(f_0(\lambda)+g_0(\lambda))^2}f(\lambda)d\lambda$$

$$+\frac{1}{2\pi}\int\limits_{-\infty}^{\infty}\frac{\left|A_s(e^{i\lambda})f_0(\lambda)-C_s^0(e^{i\lambda})\right|^2}{(f_0(\lambda)+g_0(\lambda))^2}g(\lambda)d\lambda,$$

$$C_s^0(e^{i\lambda})=\sum_{l=1}^{s}\int\limits_{-M_l-N_l}^{-M_l}((\mathbf{B}_s^0)^{-1}\mathbf{R}_s^0\mathbf{a})(t)e^{it\lambda}dt,\quad t\in S.$$

The constrained optimization problem (2.15) is equivalent to the unconstrained optimization problem (see B. N. Pshenichnyj [156])

$$\Delta_D(f,g)=-\Delta(h(f_0,g_0);f,g)+\delta((f,g)\left|D_f\times D_g\right.)\to\inf,\qquad(2.16)$$

where $\delta((f,g)\left|D_f\times D_g\right.)$ is the indicator function of the set $D=D_f\times D_g$. A solution of the problem (2.16) is characterized by the condition $0\in\partial\Delta_D(f_0,g_0)$, where $\partial\Delta_D(f_0,g_0)$ is the subdifferential of the convex functional $\Delta_D(f,g)$ at point (f_0,g_0).

The form of the functional $\Delta(h(f_0,g_0);f,g)$ is convenient for application of the Lagrange method of indefinite multipliers for finding solutions of the optimization problem. Thus, the complexity of solving the optimization problems (2.16) is determined by the complexity of calculating subdifferentials of the indicator functions of sets $D_f\times D_g$. Making use of the method of Lagrange multipliers and the form of subdifferentials of the indicator functions, we describe relations that determine the least favourable spectral densities in some special classes of spectral densities.

Lemma 2.2. *Let (f_0,g_0) be a solution to the optimization problem (2.16). The spectral densities $f_0(\lambda)$ and $g_0(\lambda)$ are the least favourable in the class $D=D_f\times D_g$, and $h^0=h(f_0,g_0)$ is the minimax spectral characteristic of the optimal linear estimate of the functional $A_s\xi$, if $h(f_0,g_0)\in H_D$.*

2.3. Least Favourable Spectral Densities in the Class $D_0\times D_\varepsilon$

Consider the problem of minimax interpolation of the functional $A_s\xi$ based on observations of the process $\xi(t)+\eta(t)$ at points $t\in\mathbb{R}\backslash S$, under the condition that the stationary processes $\xi(t)$ and $\eta(t)$ are not correlated, and the spectral densities $f(\lambda)$ and $g(\lambda)$ of the processes belong to the class of admissible spectral densities $D = D_0 \times D_\varepsilon$,

$$D_0 = \left\{ f(\lambda) \left| \frac{1}{2\pi} \int_{-\infty}^{\infty} f(\lambda) d\lambda \leq P_1 \right. \right\},$$

$$D_\varepsilon = \left\{ g(\lambda) \left| g(\lambda) = (1-\varepsilon) g_1(\lambda) + \varepsilon w(\lambda), \frac{1}{2\pi} \int_{-\infty}^{\infty} g(\lambda) d\lambda \leq P_2 \right. \right\},$$

where the spectral density $g_1(\lambda)$ is known and fixed, and $w(\lambda)$ is an unknown spectral density.

Suppose that $f_0(\lambda) \in D_0$, $g_0(\lambda) \in D_\varepsilon$. Introduce the following notations

$$h_f(f_0, g_0) = \frac{\left| A_s(e^{i\lambda}) g_0(\lambda) + C_s^0(e^{i\lambda}) \right|^2}{(f_0(\lambda) + g_0(\lambda))^2}, \tag{2.17}$$

$$h_g(f_0, g_0) = \frac{\left| A_s(e^{i\lambda}) f_0(\lambda) - C_s^0(e^{i\lambda}) \right|^2}{(f_0(\lambda) + g_0(\lambda))^2}. \tag{2.18}$$

The functional $\Delta(h(f_0, g_0); f, g)$ can be represented in the form

$$\Delta(h(f_0, g_0); f, g) = \frac{1}{2\pi} \int_{-\infty}^{\infty} h_f(f_0, g_0) f(\lambda) d\lambda + \frac{1}{2\pi} \int_{-\infty}^{\infty} h_g(f_0, g_0) g(\lambda) d\lambda.$$

Assume that the functions $h_f(f_0, g_0)$, $h_g(f_0, g_0)$, defined by formulas (2.17), (2.18) are bounded. In this case $\Delta(h(f_0, g_0); f, g)$ is a continuous linear functional in the space $L_1 \times L_1$. Therefore, the following relation is true (see B. N. Pshenichnyj [156])

$$\partial \Delta_{D_0 \times D_\varepsilon}(f_0, g_0) = -\partial \Delta(h(f_0, g_0); f_0, g_0) + \partial \delta((f_0, g_0) | D_0 \times D_\varepsilon).$$

From the condition $0 \in \partial \Delta_{D_0 \times D_\varepsilon}(f_0, g_0)$ we determine the following equations that the least favourable densities should satisfy:

$$\left| A_s(e^{i\lambda}) g_0(\lambda) + C_s^0(e^{i\lambda}) \right| = \alpha_1 (f_0(\lambda) + g_0(\lambda)), \tag{2.19}$$

$$\left|A_s(e^{i\lambda})f_0(\lambda) - C_s^0(e^{i\lambda})\right| = (f_0(\lambda) + g_0(\lambda))(\varphi(\lambda) + \alpha_2^{-1}), \tag{2.20}$$

where $\varphi(\lambda) \leq 0$, and $\varphi(\lambda) = 0$, when $g_0(\lambda) \geq (1-\varepsilon)g_1(\lambda)$, and constants $\alpha_1 \geq 0$, $\alpha_2 \geq 0$. Note that $\alpha_1 \neq 0$, if

$$\frac{1}{2\pi}\int_{-\infty}^{\infty} f_0(\lambda)d\lambda = P_1.$$

Thus, the following theorem is true.

Theorem 2.6. *Assume that the spectral densities $f_0(\lambda) \in D_0$, $g_0(\lambda) \in D_\varepsilon$ satisfy the minimality condition (2.5), and the functions, defined by formulas (2.17), (2.18), are bounded. The spectral densities $f_0(\lambda)$ and $g_0(\lambda)$ are the least favourable in the class $D_0 \times D_\varepsilon$ for the optimal interpolation of the functional $A_s\xi$, if they satisfy the system of equations (2.19), (2.20) and determine a solution of the optimization problem (2.13). The function, calculated by formula (2.9), is the minimax spectral characteristic of the optimal estimation of the functional $A_s\xi$.*

Consider the problem of minimax interpolation of the functional $A_s\xi$ under the condition that the spectral densities $f(\lambda)$ and $g(\lambda)$ of the processes belong to the class of admissible spectral densities $D = D_0^f \times D_0^g$,

$$D_0^f = \left\{ f(\lambda) \left| \frac{1}{2\pi}\int_{-\infty}^{\infty} f(\lambda)d\lambda \leq P_1 \right. \right\},$$

$$D_0^g = \left\{ g(\lambda) \left| \frac{1}{2\pi}\int_{-\infty}^{\infty} g(\lambda)d\lambda \leq P_2 \right. \right\}.$$

Suppose that densities $f_0(\lambda) \in D_0^f$, $g_0(\lambda) \in D_0^g$ and the functions, defined by formulas (2.17), (2.18) are bounded. From the condition $0 \in \partial\Delta_{D_0 \times D_\varepsilon}(f_0, g_0)$ we determine the following equations that determine the least favourable densities:

$$\left|A_s(e^{i\lambda})g_0(\lambda) + C_s^0(e^{i\lambda})\right| = \alpha_1(f_0(\lambda) + g_0(\lambda)), \tag{2.21}$$

$$\left|A_s(e^{i\lambda})f_0(\lambda) - C_s^0(e^{i\lambda})\right| = \alpha_2(f_0(\lambda) + g_0(\lambda)), \tag{2.22}$$

where $\alpha_1 \geq 0$, $\alpha_2 \geq 0$, moreover $\alpha_1 \neq 0$, if

$$\frac{1}{2\pi}\int_{-\infty}^{\infty} f_0(\lambda)d\lambda = P_1$$

and $\alpha_2 \neq 0$, if

$$\frac{1}{2\pi}\int_{-\infty}^{\infty} g_0(\lambda)d\lambda = P_2.$$

The following theorem is true.

Theorem 2.7. *Let the spectral densities $f_0(\lambda) \in D_0^f$, $g_0(\lambda) \in D_0^g$ satisfy the minimality condition (2.5). Assume and the functions, defined by formulas (2.17), (2.18), are bounded. The spectral densities $f_0(\lambda)$, $g_0(\lambda)$ are the least favourable in the class $D_0^f \times D_0^g$ for the optimal interpolation of the functional $A_s\xi$, if they satisfy the system of equations (2.21), (2.22) and the pair $(f_0(\lambda), g_0(\lambda))$ determines a solution of the optimization problem (2.13). The function, calculated by formula (2.9), is the minimax spectral characteristic of the optimal estimate of the functional $A_s\xi$.*

Corollary 2.2. *Suppose that the spectral density $g(\lambda)$ is known, and the unknown spectral density $f_0(\lambda) \in D_0$. Let the function $f_0(\lambda) + g(\lambda)$ satisfy the minimality condition (2.5), and let the function $h_f(f_0, g)$, defined by formula (2.17), be bounded. The spectral density $f_0(\lambda)$ is the least favourable in the class D_0 for the optimal interpolation of the functional $A_s\xi$, if it is of the form*

$$f_0(\lambda) = \max\{0, \alpha_1^{-1}\left|A_s(e^{i\lambda})g(\lambda) + C_s^0(e^{i\lambda})\right| - g(\lambda)\},$$

and the pair $(f_0(\lambda), g_0(\lambda))$ determines a solution of the optimization problem (2.13). The function, calculated by the formula (2.9), is the minimax spectral characteristic of the optimal estimate of the functional $A_s\xi$.

Corollary 2.3. *Let the spectral density $f_0(\lambda) \in D_0$ and satisfy the minimality condition (2.5) with $g(\lambda) = 0$. Suppose that the function, defined by (2.11), is bounded. The spectral density $f_0(\lambda)$ is the least favourable in the class D_0 for* *the optimal interpolation of the functional $A_s\xi$, based on observations of the process $\xi(t)$ at points $t \in \mathbb{R}\backslash S$, if it is of the form*

$$f_0(\lambda) = \alpha_1 \left|C_s^0(e^{i\lambda})\right|,$$

and $f_0(\lambda)$ determines a solution of the optimization problem (2.14). The function, calculated by formula (2.11), is the minimax spectral characteristic of the optimal estimate of the functional $A_s\xi$.

2.4. Least Favourable Spectral Densities in the Class $D_{\varepsilon_1} \times D^2_{\varepsilon_2}$

Consider the problem of minimax interpolation of the functional $A_s\xi$, based on observations of the process $\xi(t) + \eta(t)$ at points $t \in \mathbb{R}\backslash S$, under the condition that the stationary processes $\xi(t)$ and $\eta(t)$ are not correlated, and the spectral densities $f(\lambda)$ and $g(\lambda)$ of the processes belong to the class of admissible spectral $D_{\varepsilon_1} \times D^2_{\varepsilon_2}$,

$$D_{\varepsilon_1} = \left\{ f(\lambda) \left| f(\lambda) = (1-\varepsilon_1)f_1(\lambda) + \varepsilon_1 w(\lambda), \frac{1}{2\pi}\int_{-\infty}^{\infty} f(\lambda)d\lambda \leq P_1 \right.\right\},$$

$$D^2_{\varepsilon_2} = \left\{ g(\lambda) \left| \frac{1}{2\pi}\int_{-\infty}^{\infty} |g(\lambda) - g_1(\lambda)|^2 d\lambda \leq \varepsilon_2 \right.\right\},$$

where the spectral densities $f_1(\lambda), g_1(\lambda)$ are known and fixed, and $w(\lambda)$ is an unknown spectral density.

Densities from the class of densities D_{ε_1} form the "ε–contamination" model of a given bounded spectral density $f_1(\lambda)$.

Densities from the class of densities $D^2_{\varepsilon_2}$ form the model of "δ–neighbourhood" in the space L_2 of a given bounded spectral density $g_1(\lambda)$.

Suppose that densities $f^0(\lambda) \in D_{\varepsilon_1}$, $g^0(\lambda) \in D^2_{\varepsilon_2}$ define the bounded functions $h_f(f_0, g_0)$, $h_g(f_0, g_0)$ by formulas (2.17), (2.18). From the condition $0 \in \partial\Delta_D(f_0, g_0)$ we get the following equations that determine the least favourable spectral densities

$$\left|A_s(e^{i\lambda})g_0(\lambda) + C_s^0(e^{i\lambda})\right| = (f_0(\lambda) + g_0(\lambda))(\varphi(\lambda) + \alpha_1^{-1}), \qquad (2.23)$$

$$\left|A_s(e^{i\lambda})f_0(\lambda)-C_s^0(e^{i\lambda})\right|^2=(f_0(\lambda)+g_0(\lambda))^2(g_0(\lambda)-g_1(\lambda))\alpha_2, \tag{2.24}$$

where α_1,α_2 are constants, $\varphi(\lambda)\leq 0$, and $\varphi(\lambda)=0$, when $f_0(\lambda)\geq(1-\varepsilon_1)f_1(\lambda)$.

Equations (2.23), (2.24) together with the optimization problem (2.13) and condition

$$\frac{1}{2\pi}\int_{-\infty}^{\infty}|g(\lambda)-g_1(\lambda)|^2\,d\lambda=\varepsilon_2, \tag{2.25}$$

determine the least favourable density in the class $D_{\varepsilon_1}\times D_{\varepsilon_2}^2$.

The following statements are true.

Theorem 2.8. *Suppose that the spectral densities $f_0(\lambda)\in D_{\varepsilon_1}$, $g_0(\lambda)\in D_{\varepsilon_2}^2$ satisfy the minimality condition (2.5), and functions, calculated by formulas (2.17), (2.18) are bounded. The spectral densities $f_0(\lambda)$, $g_0(\lambda)$ are the least favourable in the class $D_{\varepsilon_1}\times D_{\varepsilon_2}^2$ for the optimal interpolation of the functional $A_s\xi$, if they satisfy equations (2.23), (2.24), condition (2.25), and determine a solution of the optimization problem (2.13). The function, calculated by formula (2.9), is the minimax spectral characteristic of the optimal estimate of the functional $A_s\xi$.*

Corollary 2.4. *Assume that we know the density $g(\lambda)$, and the unknown spectral density $f_0(\lambda)\in D_{\varepsilon_1}$. Let the function $f_0(\lambda)+g(\lambda)$ satisfy the minimality condition (2.5), and let the function $h_f(f_0,g)$, defined by (2.17), be bounded. The spectral density $f_0(\lambda)$ is the least favourable in the class D_{ε_1} for the optimal interpolation of the functional $A_s\xi$, if it satisfies equation*

$$f_0(\lambda)=\max\left\{(1-\varepsilon_1)f_1(\lambda),\alpha_1\left|A_s(e^{i\lambda})g(\lambda)+C_s^0(e^{i\lambda})\right|-g(\lambda)\right\},$$

and the pair $(f_0(\lambda),g(\lambda))$ determines a solution of the optimization problem (2.13). The function $h(f_0,g)$, calculated by formula (2.9), is the minimax spectral characteristic of the optimal estimate of the functional $A_s\xi$.

Corollary 2.5. *Suppose that the spectral density $f(\lambda)$ is known, and the unknown spectral density $g_0(\lambda)\in D_{\varepsilon_2}^2$. Let the function $f(\lambda)+g_0(\lambda)$ satisfy the minimality condition (2.5), and let the function $h_g(f,g_0)$, defined by formula (2.18), be bounded. The spectral density $g_0(\lambda)$ is the least favourable in the class $D_{\varepsilon_2}^2$ for the optimal interpolation of the functional $A_s\xi$, if it satisfies the relation*

$$\left|A_s(e^{i\lambda})f(\lambda)-C_s^0(e^{i\lambda})\right|^2=(f(\lambda)+g_0(\lambda))^2(g_0(\lambda)-g_1(\lambda)),$$

and the pair $(f(\lambda), g_0(\lambda))$ determines a solution of the optimization problem (2.13). The minimax spectral characteristic $h(f, g_0)$ of the optimal estimate of the functional $A_s\xi$ is calculated by the formula (2.9).

Corollary 2.6. *Let the spectral density $f_0(\lambda) \in D^2_{\varepsilon_2}$ satisfy the minimality condition (2.5) with $g(\lambda) = 0$, and the function, defined by (2.11), be bounded. The spectral density $f_0(\lambda)$ is the least favourable in the class $D^2_{\varepsilon_2}$ for the optimal interpolation of the functional $A_s\xi$, based on observations of the process $\xi(t)$ at points $t \in \mathbb{R}\backslash S$, if the following relation is satisfied*

$$\left|C^0_s(e^{i\lambda})\right|^2 = f_0^2(\lambda)(f_0(\lambda) - g_1(\lambda))\alpha_1,$$

and $f_0(\lambda)$ determines a solution of the optimization problem (2.14). The function, calculated by the formula (2.11), is the minimax spectral characteristic of the optimal estimate of the functional $A_s\xi$.

3. The Problem of Extrapolation of Stationary Processes Based on Observations with Missing Values

In this section, we propose results of the investigation of the problem of mean-square optimal linear estimation of the functional

$$A\xi = \int_0^\infty a(t)\xi(t)dt$$

constructed from unobserved values of a stationary stochastic process $\xi(t)$. Estimates are based on observations of the process $\xi(t) + \eta(t)$ at points $t \in \mathbb{R}^-\backslash S$,

$$S = \bigcup_{l=1}^{s} [-M_l - N_l, \ldots, -M_l],$$

$$M_l = \sum_{k=0}^{l} (N_k + K_k), \; N_0 = 0, K_0 = 0.$$

The stationary stochastic process $\eta(t)$ is uncorrelated with the process $\xi(t)$. Formulas for calculating the spectral characteristic and the mean square error of the optimal linear estimate of the functional are derived in the case of spectral certainty, where the spectral densities of the stationary processes $\xi(t)$ and $\eta(t)$ are exactly known. In the case of spectral uncertainty, where the spectral densities are not exactly known while some sets of admissible spectral densities are given, we apply the minimax method of estimation. Formulas that determine the least favourable spectral densities and the minimax-robust spectral characteristics of the optimal estimates of the functional are derived for some specific classes of admissible spectral densities.

3.1. Hilbert Space Projection Method of Extrapolation

Let $\xi(t)$ and $\eta(t)$ be uncorrelated stationary stochastic processes with zero mathematical expectations, $E\xi(t) = 0$, $E\eta(t) = 0$, and let the covariance functions $R_\xi(t) = E\xi(t+s)\overline{\xi(s)}$ and $R_\eta(t) = E\eta(t+s)\overline{\eta(s)}$ of the stationary processes $\{\xi(t), t \in \mathbb{R}\}$ and $\{\eta(t), t \in \mathbb{R}\}$ admit the spectral representation (2.2) with the spectral densities $f(\lambda)$ and $g(\lambda)$ satisfying the minimality condition

$$\int_{-\infty}^{\infty} \frac{|\gamma(\lambda)|^2}{f(\lambda)+g(\lambda)} d\lambda < \infty, \tag{2.26}$$

where

$$\gamma(\lambda) = \int_0^{\infty} \alpha(t) e^{it\lambda} dt$$

is a nonzero function of the exponential type. The processes $\xi(t)$ and $\eta(t)$ admit the spectral representation (2.3) with the orthogonal stochastic measures $Z_\xi(d\lambda)$ and $Z_\eta(d\lambda)$.

Consider the problem of mean square optimal linear estimation of the functional

$$A\xi = \int_0^{\infty} a(t)\xi(t)dt,$$

based on observations of the process $\xi(t)+\eta(t)$ at points $t\in\mathbb{R}^-\setminus S$, where

$$S=\bigcup_{l=1}^{s}[-M_l-N_l,\ldots,-M_l].$$

Let the function $a(t)$, that defines the functional $A\xi$, satisfy conditions

$$\int_0^\infty |a(t)|dt<\infty,\quad \int_0^\infty t\,|a(t)|^2\,dt<\infty. \tag{2.27}$$

The spectral representation of the stationary process $\xi(t)$ allows us to write the functional $A\xi$ in the form

$$A\xi=\int_{-\infty}^{\infty} A(e^{i\lambda})Z_\xi(d\lambda),$$

where

$$A(e^{i\lambda})=\int_0^\infty a(t)e^{it\lambda}dt.$$

We denote by $\hat{A}\xi$ a linear estimate of the functional $A\xi$ based on observations of the process $\xi(t)+\eta(t)$. It has the following form

$$\hat{A}\xi=\int_{-\infty}^{\infty} h(e^{i\lambda})(Z_\xi(d\lambda)+Z_\eta(d\lambda)), \tag{2.28}$$

where $h(e^{i\lambda})$ is the spectral characteristic of the estimate.
The mean-square error the optimal linear estimate $\hat{A}\xi$ of the functional $A\xi$ is calculated by the formula

$$\Delta(h;f,g)=\frac{1}{2\pi}\int_{-\infty}^{\infty}\left|A(e^{i\lambda})-h(e^{i\lambda})\right|^2 f(\lambda)d\lambda+ \\ +\frac{1}{2\pi}\int_{-\infty}^{\infty}\left|h(e^{i\lambda})\right|^2 g(\lambda)d\lambda. \tag{2.29}$$

We first consider the case of spectral certainty, where the spectral densities of the stationary processes $\xi(t)$ and $\eta(t)$ are known. To find the estimate $\hat{A}\xi$ we use the method of orthogonal projections in the Hilbert space proposed by A. M. Kolmogorov [1].

Denote by $H^s(\xi+\eta)$ the closed linear subspace, generated by values $\{\xi(t)+\eta(t) : t \in \mathbb{R}^- \backslash S\}$ in the Hilbert space $H = L_2(\Omega, \mathcal{F}, P)$. Denote by $L_2^s(f+g)$ the subspace, generated by functions $\{e^{it\lambda}, t \in \mathbb{R}^- \backslash S\}$ in the space $L_2(f+g)$. By the method of orthogonal projections in the Hilbert space, the optimal estimate of the functional $A\xi$ is determined by the following conditions:

$$1)\, \hat{A}\xi \in H^s(\xi+\eta),$$
$$2)\, A\xi - \hat{A}\xi \perp H^s(\xi+\eta).$$

It follows From condition 2) that the spectral characteristic $h(e^{i\lambda})$ of the optimal linear estimate $\hat{A}\xi$ for all $t \in \mathbb{R}^- \backslash S$ satisfies the relation

$$E\left[\left(A\xi - \hat{A}\xi\right)\left(\overline{\xi(t)} + \overline{\eta(t)}\right)\right] =$$
$$= \frac{1}{2\pi}\int_{-\infty}^{\infty}\left(A(e^{i\lambda}) - h(e^{i\lambda})\right)e^{-it\lambda}f(\lambda)d\lambda -$$
$$-\frac{1}{2\pi}\int_{-\infty}^{\infty} h(e^{i\lambda})e^{-it\lambda}g(\lambda)d\lambda = 0.$$

This relation can be written in the form

$$\frac{1}{2\pi}\int_{-\infty}^{\infty}\left[A(e^{i\lambda})f(\lambda) - h(e^{i\lambda})(f(\lambda)+g(\lambda))\right]e^{-it\lambda}d\lambda = 0, t \in \mathbb{R}^- \backslash S. \qquad (2.30)$$

Denote the function in the brackets by

$$C(e^{i\lambda}) = A(e^{i\lambda})f(\lambda) - h(e^{i\lambda})(f(\lambda)+g(\lambda)), \quad \lambda \in \mathbb{R},$$

and its Fourier transform denote by

$$\mathbf{c}(t) = \frac{1}{2\pi}\int_{-\infty}^{\infty} C(e^{i\lambda})e^{-it\lambda}d\lambda, \quad t \in \mathbb{R}.$$

It follows from the condition (2.30) that the function $\mathbf{c}(t)$ is nonzero on the set $T = S \cup [0,\infty)$. Hence, the function $C(e^{i\lambda})$ is of the form

$$C(e^{i\lambda}) = \sum_{l=1}^{s} \int_{-M_l-N_l}^{-M_l} \mathbf{c}(t)e^{it\lambda}dt + \int_{0}^{\infty} \mathbf{c}(t)e^{it\lambda}dt.$$

The spectral characteristic $h(e^{i\lambda})$ of the estimate $\hat{A}\xi$ is of the form

$$h(e^{i\lambda}) = A(e^{i\lambda})\frac{f(\lambda)}{f(\lambda)+g(\lambda)} - \frac{C(e^{i\lambda})}{f(\lambda)+g(\lambda)}. \tag{2.31}$$

It follows from the first condition, $\hat{A}\xi \in H^s(\xi+\eta)$, which determines the optimal estimate of the functional $A\xi$, that for some $v(t) \in L_2^s(f+g)$

$$h(e^{i\lambda}) = \frac{1}{2\pi} \int_{\mathbb{R}^- \setminus S} v(t)e^{it\lambda}d\lambda.$$

This means that

$$\int_{-\infty}^{\infty} \left(A(e^{i\lambda})\frac{f(\lambda)}{f(\lambda)+g(\lambda)} - \frac{C(e^{i\lambda})}{f(\lambda)+g(\lambda)} \right) e^{-it\lambda}d\lambda = 0, \quad t \in T. \tag{2.32}$$

We write the last relation in terms of linear operators in the space $L_2(T)$. In order to do this, we introduce the following operators

$$(\mathbf{B}\mathbf{x})(t) = \frac{1}{2\pi}\sum_{l=1}^{s} \int_{-M_l-N_l}^{-M_l} \mathbf{x}(u) \int_{-\infty}^{\infty} e^{i\lambda(u-t)} \frac{1}{f(\lambda)+g(\lambda)} d\lambda du$$
$$+ \frac{1}{2\pi}\int_{0}^{\infty} \mathbf{x}(u) \int_{-\infty}^{\infty} e^{i\lambda(u-t)} \frac{1}{f(\lambda)+g(\lambda)} d\lambda du,$$

$$(\mathbf{R}\mathbf{x})(t) = \frac{1}{2\pi}\sum_{l=1}^{s} \int_{-M_l-N_l}^{-M_l} \mathbf{x}(u) \int_{-\infty}^{\infty} e^{i\lambda(u-t)} \frac{f(\lambda)}{f(\lambda)+g(\lambda)} d\lambda du$$
$$+ \frac{1}{2\pi}\int_{0}^{\infty} \mathbf{x}(u) \int_{-\infty}^{\infty} e^{i\lambda(u-t)} \frac{f(\lambda)}{f(\lambda)+g(\lambda)} d\lambda du,$$

$$(\mathbf{Qx})(t) = \frac{1}{2\pi}\sum_{l=1}^{s}\int_{-M_l-N_l}^{-M_l}\mathbf{x}(u)\int_{-\infty}^{\infty}e^{i\lambda(u-t)}\frac{f(\lambda)g(\lambda)}{f(\lambda)+g(\lambda)}d\lambda du$$
$$+\frac{1}{2\pi}\int_{0}^{\infty}\mathbf{x}(u)\int_{-\infty}^{\infty}e^{i\lambda(u-t)}\frac{f(\lambda)g(\lambda)}{f(\lambda)+g(\lambda)}d\lambda du,$$
$$\mathbf{x}(t)\in L_2(T), \quad t\in T.$$

We represent equality (2.32) in the form

$$\int_{-\infty}^{\infty}\int_{0}^{\infty}a(u)e^{i(u-t)}\frac{f(\lambda)}{f(\lambda)+g(\lambda)}dud\lambda- \tag{2.33}$$
$$-\left(\int_{-\infty}^{\infty}\sum_{l=1}^{s}\int_{-M_l-N_l}^{-M_l}\frac{\mathbf{c}(e^{i(u-t)\lambda})}{f(\lambda)+g(\lambda)}dud\lambda+\int_{-\infty}^{\infty}\int_{0}^{\infty}\frac{\mathbf{c}(e^{i(u-t)\lambda})}{f(\lambda)+g(\lambda)}dud\lambda\right)=0.$$

and define the function $\mathbf{a}(t)$ as follows: $\mathbf{a}(t)=0, t\in S, \mathbf{a}(t)=a(t), t\geq 0$. Making use of the introduced notations, we can write equation (2.33) in the form

$$(\mathbf{Ra})(t) = (\mathbf{Bc})(t), \quad t\in T. \tag{2.34}$$

Assume that the operator $\mathbf{B}$ has an inverse operator. In this case the function $\mathbf{c}(t)$ can be calculated by the formula

$$\mathbf{c}(t) = (\mathbf{B}^{-1}\mathbf{Ra})(t), \quad t\in T.$$

Consequently, the spectral characteristic $h(e^{i\lambda})$ of the estimate $\hat{A}\xi$ is calculated by the formula

$$h(e^{i\lambda}) = A(e^{i\lambda})\frac{f(\lambda)}{f(\lambda)+g(\lambda)}-\frac{C(e^{i\lambda})}{f(\lambda)+g(\lambda)}, \tag{2.35}$$

where

$$C(e^{i\lambda}) = \sum_{l=1}^{s}\int_{-M_l-N_l}^{-M_l}(\mathbf{B}^{-1}\mathbf{Ra})(t)e^{it\lambda}dt+\int_{0}^{\infty}(\mathbf{B}^{-1}\mathbf{Ra})(t)e^{it\lambda}dt.$$

The mean-square error of the estimate $\hat{A}\xi$ is calculated by the formula

$$\Delta(h;f,g) = E\left|A\xi - \hat{A}\xi\right|^2 =$$
$$= \frac{1}{2\pi}\int_{-\infty}^{\infty} \frac{\left|A(e^{i\lambda})g(\lambda)+C(e^{i\lambda})\right|^2}{(f(\lambda)+g(\lambda))^2} f(\lambda)d\lambda +$$
$$+ \frac{1}{2\pi}\int_{-\infty}^{\infty} \frac{\left|A(e^{i\lambda})f(\lambda)-C(e^{i\lambda})\right|^2}{(f(\lambda)+g(\lambda))^2} g(\lambda)d\lambda$$
$$= \langle \mathbf{R}\vec{\mathbf{a}}, \mathbf{B}^{-1}\mathbf{R}\vec{\mathbf{a}}\rangle + \langle \mathbf{Q}\vec{\mathbf{a}}, \vec{\mathbf{a}}\rangle, \tag{2.36}$$

where

$$\langle A, C\rangle = \sum_{l=1}^{s} \int_{-M_l-N_l}^{-M_l} A(t)\overline{C(t)}dt + \int_{0}^{\infty} A(t)\overline{C(t)}dt$$

is the inner product in the space $L_2(T)$.

We can formulate the obtained results in the form of a theorem.

Theorem 2.9. *Suppose that $\xi(t)$ and $\eta(t)$ are uncorrelated stationary stochastic processes with the spectral densities $f(\lambda)$ and $g(\lambda)$ satisfying the minimality condition (2.26). The spectral characteristic $h(e^{i\lambda})$ and the value of the mean-square error $\Delta(h;f,g)$ of the optimal linear estimate of the functional $A\xi$ based on observations of the process $\xi(t)+\eta(t)$ at points $t \in \mathbb{R}^-\backslash S$ can be calculated by formulas (2.35), (2.36).*

Estimation of the Functional $A_N\xi$

Consider the problem of mean-square optimal linear estimation of the functional

$$A_N\xi = \int_0^N a(t)\xi(t)dt,$$

based on observations of the process $\xi(t)+\eta(t)$ at points $t \in \mathbb{R}^-\backslash S$.
Each linear estimate $\hat{A}_N\xi$ of the functional $A_N\xi$ can be represented in the form

$$\hat{A}_N\xi = \int_{-\infty}^{\infty} h_N(e^{i\lambda})(Z_\xi(d\lambda)+Z_\eta(d\lambda)),$$

where $h_N(e^{i\lambda}) \in L_2^s(f+g)$ is the spectral characteristic of the estimate. Introduce the function $\mathbf{a}_N(t)$ in the following way:

$$\mathbf{a}_N(t) = a(t), t \in T \cap [0,N], \mathbf{a}_N(t) = 0, t \in T \backslash [0,N].$$

The spectral characteristic $h_N(e^{i\lambda})$ of the optimal estimate $\hat{A}_N\xi$ can be calculated by the formula

$$h_N(e^{i\lambda}) = A_N(e^{i\lambda})\frac{f(\lambda)}{f(\lambda)+g(\lambda)} - \frac{C_N(e^{i\lambda})}{f(\lambda)+g(\lambda)}, \tag{2.37}$$

$$C_N(e^{i\lambda}) = \sum_{l=1}^{s} \int_{-M_l-N_l}^{-M_l} (\mathbf{B}^{-1}\mathbf{R}\mathbf{a}_N)(t)e^{it\lambda}dt + \int_0^{\infty} (\mathbf{B}^{-1}\mathbf{R}\mathbf{a}_N)(t)e^{it\lambda}dt,$$

where

$$A_N(e^{i\lambda}) = \int_0^N a(t)e^{-it\lambda}dt.$$

The mean-square error $\Delta(h_N; f,g)$ of the estimate $\hat{A}_N\xi$ is calculated by the formula

$$\begin{aligned} \Delta(h_N; f,g) &= E\left|A_N\xi - \hat{A}_N\xi\right|^2 = \\ &= \langle \mathbf{R}\vec{\mathbf{a}}_N, \mathbf{B}^{-1}\mathbf{R}\vec{\mathbf{a}}_N \rangle + \langle \mathbf{Q}\vec{\mathbf{a}}_N, \vec{\mathbf{a}}_N \rangle. \end{aligned} \tag{2.38}$$

The obtained results we formulate as the following statement.

Corollary 2.7. *Suppose that $\xi(t)$ and $\eta(t)$ are uncorrelated stationary stochastic processes with the spectral densities $f(\lambda)$ and $g(\lambda)$ satisfying the minimality condition (2.26). The spectral characteristic $h_N(e^{i\lambda})$ and the mean-square error $\Delta(h_N; f,g)$ of the optimal linear estimate of the functional $A_N\xi$ based on observations of the process $\xi(t)+\eta(t)$ at points $t \in \mathbb{R}^- \backslash S$ can be calculated by formulas (2.37), (2.38).*

Estimates Based on Observations without Noise

Consider the problem of mean-square optimal linear estimation of the functional $A\xi$ in the case where the stationary process $\xi(t)$ is observed without noise. In this case the spectral characteristic of the estimate $\hat{A}\xi$ is of the form

$$h(e^{i\lambda}) = A(e^{i\lambda}) - C(e^{i\lambda})f^{-1}(\lambda), \tag{2.39}$$

$$C(e^{i\lambda}) = \sum_{l=1}^{s} \int_{-M_l-N_l}^{-M_l} \mathbf{c}(t)e^{it\lambda}dt + \int_{0}^{\infty} \mathbf{c}(t)e^{it\lambda}dt.$$

and relation (2.34) is of the form

$$\mathbf{a}(t) = (\mathbf{Bc})(t), \quad t \in T.$$

If the operator $\mathbf{B}$ is invertible, then the unknown function $\mathbf{c}(t)$ is determined by the formula

$$\mathbf{c}(t) = (\mathbf{B}^{-1}\mathbf{a})(t), \quad t \in T,$$

and the spectral characteristic $\hat{A}\xi$ is of the form

$$h(e^{i\lambda}) = A(e^{i\lambda}) - C(e^{i\lambda})f^{-1}(\lambda), \tag{2.40}$$

$$C(e^{i\lambda}) = \sum_{l=1}^{s} \int_{-M_l-N_l}^{-M_l} (\mathbf{B}^{-1}\mathbf{a})(t)e^{it\lambda}dt + \int_{0}^{\infty} (\mathbf{B}^{-1}\mathbf{a})(t)e^{it\lambda}dt.$$

The mean-square error is calculated by the formula

$$\Delta(h; f) = \langle \mathbf{B}^{-1}\mathbf{a}, \mathbf{a} \rangle. \tag{2.41}$$

The following theorem is true.

Theorem 2.10. *Let $\xi(t)$ be a stationary stochastic process with the spectral density $f(\lambda)$, which satisfies the minimality condition (2.26) with $g(\lambda) = 0$. The spectral characteristic $h(e^{i\lambda})$ and the mean-square error $\Delta(h, f)$ of the optimal linear estimate $\hat{A}\xi$ of the functional $A\xi$ based on observations of the process $\xi(t)$ at points $t \in \mathbb{R}^- \setminus S$, $S = \bigcup_{l=1}^{s} [-M_l - N_l, \dots, -M_l]$, can be calculated by formulas (2.40), (2.41).*

3.2. Minimax Robust Method of Extrapolation

The obtained theorems and corollaries can be applied to calculate the spectral characteristic and the mean-square error of the optimal linear estimate of the functional $A\xi$ only in the case, where the spectral densities of the investigated processes are known. In the case, where the spectral densities are unknown while some sets of admissible spectral densities are given, it is reasonable to apply the minimax robust method of estimation. According to this method, we find an estimate which minimizes the maximal values of the mean square errors for all spectral densities from a given class D of admissible spectral densities simultaneously.

To describe the minimax method of estimation, we introduce the following definitions.

Definition 2.4. *For a given set of spectral densities $D = D_f \times D_g$ the spectral densities $f_0(\lambda) \in D_f$, $g_0(\lambda) \in D_g$ are called the least favourable for the optimal extrapolation of the functional $A\xi$, if the following relation is true*

$$\Delta(f_0, g_0) = \Delta(h(f_0, g_0); f_0, g_0) = \max_{(f,g) \in D_f \times D_g} \Delta(h(f,g); f,g).$$

Definition 2.5. *For a given set of spectral densities $D = D_f \times D_g$ the spectral characteristic $h^0(e^{i\lambda})$ of the optimal linear estimate $A\xi$ is called the minimax robust, if the following relations are true*

$$h^0(e^{i\lambda}) \in H_D = \bigcap_{(f,g) \in D} L_2^s(f+g),$$

$$\min_{h \in H_D} \max_{(f,g) \in D} \Delta(h; f,g) = \max_{(f,g) \in D} \Delta\left(h^0; f,g\right).$$

Using the obtained results and introduced definitions, we can formulate the following statements.

Lemma 2.3. *The spectral densities $f_0(\lambda) \in D_f$ and $g_0(\lambda) \in D_g$, that satisfy the minimality condition (2.26), are the least favourable spectral densities in the set $D = D_f \times D_g$ for the optimal linear extrapolation of the functional $A\xi$, if the Fourier transforms of the functions*

$$(f_0(\lambda) + g_0(\lambda))^{-1}, \quad f_0(\lambda)(f_0(\lambda) + g_0(\lambda))^{-1}, \quad f_0(\lambda) g_0(\lambda)(f_0(\lambda) + g_0(\lambda))^{-1}$$

define operators $\mathbf{B}^0, \mathbf{R}^0, \mathbf{Q}^0$, *which determine a solution of the optimization problem*

$$\max_{(f,g)\in D_f\times D_g} \langle \mathbf{R}\vec{\mathbf{a}}, \mathbf{B}^{-1}\mathbf{R}\vec{\mathbf{a}}\rangle + \langle \mathbf{Q}\vec{\mathbf{a}}, \vec{\mathbf{a}}\rangle = \\ = \langle \mathbf{R}^0\vec{\mathbf{a}}, (\mathbf{B}^0)^{-1}\mathbf{R}^0\vec{\mathbf{a}}\rangle + \langle \mathbf{Q}^0\vec{\mathbf{a}}, \vec{\mathbf{a}}\rangle. \tag{2.42}$$

The minimax spectral characteristic $h^0 = h(f_0, g_0)$ *is calculated by formula (2.35) if* $h(f_0, g_0) \in H_D$.

Corollary 2.8. *Suppose that the spectral density* $f_0(\lambda) \in D_f$ *satisfy the minimality condition (2.26) with* $g(\lambda) = 0$. *The density* $f_0(\lambda) \in D_f$ *is the least favourable in the class* D_f *for the optimal extrapolation of the functional* $A\xi$, *based on observations of the process* $\xi(t)$ *at points* $t \in \mathbb{R}^-\backslash S$, *if the Fourier transform of the function* $f_0^{-1}(\lambda)$ *defines the operator* $\mathbf{B}^0$, *which determines a solution of the optimization problem*

$$\max_{f\in D_f} \langle \mathbf{B}^{-1}\vec{\mathbf{a}}, \vec{\mathbf{a}}\rangle = \langle (\mathbf{B}^0)^{-1}\vec{\mathbf{a}}, \vec{\mathbf{a}}\rangle. \tag{2.43}$$

The minimax spectral characteristic $h^0 = h(f_0)$ *is calculated by formula (2.40) if* $h(f_0) \in H_{D_f}$.

For more detailed analysis of properties of the least favourable spectral densities and the minimax-robust spectral characteristics we observe that the least favourable spectral densities (f_0, g_0) and the minimax spectral characteristic $h^0 = h(f_0, g_0)$ form a saddle point of the function $\Delta(h; f, g)$ on the set $H_D \times D$. The saddle point inequalities

$$\Delta\left(h^0; f, g\right) \leq \Delta\left(h^0; f_0, g_0\right) \leq \Delta(h; f_0, g_0), \forall h \in H_D, \forall f \in D_f, \forall g \in D_g,$$

are true if $h^0 = h(f_0, g_0)$ and $h(f_0, g_0) \in H_D$, where (f_0, g_0) is a solution of the optimization problem

$$\sup_{(f,g)\in D_f\times D_g} \Delta(h(f_0, g_0); f, g) = \Delta(h(f_0, g_0); f_0, g_0), \tag{2.44}$$

$$\Delta(h(f_0, g_0); f, g) = \frac{1}{2\pi}\int_{-\infty}^{\infty} \frac{\left|A(e^{i\lambda})g_0(\lambda) + C^0(e^{i\lambda})\right|^2}{(f_0(\lambda) + g_0(\lambda))^2} f(\lambda)d\lambda \\ + \frac{1}{2\pi}\int_{-\infty}^{\infty} \frac{\left|A(e^{i\lambda})f_0(\lambda) - C^0(e^{i\lambda})\right|^2}{(f_0(\lambda) + g_0(\lambda))^2} g(\lambda)d\lambda,$$

$$C^0(e^{i\lambda}) = \sum_{l=1}^{s} \int_{-M_l-N_l}^{-M_l} ((\mathbf{B}^0)^{-1}\mathbf{R}^0\mathbf{a})(t)e^{it\lambda}dt +$$

$$+ \int_0^{\infty} ((\mathbf{B}^0)^{-1}\mathbf{R}^0\mathbf{a})(t)e^{it\lambda}dt, \quad t \in S.$$

The constrained optimization problem (2.44) is equivalent to the unconstrained optimization problem (see B. N. Pshenichnyj [156])

$$\Delta_D(f,g) = -\Delta(h(f_0,g_0);f,g) + \delta((f,g)\left|D_f \times D_g\right.) \to \inf, \tag{2.45}$$

where $\delta((f,g)\left|D_f \times D_g\right.)$ is the indicator function of the set $D = D_f \times D_g$. A solution of the problem (2.45) is determined by the condition $0 \in \partial\Delta_D(f_0,g_0)$, where $\partial\Delta_D(f_0)$ is a subdifferential of the convex functional $\Delta_D(f,g)$ at point (f_0,g_0).

The form of the functional $\Delta(h(f_0,g_0);f,g)$ is convenient for application of the Lagrange method of indefinite multipliers for finding a solution of the optimization problem. Thus the complexity of solving the optimization problems (2.45) is determined by the complexity of calculating the subdifferentials of the indicator functions of sets $D_f \times D_g$. Making use of the method of Lagrange multipliers and the form of subdifferentials of the indicator functions we describe relations that determine the least favourable spectral densities in some special classes of spectral densities.

Lemma 2.4. *Let (f_0,g_0) be a solution to the optimization problem (2.45). The spectral densities $f_0(\lambda)$ and $g_0(\lambda)$ are the least favourable in the class $D = D_f \times D_g$, and $h^0 = h(f_0,g_0)$ is the minimax spectral characteristic of the optimal linear estimate of the functional $A_s\xi$, if $h(f_0,g_0) \in H_D$.*

3.3. Least Favourable Spectral Densities in the Class $D_0 \times D_\varepsilon^1$

Consider the problem of minimax extrapolation of the functional $A\xi$, based on observations of the process $\xi(t)+\eta(t)$ at points $t \in \mathbb{R}^-\backslash S$, under the condition that the stationary processes $\xi(t)$ and $\eta(t)$ are not correlated, and the spectral densities $f(\lambda)$ and $g(\lambda)$ of the processes belong to the class of admissible spectral densities $D = D_0 \times D_\varepsilon^1$,

$$D_0 = \left\{ f(\lambda) \left| \frac{1}{2\pi} \int_{-\infty}^{\infty} f(\lambda) d\lambda \leq P_1 \right. \right\},$$

$$D_\varepsilon^1 = \left\{ g(\lambda) \left| \frac{1}{2\pi} \int_{-\infty}^{\infty} |g(\lambda) - g_1(\lambda)| d\lambda \leq \varepsilon \right. \right\},$$

where the spectral density $g_1(\lambda)$ is known and fixed.
Densities from the class of admissible spectral densities D_ε^1 form the "ε-neighbourhood" model in the space L_1 of a given bounded spectral density $g_1(\lambda)$.

Suppose that the spectral densities $f_0(\lambda) \in D_0$, $g_0(\lambda) \in D_\varepsilon^1$ determine the bounded functions

$$h_f(f_0, g_0) = \frac{\left|A(e^{i\lambda})g_0(\lambda) + C^0(e^{i\lambda})\right|^2}{(f_0(\lambda) + g_0(\lambda))^2}, \tag{2.46}$$

$$h_g(f_0, g_0) = \frac{\left|A(e^{i\lambda})f_0(\lambda) - C^0(e^{i\lambda})\right|^2}{(f_0(\lambda) + g_0(\lambda))^2}. \tag{2.47}$$

In this case the functional $\Delta(h(f_0, g_0); f, g)$ can be written in the form

$$\Delta(h(f_0, g_0); f, g) = \frac{1}{2\pi} \int_{-\infty}^{\infty} h_f(f_0, g_0) f(\lambda) d\lambda + \\ + \frac{1}{2\pi} \int_{-\infty}^{\infty} h_g(f_0, g_0) g(\lambda) d\lambda.$$

Since the functions $h_f(f_0, g_0)$, $h_g(f_0, g_0)$, defined by formulas (2.46), (2.47), are supposed to be bounded, the functional $\Delta(h(f_0, g_0); f, g)$ is a continuous linear functional in the space $L_1 \times L_1$.
From the condition $0 \in \partial \Delta_{D_0 \times D_\varepsilon^1}(f_0, g_0)$ we can derive the following equations which determine the least favourable densities

$$\left|A(e^{i\lambda})g_0(\lambda) + C^0(e^{i\lambda})\right| = \alpha_1 (f_0(\lambda) + g_0(\lambda)), \tag{2.48}$$

$$\left|A(e^{i\lambda})f_0(\lambda)-C^0(e^{i\lambda})\right|=(f_0(\lambda)+g_0(\lambda))\Psi(\lambda)\alpha_2. \tag{2.49}$$

Here $|\Psi(\lambda)|\leq 1$ and $\Psi(\lambda)=\text{sign}(g_0(\lambda)-g_1(\lambda))$, when $g_0(\lambda)\neq g_1(\lambda)$, and constants $\alpha_1\geq 0$, $\alpha_2\geq 0$.
Note that $\alpha_1\neq 0$, if

$$\frac{1}{2\pi}\int_{-\infty}^{\infty} f_0(\lambda)d\lambda=P_1.$$

Equations (2.48), (2.49) and the normalization condition

$$\frac{1}{2\pi}\int_{-\infty}^{\infty}|g(\lambda)-g_1(\lambda)|d\lambda=\varepsilon \tag{2.50}$$

together with the optimization problem (2.42) determine the least favourable spectral density in the class $D=D_0\times D^1_{\varepsilon}$.

Theorem 2.11. *Suppose that the spectral densities $f_0(\lambda)\in D_0$, $g_0(\lambda)\in D^1_{\varepsilon}$ satisfy the minimality condition (2.26), and functions, calculated by formulas (2.46), (2.47), are bounded. The spectral densities $f_0(\lambda)$, $g_0(\lambda)$ are the least favourable spectral densities in the class $D_0\times D^1_{\varepsilon}$ for the optimal linear extrapolation of the functional $A\xi$, if they satisfy the system of equations (2.48)–(2.50) and determine a solution of the optimization problem (2.42). The function, calculated by formula (2.35), is the minimax spectral characteristic of the optimal estimate of the functional $A\xi$.*

Theorem 2.12. *Assume that densities $f_0(\lambda)\in D^1_{\varepsilon_1}$, $g_0(\lambda)\in D^1_{\varepsilon_2}$,*

$$D^1_{\varepsilon_1}=\left\{f(\lambda)\left|\frac{1}{2\pi}\int_{-\infty}^{\infty}|f(\lambda)-f_1(\lambda)|d\lambda\leq\varepsilon_1\right.\right\},$$

$$D^1_{\varepsilon_2}=\left\{g(\lambda)\left|\frac{1}{2\pi}\int_{-\infty}^{\infty}|g(\lambda)-g_1(\lambda)|d\lambda\leq\varepsilon_2\right.\right\},$$

where the spectral densities $f_1(\lambda)$, $g_1(\lambda)$ are known and fixed. We also assume that the spectral densities $f_0(\lambda),g_0(\lambda)$ satisfy the minimality condition (2.26), *and the functions, calculated by formulas (2.46), (2.47), are bounded. The spectral densities $f_0(\lambda)$, $g_0(\lambda)$ are the least favourable in the class $D^1_{\varepsilon_1} \times D^1_{\varepsilon_2}$ for the optimal linear extrapolation of the functional $A\xi$, if they satisfy the system of equations*

$$\left|A(e^{i\lambda})g_0(\lambda)+C^0(e^{i\lambda})\right| = (f_0(\lambda)+g_0(\lambda))\Psi_1(\lambda)\alpha_1,$$

$$\left|A(e^{i\lambda})f_0(\lambda)-C^0(e^{i\lambda})\right| = (f_0(\lambda)+g_0(\lambda))\Psi_2(\lambda)\alpha_2,$$

where $|\Psi_1(\lambda)| \leq 1$ *and* $\Psi_1(\lambda) = \mathrm{sign}(f_0(\lambda) - f_1(\lambda))$, *when* $f_0(\lambda) \neq f_1(\lambda)$, $|\Psi_2(\lambda)| \leq 1$ *and* $\Psi_2(\lambda) = \mathrm{sign}(g_0(\lambda) - g_1(\lambda))$, *when* $g_0(\lambda) \neq g_1(\lambda)$, *constants* $\alpha_1 \geq 0$, $\alpha_2 \geq 0$, *satisfy the normalization conditions*

$$\frac{1}{2\pi}\int_{-\infty}^{\infty} |f(\lambda)-f_1(\lambda)|\,d\lambda = \varepsilon_1,$$

$$\frac{1}{2\pi}\int_{-\infty}^{\infty} |g(\lambda)-g_1(\lambda)|\,d\lambda = \varepsilon_2,$$

and determine a solution of the optimization problem (2.42). The function, calculated by formula (2.35), is the minimax spectral characteristic of the optimal estimate of the functional $A\xi$.

Corollary 2.9. *Suppose that the spectral density $g(\lambda)$ is known, and the unknown spectral density $f_0(\lambda) \in D^1_{\varepsilon_1}$. Suppose that the function $f_0(\lambda) + g(\lambda)$ satisfy the minimality condition (2.26), and the function $h_f(f_0, g)$, defined by formula (2.46), is bounded. The spectral density $f_0(\lambda)$ is the least favourable in the class $D^1_{\varepsilon_1}$ for the optimal linear extrapolation of the functional $A\xi$, if it is of the form*

$$f_0(\lambda) = \max\left\{f_1(\lambda), \alpha_1\left|A(e^{i\lambda})g(\lambda)+C^0(e^{i\lambda})\right| - g(\lambda)\right\},$$

and the pair $(f_0(\lambda), g(\lambda))$ determines a solution of the optimization problem (2.42). The minimax spectral characteristic of the optimal estimate of the functional $A\xi$ is calculated by formula (2.35).

Corollary 2.10. *Let the spectral density $f_0(\lambda) \in D^1_{\varepsilon_1}$ satisfy the minimality condition ((2.26)) with $g(\lambda) = 0$, and let the function, defined by (2.40), be bounded. The spectral density $f_0(\lambda)$ is the least favourable in the class $D^1_{\varepsilon_1}$ for the optimal linear extrapolation of the functional $A\xi$, based on observations of the process $\xi(t)$ at points $t \in \mathbb{R}^- \setminus S$, if it satisfies the relation*

$$\left|C^0(e^{i\lambda})\right| = f_0(\lambda)\Psi_1(\lambda)\alpha_1,$$

and $f_0(\lambda)$ determines a solution of the optimization problem (2.43). The minimax spectral characteristic of the optimal estimate of the functional $A\xi$ is calculated by formula (2.40).

3.4. Least Favourable Spectral Densities in the Class $D^u_v \times D_0$

Consider the problem of minimax extrapolation of the functional $A\xi$ based on observations of the process $\xi(t) + \eta(t)$ under the condition that the stationary processes $\xi(t)$ and $\eta(t)$ are not correlated, and the spectral densities $f(\lambda)$ and $g(\lambda)$ of the processes belong to the class of admissible spectral densities $D^u_v \times D_0$,

$$D^u_v = \left\{ f(\lambda) \left| v(\lambda) \leq f(\lambda) \leq u(\lambda), \frac{1}{2\pi}\int_{-\pi}^{\pi} f(\lambda)d\lambda \leq P_1 \right. \right\},$$

$$D_0 = \left\{ g(\lambda) \left| \frac{1}{2\pi}\int_{-\infty}^{\infty} g(\lambda)d\lambda \leq P_2 \right. \right\},$$

where the spectral densities $u(\lambda), v(\lambda)$ are known and fixed.
Densities from the class of admissible spectral densities D^u_v form the "stripe" model of spectral densities.

Suppose that the spectral densities $f^0(\lambda) \in D^u_v$, $g^0(\lambda) \in D_0$ satisfy the minimality condition (2.26), and determine the bounded functions $h_f(f_0, g_0)$, $h_g(f_0, g_0)$ by formulas (2.46), (2.47). From the condition $0 \in \partial\Delta_D(f_0, g_0)$ we derive equations that determine the least favourable spectral densities

$$\left|A(e^{i\lambda})g_0(\lambda) + C^0(e^{i\lambda})\right| = (f_0(\lambda) + g_0(\lambda))(\gamma_1(\lambda) + \gamma_2(\lambda) + \alpha_1^{-1}), \quad (2.51)$$

$$\left|A(e^{i\lambda})f_0(\lambda) - C^0(e^{i\lambda})\right|^2 = \alpha_2(f_0(\lambda) + g_0(\lambda)). \qquad (2.52)$$

Here α_1, α_2 are constants, $\gamma_1 \leq 0$ and $\gamma_1 = 0$, when $f_0(\lambda) \geq v(\lambda)$, $\gamma_2 \geq 0$ and $\gamma_2 = 0$, when $f_0(\lambda) \leq u(\lambda)$.
Note, that the constant $\alpha_2 \neq 0$, if

$$\frac{1}{2\pi}\int_{-\infty}^{\infty} g_0(\lambda)d\lambda = P_2.$$

The following statements are true.

Theorem 2.13. *Let the spectral densities $f^0(\lambda) \in D_v^u$, $g^0(\lambda) \in D_0$ satisfy the minimality condition (2.26) and determine the bounded functions $h_f(f_0, g_0)$, $h_g(f_0, g_0)$ by formulas (2.46), (2.47). The spectral densities $f_0(\lambda)$, $g_0(\lambda)$ are the least favourable spectral densities in the class $D_v^u \times D_0$ for the optimal linear extrapolation of the functional $A\xi$, if they satisfy equations (2.51) – (2.52) and determine a solution of the optimization problem (2.42). The function, calculated by formula (2.35), is the minimax spectral characteristic of the optimal estimate of the functional $A\xi$.*

Corollary 2.11. *Let the spectral density $g(\lambda)$ be known, and the unknown spectral density $f_0(\lambda) \in D_v^u$. Suppose that the function $f_0(\lambda) + g(\lambda)$ satisfy the minimality condition (2.26) and the function $h_f(f_0, g)$, defined by formula (2.46), is bounded. The spectral density $f_0(\lambda)$ is the least favourable spectral density in the class D_v^u for the optimal linear extrapolation of the functional $A\xi$, if it satisfies the relation*

$$f_0(\lambda) = \min\left\{\max\left\{f_1(\lambda), v(\lambda)\right\}, u(\lambda)\right\},$$

$$f_1(\lambda) = \alpha_1\left|A(e^{i\lambda})g(\lambda) + C^0(e^{i\lambda})\right| - g(\lambda),$$

and the pair $(f_0(\lambda), g(\lambda))$ determines a solution of the optimization problem (2.42). The function $h(f_0, g)$, calculated by formula (2.35), is the minimax spectral characteristic of the optimal estimate of the functional $A\xi$.

Corollary 2.12. *Assume that a spectral density $f_0(\lambda) \in D_v^u$ satisfy the minimality condition ((2.26)) with $g(\lambda) = 0$ and the function, defined by (2.40), is* *bounded. The spectral density* $f_0(\lambda)$ *is the least favourable density in the class* D_v^u *for the optimal linear extrapolation of the functional* $A\xi$*, based on observations of the process* $\xi(t)$ *at points* $t \in \mathbb{R}^- \backslash S$*, if it is of the form*

$$f_0(\lambda) = \max\left\{v(\lambda), \min\left\{u(\lambda), \alpha_1 \left|C^0(e^{i\lambda})\right|\right\}\right\}$$

and the function $f_0(\lambda)$ *determines a solution of the optimization problem (2.43). The minimax spectral characteristic of the optimal estimate of the functional* $A\xi$ *is calculated by formula (2.40).*

4. The Problem of Filtering of Stationary Processes Based on Observations with Missing Values

In this section, we propose results of the investigation of the problem of mean-square optimal linear estimation of the functional

$$A\xi = \int_{R^s} a(t)\xi(-t)dt$$

constructed from unobserved values of a stationary stochastic process $\xi(t)$. Estimates are based on observations of the process $\xi(t)+\eta(t)$ at points

$$t \in \mathbb{R}^- \backslash S, \quad S = \bigcup_{l=1}^{s}[-M_l - N_l, \ldots, -M_l],$$

$$R^s = [0,\infty)\backslash S^+, \quad S^+ = \bigcup_{l=1}^{s}[M_l, \ldots, M_l + N_l],$$

$$M_l = \sum_{k=0}^{l}(N_k + K_k),\ N_0 = 0, K_0 = 0.$$

The stationary stochastic process $\eta(t)$ is uncorrelated with the process $\xi(t)$. The problem is investigated in the case of spectral certainty as well as in the case of spectral uncertainty. Formulas for calculating the spectral characteristic and the mean square error of the optimal linear estimate of the functional are derived in the case of spectral certainty, where the spectral densities of the stationary processes $\xi(t)$ and $\eta(t)$ are exactly known. Formulas that determine the least favourable spectral densities and the minimax-robust spectral characteristics of the optimal estimates of the functional are derived in the case of spectral uncertainty, where the spectral densities are not exactly known while some sets of admissible spectral densities are given.

4.1. Hilbert Space Projection Method of Filtering

Consider two uncorrelated stationary stochastic processes $\xi(t)$ and $\eta(t)$ with zero mathematical expectations, $E\xi(t) = 0$, $E\eta(t) = 0$, and covariance functions $R_\xi(t) = E\xi(t+s)\overline{\xi(s)}$ and $R_\eta(t) = E\eta(t+s)\overline{\eta(s)}$, which admit the spectral representation (2.2). We will assume that the spectral densities $f(\lambda)$ and $g(\lambda)$ satisfy the minimality condition

$$\int_{-\infty}^{\infty} \frac{|\gamma(\lambda)|^2}{f(\lambda)+g(\lambda)} d\lambda < \infty, \tag{2.53}$$

where

$$\gamma(\lambda) = \int_{R^s} \alpha(t) e^{it\lambda} dt$$

is a nonzero function of an exponential type.

The stationary stochastic processes $\xi(t)$ and $\eta(t)$ admit the spectral representation (2.3) with the orthogonal stochastic measures $Z_\xi(d\lambda)$ and $Z_\eta(d\lambda)$ correspondingly.

Consider the problem of mean-square optimal linear estimation of the functional

$$A\xi = \int_{R^s} a(t)\xi(-t)dt$$

based on observations of the process $\xi(t) + \eta(t)$ at points $t \in \mathbb{R}^- \setminus S$,

$$S = \bigcup_{l=1}^{s} [-M_l - N_l, \ldots, -M_l],$$

$$R^s = [0, \infty) \setminus S^+, \quad S^+ = \bigcup_{l=1}^{s} [M_l, \ldots, M_l + N_l].$$

Assume that the function $a(t)$, which determines the functional $A\xi$, satisfy conditions

$$\int_{R^s} |a(t)|dt < \infty, \quad \int_{R^s} t\,|a(t)|^2\,dt < \infty. \tag{2.54}$$

We use the spectral representation (2.3) of the stationary process $\xi(t)$ to write the functional $A\xi$ in the form

$$A\xi = \int_{-\infty}^{\infty} A(e^{i\lambda}) Z_{\xi}(d\lambda),$$

where

$$A(e^{i\lambda}) = \int_{R^s} a(t)e^{-it\lambda}dt.$$

The optimal linear estimate $\hat{A}\xi$ of the functional $A\xi$ based on observations of the process $\xi(t)+\eta(t)$ is of the form (2.28). We first assume that the spectral densities of the stationary processes $\xi(t)$ and $\eta(t)$ are known and use the method of orthogonal projections in the Hilbert space to find the optimal estimate $\hat{A}\xi$ of the functional $A\xi$.

Denote by $H^s(\xi+\eta)$ the closed linear subspace, generated by values $\{\xi(t)+\eta(t) : t \in \mathbb{R}^- \setminus S\}$ in the Hilbert space $H = L_2(\Omega, \mathcal{F}, P)$. Denote by $L_2^s(f+g)$ the subspace of the space $L_2(f+g)$, generated by functions $\{e^{it\lambda} : t \in \mathbb{R}^- \setminus S\}$.

The optimal linear estimate of the functional $A\xi$ is a projection of the element $A\xi$ of the space $H = L_2(\Omega, \mathcal{F}, P)$ onto the subspace $H^s(\xi+\eta)$. It is determined by the following conditions:

$$1)\, \hat{A}\xi \in H^s(\xi+\eta),$$
$$2)\, A\xi - \hat{A}\xi \perp H^s(\xi+\eta).$$

It follows from the second condition that the spectral characteristic $h(e^{i\lambda})$ of the optimal linear estimate $\hat{A}\xi$ satisfies relation (2.30) for all $t \in \mathbb{R}^- \setminus S$. This relation can be written in the form

$$\frac{1}{2\pi}\int_{-\infty}^{\infty} \Big(A(e^{i\lambda})f(\lambda) - h(e^{i\lambda})(f(\lambda)+g(\lambda))\Big)e^{-it\lambda}d\lambda = 0, t \in \mathbb{R}^- \setminus S. \tag{2.55}$$

Denote the function in the brackets by

$$C(e^{i\lambda}) = A(e^{i\lambda})f(\lambda) - h(e^{i\lambda})(f(\lambda)+g(\lambda)), \quad \lambda \in \mathbb{R},$$

and its Fourier transform denote by

$$\mathbf{c}(t) = \frac{1}{2\pi}\int_{-\infty}^{\infty} C(e^{i\lambda})e^{-it\lambda}d\lambda, \quad t \in \mathbb{R}.$$

It follows from condition (2.55), that the function $\mathbf{c}(t)$ is nonzero on the set $T = S \cup [0,\infty)$. Hence, the function $C(e^{i\lambda})$ is of the form

$$C(e^{i\lambda}) = \sum_{l=1}^{s}\int_{-M_l-N_l}^{-M_l} \mathbf{c}(t)e^{it\lambda}dt + \int_{0}^{\infty}\mathbf{c}(t)e^{it\lambda}dt,$$

and the spectral characteristic of the optimal estimate $\hat{A}\xi$ of the functional $A\xi$ is of the form

$$h(e^{i\lambda}) = A(e^{i\lambda})\frac{f(\lambda)}{f(\lambda)+g(\lambda)} - \frac{C(e^{i\lambda})}{f(\lambda)+g(\lambda)}. \tag{2.56}$$

From the first condition, $\hat{A}\xi \in H^s(\xi+\eta)$, which determines the optimal estimate of the functional $A\xi$, we can derive that the spectral characteristic $h(e^{i\lambda})$ is of the form

$$h(e^{i\lambda}) = \frac{1}{2\pi}\int_{\mathbb{R}^-\setminus S} v(t)e^{it\lambda}d\lambda$$

for some function $v(t) \in L_2^s(f+g)$. Hence,

$$\int_{-\infty}^{\infty}\left(A(e^{i\lambda})\frac{f(\lambda)}{f(\lambda)+g(\lambda)} - \frac{C(e^{i\lambda})}{f(\lambda)+g(\lambda)}\right)e^{-it\lambda}d\lambda = 0,\ t \in T. \tag{2.57}$$

We write the last relation in terms of linear operators in the space $L_2(T)$. To do this, we introduce the following linear operators:

$$(\mathbf{B}\mathbf{x})(t) = \frac{1}{2\pi}\sum_{l=1}^{s}\int\limits_{-M_l-N_l}^{-M_l}\mathbf{x}(u)\int\limits_{-\infty}^{\infty}e^{i\lambda(u-t)}\frac{1}{f(\lambda)+g(\lambda)}d\lambda du$$
$$+\frac{1}{2\pi}\int\limits_{0}^{\infty}\mathbf{x}(u)\int\limits_{-\infty}^{\infty}e^{i\lambda(u-t)}\frac{1}{f(\lambda)+g(\lambda)}d\lambda du,$$

$$(\mathbf{R}\mathbf{x})(t) = \frac{1}{2\pi}\sum_{l=1}^{s}\int\limits_{-M_l-N_l}^{-M_l}\mathbf{x}(u)\int\limits_{-\infty}^{\infty}e^{-i\lambda(u+t)}\frac{f(\lambda)}{f(\lambda)+g(\lambda)}d\lambda du$$
$$+\frac{1}{2\pi}\int\limits_{0}^{\infty}\mathbf{x}(u)\int\limits_{-\infty}^{\infty}e^{i\lambda(u+t)}\frac{f(\lambda)}{f(\lambda)+g(\lambda)}d\lambda du,$$

$$(\mathbf{Q}\mathbf{x})(t) = \frac{1}{2\pi}\sum_{l=1}^{s}\int\limits_{-M_l-N_l}^{-M_l}\mathbf{x}(u)\int\limits_{-\infty}^{\infty}e^{i\lambda(u-t)}\frac{f(\lambda)g(\lambda)}{f(\lambda)+g(\lambda)}d\lambda du$$
$$+\frac{1}{2\pi}\int\limits_{0}^{\infty}\mathbf{x}(u)\int\limits_{-\infty}^{\infty}e^{i\lambda(u-t)}\frac{f(\lambda)g(\lambda)}{f(\lambda)+g(\lambda)}d\lambda du,$$

$\mathbf{x}(t) \in L_2(T), \quad t \in T.$
Note, that equality (2.57) can be written in the form

$$\begin{aligned}
&\int\limits_{-\infty}^{\infty}\int\limits_{R^s} a(u)e^{i(u-t)}\frac{f(\lambda)}{f(\lambda)+g(\lambda)}dud\lambda \\
&-\left[\int\limits_{-\infty}^{\infty}\left(\sum_{l=1}^{s}\int\limits_{-M_l-N_l}^{-M_l}\frac{\mathbf{c}(e^{i(u-t)\lambda})}{f(\lambda)+g(\lambda)}du\right)d\lambda\right. \\
&\left.+\int\limits_{-\infty}^{\infty}\int\limits_{0}^{\infty}\frac{\mathbf{c}(e^{i(u-t)\lambda})}{f(\lambda)+g(\lambda)}dud\lambda\right] = 0.
\end{aligned} \tag{2.58}$$

Denote by $\mathbf{a}(t)$ a function of the form

$$\mathbf{a}(t)=0, t\in S, \quad \mathbf{a}(t)=a(t), t\in R^s \quad \mathbf{a}(t)=0, t\in S^+.$$

Making use of the introduced above notations, equality (2.58) can be written in the form

$$(\mathbf{Ra})(t)=(\mathbf{Bc})(t), \quad t\in T.$$

Assume that the operator $\mathbf{B}$ is invertible. In this case the function $\mathbf{c}(t)$ can be calculated by the formula

$$\mathbf{c}(t)=(\mathbf{B}^{-1}\mathbf{Ra})(t), \quad t\in T.$$

Consequently, the spectral characteristic $h(e^{i\lambda})$ of the optimal estimate $\hat{A}\xi$ is calculated by the formula

$$h(e^{i\lambda})=A(e^{i\lambda})\frac{f(\lambda)}{f(\lambda)+g(\lambda)}-\frac{C(e^{i\lambda})}{f(\lambda)+g(\lambda)}, \tag{2.59}$$

where

$$C(e^{i\lambda})=\sum_{l=1}^{s}\int_{-M_l-N_l}^{-M_l}(\mathbf{B}^{-1}\mathbf{Ra})(t)e^{it\lambda}dt+\int_{0}^{\infty}(\mathbf{B}^{-1}\mathbf{Ra})(t)e^{it\lambda}dt.$$

Formula (2.29), the form of spectral characteristic (2.59) and the introduced operators $\mathbf{B}$, $\mathbf{R}$, $\mathbf{Q}$ allows us to derive the following formula for calculating the value of the mean-square error of the estimate $\hat{A}\xi$

$$\begin{aligned}\Delta(h;f,g)&=\frac{1}{2\pi}\int_{-\infty}^{\infty}\frac{\left|A(e^{i\lambda})g(\lambda)+C(e^{i\lambda})\right|^2}{(f(\lambda)+g(\lambda))^2}f(\lambda)d\lambda\\&+\frac{1}{2\pi}\int_{-\infty}^{\infty}\frac{\left|A(e^{i\lambda})f(\lambda)-C(e^{i\lambda})\right|^2}{(f(\lambda)+g(\lambda))^2}g(\lambda)d\lambda\\&=\langle\mathbf{R}\vec{\mathbf{a}},\mathbf{B}^{-1}\mathbf{R}\vec{\mathbf{a}}\rangle+\langle\mathbf{Q}\vec{\mathbf{a}},\vec{\mathbf{a}}\rangle.\end{aligned} \tag{2.60}$$

Here

$$\langle A,C\rangle=\sum_{l=1}^{s}\int_{-M_l-N_l}^{-M_l}A(t)\overline{C(t)}dt+\int_{0}^{\infty}A(t)\overline{C(t)}dt$$

is the inner product in the space $L_2(T)$.

The following statement is true.

Theorem 2.14. *Let $\xi(t)$ and $\eta(t)$ be uncorrelated stationary stochastic processes with the spectral densities $f(\lambda)$ and $g(\lambda)$ that satisfy the minimality condition (2.53). The spectral characteristic $h(e^{i\lambda})$ and the mean-square error $\Delta(h;f,g)$ of the optimal linear estimate of the functional $A\xi$ based on observations of the process $\xi(t)+\eta(t)$ at points $t\in\mathbb{R}^-\backslash S$, can be calculated by formulas (2.59), (2.60).*

Estimation of the Functional $A_N\xi$

Consider the problem of mean-square optimal linear estimation of the functional

$$A_N\xi=\int_{N^s} a(t)\xi(-t)dt, \quad N^s=[0,N]\cap R^s,$$

based on observations of the process $\xi(t)+\eta(t)$ at points $t\in\mathbb{R}^-\backslash S$. Each linear estimate $\hat{A}_N\xi$ of the functional $A_N\xi$ is of the form

$$\hat{A}_N\xi=\int_{-\infty}^{\infty} h_N(e^{i\lambda})(Z_\xi(d\lambda)+Z_\eta(d\lambda)),$$

where $h_N(e^{i\lambda})\in L_2^s(f+g)$ is the spectral characteristic of the estimate. We introduce the function $\mathbf{a}_N(t)$ as follows

$$\mathbf{a}_N(t)=a(t), t\in T\cap[0,N], \mathbf{a}_N(t)=0, t\in T\backslash[0,N].$$

Making use of the introduced function and the derived above formulas we come to the conclusion that the spectral characteristic $h_N(e^{i\lambda})$ and the mean-square error $\Delta(h_N;f,g)$ of the estimate $\hat{A}_N\xi$ are calculated by formulas

$$h_N(e^{i\lambda})=A_N(e^{i\lambda})\frac{f(\lambda)}{f(\lambda)+g(\lambda)}-\frac{C_N(e^{i\lambda})}{f(\lambda)+g(\lambda)}, \tag{2.61}$$

$$C_N(e^{i\lambda})=\sum_{l=1}^{s}\int_{-M_l-N_l}^{-M_l}(\mathbf{B}^{-1}\mathbf{R}\mathbf{a}_N)(t)e^{it\lambda}dt+\int_0^{\infty}(\mathbf{B}^{-1}\mathbf{R}\mathbf{a}_N)(t)e^{it\lambda}dt,$$

$$A_N(e^{i\lambda}) = \int\limits_{N^s} a(t)e^{-it\lambda}dt,$$

$$\Delta(h_N; f, g) = E\left|A_N\xi - \hat{A}_N\xi\right|^2 = \\ = \langle \mathbf{R}\vec{\mathbf{a}}_N, \mathbf{B}^{-1}\mathbf{R}\vec{\mathbf{a}}_N\rangle + \langle \mathbf{Q}\vec{\mathbf{a}}_N, \vec{\mathbf{a}}_N\rangle. \tag{2.62}$$

The obtained results we formulate in the form of a corollary.

Corollary 2.13. *Suppose that $\xi(t)$ and $\eta(t)$ are uncorrelated stationary stochastic processes with the spectral densities $f(\lambda)$ and $g(\lambda)$ satisfying the minimality condition (2.53). The spectral characteristic $h_N(e^{i\lambda})$ and the mean-square error $\Delta(h_N; f, g)$ of the optimal linear estimate of the functional $A_N\xi$ based on observations of the process $\xi(t) + \eta(t)$ at points $t \in \mathbb{R}^- \backslash S$, can be calculated by formulas (2.61), (2.62).*

4.2. Minimax Robust Method of Filtering

The derived in the previous section formulas for calculating values of the mean square error and the spectral characteristic of the optimal estimate $\hat{A}\xi$ of the functional $A\xi$, can be applied in the case, where the spectral densities $f(\lambda)$, $g(\lambda)$ of the processes are exactly known. In the case of spectral uncertainty, where information about the exact form of densities is not complete, while it is known that densities belong to a certain class of admissible spectral densities, the minimax robust method of estimation of functionals may be applied. According to this method, we find an estimate which minimizes the maximal values of the mean square errors for all spectral densities from a given class D of admissible spectral densities simultaneously.

For a description of the minimax method of estimation, we introduce the following definitions.

Definition 2.6. *For a given set of spectral densities $D = D_f \times D_g$, the spectral densities $f_0(\lambda) \in D_f$, $g_0(\lambda) \in D_g$ are called the least favourable for the optimal filtering of the functional $A\xi$, if the following relation is satisfied*

$$\Delta(f_0, g_0) = \Delta(h(f_0, g_0); f_0, g_0) = \max_{(f,g) \in D_f \times D_g} \Delta(h(f, g); f, g).$$

Definition 2.7. *For a given set of spectral densities $D = D_f \times D_g$, the spectral characteristic $h^0(e^{i\lambda})$ of the optimal estimate of the functional $A\xi$ is called minimax-robust, if the following relations are satisfied*

$$h^0(e^{i\lambda}) \in H_D = \bigcap_{(f,g)\in D} L_2^s(f+g),$$

$$\min_{h\in H_D} \max_{(f,g)\in D} \Delta(h; f,g) = \max_{(f,g)\in D} \Delta\left(h^0; f,g\right).$$

By using the obtained results and the introduced definitions, the following statement can be verified.

Lemma 2.5. *The spectral densities $f_0(\lambda) \in D_f$, $g_0(\lambda) \in D_g$, that satisfy the minimality condition (2.53), are the least favourable in the class $D = D_f \times D_g$ for the optimal linear filtering of the functional $A\xi$, if the Fourier transforms of the functions*

$$(f_0(\lambda)+g_0(\lambda))^{-1}, \quad f_0(\lambda)(f_0(\lambda)+g_0(\lambda))^{-1}, \quad f_0(\lambda)g_0(\lambda)(f_0(\lambda)+g_0(\lambda))^{-1}$$

define operators $\mathbf{B}^0, \mathbf{R}^0, \mathbf{Q}^0$, which determine a solution of the optimization problem

$$\begin{aligned}\max_{(f,g)\in D_f\times D_g} \langle \mathbf{R}\vec{\mathbf{a}}, \mathbf{B}^{-1}\mathbf{R}\vec{\mathbf{a}}\rangle + \langle \mathbf{Q}\vec{\mathbf{a}}, \vec{\mathbf{a}}\rangle = \\ = \langle \mathbf{R}^0\vec{\mathbf{a}}, (\mathbf{B}^0)^{-1}\mathbf{R}^0\vec{\mathbf{a}}\rangle + \langle \mathbf{Q}^0\vec{\mathbf{a}}, \vec{\mathbf{a}}\rangle.\end{aligned} \tag{2.63}$$

The minimax spectral characteristic $h^0 = h(f_0, g_0)$ is calculated by formula (2.59) if $h(f_0, g_0) \in H_D$.

The least favourable spectral densities $f_0(\lambda)$, $g_0(\lambda)$ and the minimax spectral characteristic $h^0 = h(f_0, g_0)$ form a saddle point of the function $\Delta(h; f,g)$ on the set $H_D \times D$. The saddle point inequalities

$$\Delta\left(h^0; f,g\right) \leq \Delta\left(h^0; f_0,g_0\right) \leq \Delta(h; f_0,g_0), \quad \forall h \in H_D, \forall f \in D_f, \forall g \in D_g,$$

are true if $h^0 = h(f_0, g_0)$ and $h(f_0, g_0) \in H_D$, where (f_0, g_0) is a solution of the constrained optimization problem

$$\sup_{(f,g)\in D_f\times D_g} \Delta(h(f_0,g_0); f,g) = \Delta(h(f_0,g_0); f_0,g_0), \tag{2.64}$$

$$\Delta(h(f_0,g_0);f,g) = \frac{1}{2\pi}\int_{-\infty}^{\infty}\frac{\left|A(e^{i\lambda})g_0(\lambda)+C^0(e^{i\lambda})\right|^2}{(f_0(\lambda)+g_0(\lambda))^2}f(\lambda)d\lambda$$

$$+\frac{1}{2\pi}\int_{-\infty}^{\infty}\frac{\left|A(e^{i\lambda})f_0(\lambda)-C^0(e^{i\lambda})\right|^2}{(f_0(\lambda)+g_0(\lambda))^2}g(\lambda)d\lambda,$$

$$C^0(e^{i\lambda}) = \sum_{l=1}^{s}\int_{-M_l-N_l}^{-M_l}((\mathbf{B}^0)^{-1}\mathbf{R}^0\mathbf{a})(t)e^{it\lambda}dt+$$

$$+\int_0^{\infty}((\mathbf{B}^0)^{-1}\mathbf{R}^0\mathbf{a})(t)e^{it\lambda}dt, \quad t\in S.$$

The constrained optimization problem (2.64) is equivalent to the unconstrained optimization problem

$$\Delta_D(f,g) = -\Delta(h(f_0,g_0);f,g)+\delta((f,g)\left|D_f\times D_g\right.)\to\inf, \tag{2.65}$$

where $\delta((f,g)\left|D_f\times D_g\right.)$ is the indicator function of the set $D=D_f\times D_g$. A solution of the problem (2.65) is determined by the condition $0\in\partial\Delta_D(f_0,g_0)$, where $\partial\Delta_D(f_0)$ is the subdifferential of the convex functional $\Delta_D(f,g)$ at point (f_0,g_0) (see the book by R. T. Rockafellar [158] for additional information).

The following statement is true.

Lemma 2.6. *Let (f_0,g_0) be a solution of the optimization problem (2.65). The spectral densities $f_0(\lambda)$, $g_0(\lambda)$ are the least favourable in the class $D=D_f\times D_g$, and $h^0=h(f_0,g_0)$ is the minimax spectral characteristic of the optimal linear estimate of the functional $A\xi$, if $h(f_0,g_0)\in H_D$.*

4.3. Least Favourable Spectral Densities in the Class $D_{\varepsilon_1}^1\times D_{\varepsilon_2}^2$

Consider the problem of minimax filtering of the functional $A\xi$, based on observations of the process $\xi(t)+\eta(t)$ at points $t\in\mathbb{R}\backslash S$, under the condition that the spectral densities $f(\lambda)$ and $g(\lambda)$ of the processes belong to the class of admissible spectral densities $D=D_{\varepsilon_1}^1\times D_{\varepsilon_2}^2$,

$$D_{\varepsilon_1}^1 = \left\{f(\lambda)\left|\frac{1}{2\pi}\int_{-\infty}^{\infty}|f(\lambda)-f_1(\lambda)|d\lambda\le\varepsilon_1\right.\right\},$$

$$D_{\varepsilon_2}^2 = \left\{ g(\lambda) \left| \frac{1}{2\pi} \int_{-\infty}^{\infty} |g(\lambda) - g_1(\lambda)|^2 d\lambda \leq \varepsilon_2 \right. \right\},$$

where densities $f_1(\lambda)$, $g_1(\lambda)$ are known and fixed.

Suppose that the spectral densities $f_0(\lambda) \in D_{\varepsilon_1}^1$, $g_0(\lambda) \in D_{\varepsilon_2}^2$ and introduce the following functions

$$h_f(f_0, g_0) = \frac{\left|A(e^{i\lambda})g_0(\lambda) + C^0(e^{i\lambda})\right|^2}{(f_0(\lambda) + g_0(\lambda))^2}, \tag{2.66}$$

$$h_g(f_0, g_0) = \frac{\left|A(e^{i\lambda})f_0(\lambda) - C^0(e^{i\lambda})\right|^2}{(f_0(\lambda) + g_0(\lambda))^2}, \tag{2.67}$$

The functional $\Delta(h(f_0, g_0); f, g)$ is calculated by the formula

$$\Delta(h(f_0, g_0); f, g) = \frac{1}{2\pi} \int_{-\infty}^{\infty} h_f(f_0, g_0) f(\lambda) d\lambda + \frac{1}{2\pi} \int_{-\infty}^{\infty} h_g(f_0, g_0) g(\lambda) d\lambda.$$

Assume that the functions $h_f(f_0, g_0)$, $h_g(f_0, g_0)$, defined by formulas (2.66), (2.67), are bounded. In this case $\Delta(h(f_0, g_0); f, g)$ is a continuous linear functional in the space $L_1 \times L_1$ and the following relation is true

$$\partial \Delta_{D_{\varepsilon_1}^1 \times D_{\varepsilon_2}^2}(f_0, g_0) = -\partial \Delta(h(f_0, g_0); f_0, g_0) + \partial \delta((f_0, g_0) \left| D_{\varepsilon_1}^1 \times D_{\varepsilon_2}^2 \right.).$$

Making use of the condition $0 \in \partial \Delta_D(f_0, g_0)$, we derive the following equations which determine the least favourable densities

$$\left|A(e^{i\lambda})g_0(\lambda) + C^0(e^{i\lambda})\right| = (f_0(\lambda) + g_0(\lambda))\Psi(\lambda)\alpha_1, \tag{2.68}$$

$$\left|A(e^{i\lambda})f_0(\lambda) - C^0(e^{i\lambda})\right| = (f_0(\lambda) + g_0(\lambda))^2 (g_0(\lambda) - g_1(\lambda))\alpha_2. \tag{2.69}$$

Here $|\Psi(\lambda)| \leq 1$ and $\Psi(\lambda) = \text{sign}(f_0(\lambda) - f_1(\lambda))$, when $f_0(\lambda) \neq f_1(\lambda)$, and constants $\alpha_1 \geq 0$, $\alpha_2 \geq 0$.

Equations (2.68), (2.69) together with the optimization problem (2.63) and the normalization conditions

$$\frac{1}{2\pi} \int_{-\infty}^{\infty} |f(\lambda) - f_1(\lambda)| d\lambda = \varepsilon_1, \tag{2.70}$$

$$\frac{1}{2\pi}\int_{-\infty}^{\infty}|g(\lambda)-g_1(\lambda)|^2\,d\lambda=\varepsilon_2 \tag{2.71}$$

determine the least favourable spectral densities in the class $D^1_{\varepsilon_1}\times D^2_{\varepsilon_2}$.

Theorem 2.15. *Assume that the spectral densities $f_0(\lambda)\in D^1_{\varepsilon_1}$, $g_0(\lambda)\in D^2_{\varepsilon_2}$ satisfy the minimality condition (2.53), and the functions $h_f(f_0,g_0)$, $h_g(f_0,g_0)$, defined by formulas (2.66), (2.67), are bounded. The spectral densities $f_0(\lambda)$, $g_0(\lambda)$ are the least favourable in the class $D^1_{\varepsilon_1}\times D^2_{\varepsilon_2}$ for the optimal filtering of the functional $A\xi$, if they satisfy the system of equations (2.68) – (2.71) and determine a solution of the optimization problem (2.63). The minimax spectral characteristic of the optimal estimate of the functional $A\xi$ is calculated by formula (2.59).*

Theorem 2.16. *Assume that the spectral densities $f_0(\lambda)\in D^2_{\varepsilon_1}$, $g_0(\lambda)\in D^2_{\varepsilon_2}$,*

$$D^2_{\varepsilon_1}=\left\{f(\lambda)\left|\frac{1}{2\pi}\int_{-\infty}^{\infty}|f(\lambda)-f_1(\lambda)|^2\,d\lambda\le\varepsilon_1\right.\right\},$$

$$D^2_{\varepsilon_2}=\left\{g(\lambda)\left|\frac{1}{2\pi}\int_{-\infty}^{\infty}|g(\lambda)-g_1(\lambda)|^2\,d\lambda\le\varepsilon_2\right.\right\},$$

and satisfy the minimality condition (2.53). Assume that the functions $h_f(f_0,g_0)$, $h_g(f_0,g_0)$, defined by formulas (2.66), (2.67), are bounded. The spectral densities $f_0(\lambda)$, $g_0(\lambda)$ are the least favourable in the class $D^2_{\varepsilon_1}\times D^2_{\varepsilon_2}$ for the optimal filtering of the functional $A\xi$, if they satisfy the system of equations

$$\left|A(e^{i\lambda})g_0(\lambda)+C^0(e^{i\lambda})\right|=(f_0(\lambda)+g_0(\lambda))^2(f_0(\lambda)-f_1(\lambda))\alpha_1,$$

$$\left|A(e^{i\lambda})f_0(\lambda)-C^0(e^{i\lambda})\right|=(f_0(\lambda)+g_0(\lambda))^2(g_0(\lambda)-g_1(\lambda))\alpha_2,$$

$(\alpha_1\ge 0,\alpha_2\ge 0)$, *satisfy the normalization conditions*

$$\frac{1}{2\pi}\int_{-\infty}^{\infty}|f(\lambda)-f_1(\lambda)|^2\,d\lambda=\varepsilon_1,$$

$$\frac{1}{2\pi}\int_{-\infty}^{\infty}|g(\lambda)-g_1(\lambda)|^2 d\lambda=\varepsilon_2,$$

and determine a solution of the optimization problem (2.63) The function, defined by (2.59), is the minimax spectral characteristic of the optimal estimate of the functional $A\xi$.

Corollary 2.14. *Let the spectral density $g(\lambda)$ be known, and let the unknown spectral density $f_0(\lambda)\in D^2_{\varepsilon_1}$. Suppose that the function $f_0(\lambda)+g(\lambda)$ satisfies the minimality condition (2.53), and the function $h_f(f_0,g)$, defined by formula(2.66), is bounded. The spectral density $f_0(\lambda)$ is the least favourable in the class $D^2_{\varepsilon_1}$ for the optimal filtering of the functional $A\xi$, if it satisfies the relation*

$$\left|A(e^{i\lambda})g(\lambda)+C^0(e^{i\lambda})\right|=(f_0(\lambda)+g(\lambda))^2(f_0(\lambda)-f_1(\lambda))\alpha_1,$$

and the pair $(f_0(\lambda),g(\lambda))$ determines a solution of the optimization problem (2.63). The function, calculated by formula (2.59), is the minimax spectral characteristic of the optimal estimate of the functional $A\xi$.

4.4. Least Favourable Spectral Densities in the Class $D_{\varepsilon_1}\times D^1_{\varepsilon_2}$

Consider the problem of minimax filtering of the functional $A\xi$, based on observations of the process $\xi(t)+\eta(t)$ at points $t\in\mathbb{R}\backslash S$, under the condition that the spectral densities $f(\lambda)$ and $g(\lambda)$ of the processes belong to the class of admissible spectral densities $D_{\varepsilon_1}\times D^1_{\varepsilon_2}$,

$$D_{\varepsilon_1}=\left\{f(\lambda)\left|f(\lambda)=(1-\varepsilon_1)f_1(\lambda)+\varepsilon_1 w(\lambda),\frac{1}{2\pi}\int_{-\infty}^{\infty}f(\lambda)d\lambda\le P_1\right.\right\},$$

$$D^1_{\varepsilon_2}=\left\{g(\lambda)\left|\frac{1}{2\pi}\int_{-\infty}^{\infty}|g(\lambda)-g_1(\lambda)|d\lambda\le\varepsilon_2\right.\right\},$$

where the spectral densities $f_1(\lambda),g_1(\lambda)$ are known and fixed, and the density $w(\lambda)$ is unknown.

Densities from the class of admissible spectral densities D_{ε_1} form the "ε–contamination" model of a given bounded spectral density $f_1(\lambda)$.

Densities from the class of admissible spectral densities $D^1_{\varepsilon_2}$ form the "ε–neighbourhood" model in the space L_1 of a given bounded spectral density $g_1(\lambda)$.

Suppose that the spectral densities $f^0(\lambda) \in D_{\varepsilon_1}$, $g^0(\lambda) \in D^1_{\varepsilon_2}$ determine the bounded functions $h_f(f_0, g_0)$, $h_g(f_0, g_0)$ by formulas (2.66), (2.67). From the condition $0 \in \partial\Delta_D(f_0, g_0)$ we get the following equations that determine the least favourable spectral densities

$$\left|A(e^{i\lambda})g_0(\lambda) + C^0(e^{i\lambda})\right| = (f_0(\lambda) + g_0(\lambda))(\varphi(\lambda) + \alpha_1^{-1}), \tag{2.72}$$

$$\left|A(e^{i\lambda})f_0(\lambda) - C^0(e^{i\lambda})\right|^2 = (f_0(\lambda) + g_0(\lambda))\Psi(\lambda)\alpha_2, \tag{2.73}$$

α_1, α_2 are constants, the function $\varphi(\lambda) \leq 0$, and $\varphi(\lambda) = 0$, when $f_0(\lambda) \geq (1-\varepsilon_1)f_1(\lambda)$, the function $|\Psi(\lambda)| \leq 1$ and $\Psi(\lambda) = \text{sign}(g_0(\lambda) - g_1(\lambda))$, when $g_0(\lambda) \neq g_1(\lambda)$. Equations (2.66), (2.67) together with the normalization condition

$$\frac{1}{2\pi}\int_{-\infty}^{\infty} |g(\lambda) - g_1(\lambda)|\, d\lambda = \varepsilon_2,$$

and the optimization problem (2.63) determine the least favourable spectral densities in the class D.

The following statements are true.

Theorem 2.17. *Let the spectral densities $f^0(\lambda) \in D_{\varepsilon_1}$, $g^0(\lambda) \in D^1_{\varepsilon_2}$ satisfy the minimality condition (2.53), and let the functions, defined by formulas (2.66), (2.67), be bounded. The spectral densities $f_0(\lambda)$, $g_0(\lambda)$ are the least favourable in the class $D_{\varepsilon_1} \times D^1_{\varepsilon_2}$ for the optimal filtering of the functional $A\xi$, if they satisfy equations (2.72) – (2.73) and determine a solution of the optimization problem (2.63). The minimax spectral characteristic of the optimal estimate of the functional $A\xi$ is calculated by formula (2.59).*

Corollary 2.15. *Suppose that the density $g(\lambda)$ is known, and the unknown spectral density $f_0(\lambda) \in D_{\varepsilon_1}$. Let the function $f_0(\lambda) + g(\lambda)$ satisfy the minimality condition (2.53), and let the function $h_f(f_0, g)$, defined by (2.66), be bounded. The spectral density $f_0(\lambda)$ is the least favourable in the class D_{ε_1} for the optimal filtering of the functional $A\xi$, if it is of the form*

$$f_0(\lambda) = \max\left\{(1-\varepsilon_1)f_1(\lambda), \alpha_1\left|A(e^{i\lambda})g(\lambda) + C^0(e^{i\lambda})\right| - g(\lambda)\right\},$$

and the pair $(f_0(\lambda), g(\lambda))$ determine a solution of the optimization problem (2.63). The function, calculated by ormula (2.59), is the minimax spectral characteristic of the optimal estimate of the functional $A\xi$.

Corollary 2.16. *Suppose that that the form of the spectral density $f(\lambda)$ is known, and the unknown spectral density $g_0(\lambda) \in D^1_{\varepsilon_2}$. Suppose that the function $f(\lambda) + g_0(\lambda)$ satisfies the minimality condition (2.53), and the function $h_g(f, g_0)$, defined by formula(2.67), is bounded. The spectral density $g_0(\lambda)$ is the least favourable in the class $D^1_{\varepsilon_2}$ for the optimal filtering of the functional $A\xi$, if it is of the form*

$$g_0(\lambda) = \max\left\{g_1(\lambda), \alpha_2 \left|A(e^{i\lambda})f(\lambda) - C^0(e^{i\lambda})\right| - f(\lambda)\right\},$$

and the pair $(f(\lambda), g_0(\lambda))$ determine a solution of the optimization problem (2.63). The function, calculated by formula (2.59), is the minimax spectral characteristic of the optimal estimate of the functional $A\xi$.

Conclusion

In this chapter, the problems of interpolation, extrapolation, and filtering of linear functionals constructed from the unobserved values of the stationary processes are investigated. We, first, investigate the problems under the condition of spectral certainty, where the form of spectral densities of the processes are exactly known.

In the case of interpolation problem, formulas for calculating values of the mean-square error and the spectral characteristic of the interpolation of the functional

$$A_s\xi = \sum_{l=1}^{s} \int_{-M_l-N_l}^{-M_l} a(t)\xi(t)dt,$$

are established based on observations of the process $\xi(t) + \eta(t)$ at points $t \in \mathbb{R} \backslash S$,

$$S = \bigcup_{l=1}^{s} [-M_l - N_l, \ldots, -M_l],$$

$$M_l = \sum_{k=0}^{l} (N_k + K_k), \quad N_0 = 0, K_0 = 0.$$

The stationary process $\eta(t)$ is uncorrelated with the process $\xi(t)$.

In the case of extrapolation problem, formulas for calculating values of the mean-square error and the spectral characteristic of the optimal estimate of the functional

$$A\xi = \int_0^{\infty} a(t)\xi(t)dt,$$

are found based on observations of the process $\xi(t) + \eta(t)$ at points $t \in \mathbb{R}^- \backslash S$. The stationary process $\eta(t)$ is uncorrelated with the process $\xi(t)$.

In the case of filtering problem, formulas for calculating values of the mean-square error and the spectral characteristic of the optimal filtering of the functional

$$A\xi = \int_{R^s} a(t)\xi(-t)dt,$$

are obtained based on observations of the process $\xi(t) + \eta(t)$ at points

$$t \in \mathbb{R}^- \backslash S, \quad R^s = [0, \infty) \backslash S^+,$$

$$S^+ = \bigcup_{l=1}^{s} [M_l, \ldots, M_l + N_l],$$

$$M_l = \sum_{k=0}^{l} (N_k + K_k), \quad N_0 = 0, K_0 = 0.$$

Next, we investigate the problems in the case of spectral uncertainty, where the spectral densities of the stationary processes are not exactly known while some special sets of admissible spectral densities are given. In this case, we apply the minimax-robust method of estimation. This method allows us to find estimates that minimize the maximal values of the mean-square errors of the estimates for all spectral densities from a given class of admissible spectral densities. We derive formulas determining the least favourable spectral densities and the minimax spectral characteristics for some special classes of admissible spectral densities.

Chapter 3

Estimation of Functionals from Multidimensional Stationary Stochastic Sequences with Missing Observations

1. Extrapolation Problem for Multidimensional Stationary Stochastic Sequenceswith Missing Observations

In this section we present results of the investigation of the problem of mean-square optimal estimation of the linear functional

$$A\vec{\xi} = \sum_{j=0}^{\infty} \vec{a}(j)^{\top}\vec{\xi}(j)$$

constructed from unobserved values of a multidimensional stationary stochastic sequence $\{\vec{\xi}(j), j \in \mathbb{Z}\}$. Estimates are based on observations of the sequence with a stationary stochastic noise sequence $\{\vec{\xi}(j) + \vec{\eta}(j)\}$ at points $j \in \mathbb{Z}_{-} \backslash S = \{\ldots, -2, -1\} \backslash S$, where $S = \bigcup_{l=1}^{s} \{-M_l - N_l, -M_l - N_l + 1, \ldots, -M_l\}$. The problem is investigated in the case of spectral certainty as well as in the case of spectral uncertainty. We first propose results of the investigation of the mean-square optimal linear estimation of the linear functional in the case of spectral certainty, where the spectral densities of the signal and the noise sequences $\{\vec{\xi}(j), j \in \mathbb{Z}\}$ and $\{\vec{\eta}(j), j \in \mathbb{Z}\}$ are exactly known. To find the optimal solution of the estimation problem in this case we apply an approach based on the Hilbert space projection method proposed by A. N. Kolmogorov [1] and developed in the papers by M. P. Moklyachuk [55, 56], and M. P. Moklyachuk and O. Yu. Masytka [90],[91],[92],[93],[94],[95], [96]. We derive formulas for calculating the spectral characteristic and the mean-square error of the optimal estimate of the functional. Next, in the case of spectral uncertainty, where full information on the spectral densities is impossible, while it is known that the spectral densities of the sequences belong to some specified classes of admissible densities, the minimax-robust method of estimation is applied. This method gives us a procedure of finding estimates which minimize the maximal values of the mean-square errors of the estimates for all spectral densities from a given class of admissible spectral densities. Formulas that determine the least favourable spectral densities and the minimax-robust spectral characteristics of the optimal estimates of the functional are proposed for some specific classes of admissible spectral densities.

1.1. Hilbert Space Projection Method of Extrapolation

Let $\vec{\xi}(j) = \{\xi_k(j)\}_{k=1}^{T}$, $j \in \mathbb{Z}$ and $\vec{\eta}(j) = \{\eta_k(j)\}_{k=1}^{T}$, $j \in \mathbb{Z}$ be multidimensional stationary stochastic sequences with zero mean values, $E\vec{\xi}(j) = \vec{0}$, $E\vec{\eta}(j) = \vec{0}$ and the correlation functions which admit the spectral representation (see the book by I. I. Gikhman and A. V. Skorokhod [153] for more details)

$$R_{\xi}(n) = E\vec{\xi}(j+n)(\vec{\xi}(j))^{*} = \frac{1}{2\pi}\int_{-\pi}^{\pi} e^{in\lambda}F(\lambda)d\lambda,$$

$$R_{\xi\eta}(n) = E\vec{\xi}(j+n)(\vec{\eta}(j))^{*} = \frac{1}{2\pi}\int_{-\pi}^{\pi} e^{in\lambda}F_{\xi\eta}(\lambda)d\lambda,$$

$$R_{\eta\xi}(n) = E\vec{\eta}(j+n)(\vec{\xi}(j))^{*} = \frac{1}{2\pi}\int_{-\pi}^{\pi} e^{in\lambda}F_{\eta\xi}(\lambda)d\lambda,$$

$$R_\eta(n) = E\vec{\eta}(j+n)(\vec{\eta}(j))^* = \frac{1}{2\pi}\int_{-\pi}^{\pi} e^{in\lambda} G(\lambda)d\lambda,$$

where $F(\lambda) = \{f_{kl}(\lambda)\}_{k,l=1}^T$, $F_{\xi\eta}(\lambda) = \left\{f_{kl}^{\xi\eta}(\lambda)\right\}_{k,l=1}^T$, $F_{\eta\xi}(\lambda) = \left\{f_{kl}^{\eta\xi}(\lambda)\right\}_{k,l=1}^T$, $G(\lambda) = \{g_{kl}(\lambda)\}_{k,l=1}^T$ are spectral densities of the stationary sequences satisfying the minimality condition

$$\int_{-\pi}^{\pi} \mathrm{Tr}\,\left(F(\lambda) + F_{\xi\eta}(\lambda) + F_{\eta\xi}(\lambda) + G(\lambda)\right)^{-1} d\lambda < \infty. \tag{3.1}$$

Under this condition the mean-square error of the optimal estimate of unobserved value of the stationary sequence is nonzero (see Yu. A. Rozanov [10] for more details).

The stationary sequences $\vec{\xi}(j)$ and $\vec{\eta}(j)$ admit the spectral representations (see I. I. Gikhman and A. V. Skorokhod [153])

$$\vec{\xi}(j) = \int_{-\pi}^{\pi} e^{ij\lambda} Z_\xi(d\lambda), \qquad \vec{\eta}(j) = \int_{-\pi}^{\pi} e^{ij\lambda} Z_\eta(d\lambda), \tag{3.2}$$

where $Z_\xi(d\lambda)$ and $Z_\eta(d\lambda)$ are orthogonal stochastic measures such that the following relations are true

$$EZ_\xi(\Delta_1)(Z_\xi(\Delta_2))^* = \frac{1}{2\pi}\int_{\Delta_1\cap\Delta_2} F(\lambda)d\lambda,$$

$$EZ_\xi(\Delta_1)(Z_\eta(\Delta_2))^* = \frac{1}{2\pi}\int_{\Delta_1\cap\Delta_2} F_{\xi\eta}(\lambda)d\lambda,$$

$$EZ_\eta(\Delta_1)(Z_\xi(\Delta_2))^* = \frac{1}{2\pi}\int_{\Delta_1\cap\Delta_2} F_{\eta\xi}(\lambda)d\lambda,$$

$$EZ_\eta(\Delta_1)(Z_\eta(\Delta_2))^* = \frac{1}{2\pi}\int_{\Delta_1\cap\Delta_2} G(\lambda)d\lambda.$$

Consider the problem of mean-square optimal linear estimation of the functional

$$A\vec{\xi} = \sum_{j=0}^{\infty} \vec{a}(j)^\top \vec{\xi}(j)$$

constructed from unobserved values of the sequence $\{\vec{\xi}(j), j \in \mathbb{Z}\}$ from observations of the sequence $\vec{\xi}(j)+\vec{\eta}(j)$ at points $j \in \mathbb{Z}_-\backslash S$, $S=\bigcup_{l=1}^{s}\{-M_l-N_l,\dots,-M_l\}$.

Making use of the spectral representation (3.2) of the sequence $\vec{\xi}(j)$ we can represent the functional $A\vec{\xi}$ in the form

$$A\vec{\xi}=\int_{-\pi}^{\pi}(A(e^{i\lambda}))^{\top}Z_{\xi}(d\lambda), \quad A(e^{i\lambda})=\sum_{j=0}^{\infty}\vec{a}(j)e^{ij\lambda}.$$

We will suppose that the coefficients $\{\vec{a}(j), j=0,1,\dots\}$, which determine the functional $A\vec{\xi}$, are such that the condition

$$\sum_{j=0}^{\infty}\sum_{k=1}^{T}|a_k(j)|<\infty \tag{3.3}$$

is satisfied. This condition ensures that the functional $A\vec{\xi}$ has a finite second moment.

Consider values $\xi_k(j), k=1,\dots,T, j\in\mathbb{Z}$, and $\eta_k(j), k=1,\dots,T, j\in\mathbb{Z}$, of the sequences as elements of the Hilbert space $H=L_2(\Omega,\mathcal{F},P)$ generated by random variables ξ with zero mathematical expectations, $E\xi=0$, finite variations, $E|\xi|^2<\infty$, and the inner product $(\xi,\eta)=E(\xi\overline{\eta})$. Denote by $H^s(\xi+\eta)$ the closed linear subspace generated by elements $\{\xi_k(j)+\eta_k(j): j\in\mathbb{Z}_-\backslash S, k=1,\dots,T\}$ in the Hilbert space $H=L_2(\Omega,\mathcal{F},P)$.

Denote by $L_2(F+G)$ the Hilbert space of vector-valued functions $\vec{a}(\lambda)=\{a_k(\lambda)\}_{k=1}^{T}$ such that

$$\int_{-\pi}^{\pi}\vec{a}(\lambda)^{\top}\left(F(\lambda)+F_{\xi\eta}(\lambda)+F_{\eta\xi}(\lambda)+G(\lambda)\right)\overline{\vec{a}(\lambda)}d\lambda<\infty.$$

Denote by $L_2^s(F+G)$ the subspace of $L_2(F+G)$ generated by functions of the form

$$e^{in\lambda}\delta_k,\ \delta_k=\{\delta_{kl}\}_{l=1}^{T},\ k=1,\dots,T,\ n\in Z_-\backslash S.$$

Each linear estimate $\hat{A}\vec{\xi}$ of the functional $A\vec{\xi}$, based on observations of the sequence $\vec{\xi}(j)+\vec{\eta}(j)$, is of the form

$$\hat{A}\vec{\xi}=\int_{-\pi}^{\pi}(h(e^{i\lambda}))^{\top}(Z_{\xi}(d\lambda)+Z_{\eta}(d\lambda)), \tag{3.4}$$

where $h(e^{i\lambda}) = \left\{h_k(e^{i\lambda})\right\}_{k=1}^{T} \in L_2^s(F+G)$ is the spectral characteristic of the estimate.

The mean-square error of the estimate $\hat{A}\vec{\xi}$ is given by the formula

$$
\begin{aligned}
\Delta(h;F,G) =& E\left|A\vec{\xi}-\hat{A}\vec{\xi}\right|^2 = \\
=& \frac{1}{2\pi}\int_{-\pi}^{\pi}\left(A(e^{i\lambda})-h(e^{i\lambda})\right)^{\top} F(\lambda)\overline{\left(A(e^{i\lambda})-h(e^{i\lambda})\right)}d\lambda+ \\
&+\frac{1}{2\pi}\int_{-\pi}^{\pi}\left(h(e^{i\lambda})\right)^{\top} G(\lambda)\overline{\left(h(e^{i\lambda})\right)}d\lambda- \\
&-\frac{1}{2\pi}\int_{-\pi}^{\pi}\left(A(e^{i\lambda})-h(e^{i\lambda})\right)^{\top} F_{\xi\eta}(\lambda)\overline{h(e^{i\lambda})}d\lambda \\
&-\frac{1}{2\pi}\int_{-\pi}^{\pi}\left(h(e^{i\lambda})\right)^{\top} F_{\eta\xi}(\lambda)\overline{\left(A(e^{i\lambda})-h(e^{i\lambda})\right)}d\lambda.
\end{aligned}
\tag{3.5}
$$

According to the Hilbert space orthogonal projection method the optimal linear estimate of the functional $A\vec{\xi}$ is a projection of the element $A\vec{\xi}$ of the space $H = L_2(\Omega,\mathcal{F},P)$ on the subspace $H^s(\xi+\eta)$. The projection is determined by the following two conditions:

$$
\begin{aligned}
&1)\hat{A}\vec{\xi} \in H^s(\xi+\eta), \\
&2)A\vec{\xi}-\hat{A}\vec{\xi}\perp H^s(\xi+\eta).
\end{aligned}
$$

It follows from the second condition that the spectral characteristic $h(e^{i\lambda}) = \left\{h_k(e^{i\lambda})\right\}_{k=1}^{T}$ of the optimal linear estimate $\hat{A}\vec{\xi}$ for any $j \in \mathbb{Z}_-\setminus S$ satisfies equations

$$\frac{1}{2\pi}\int_{-\pi}^{\pi}\left(A(e^{i\lambda})-h(e^{i\lambda})\right)^{\top}F(\lambda)e^{-ij\lambda}d\lambda-$$
$$-\frac{1}{2\pi}\int_{-\pi}^{\pi}(h(e^{i\lambda}))^{\top}F_{\eta\xi}(\lambda)e^{-ij\lambda}d\lambda+$$
$$+\frac{1}{2\pi}\int_{-\pi}^{\pi}\left(A(e^{i\lambda})-h(e^{i\lambda})\right)^{\top}F_{\xi\eta}(\lambda)e^{-ij\lambda}d\lambda-$$
$$-\frac{1}{2\pi}\int_{-\pi}^{\pi}(h(e^{i\lambda}))^{\top}G(\lambda)e^{-ij\lambda}d\lambda=\vec{0}.$$

The last relation can be written in the form

$$\frac{1}{2\pi}\int_{-\pi}^{\pi}\big[(A(e^{i\lambda}))^{\top}(F(\lambda)+F_{\xi\eta}(\lambda))-$$
$$-(h(e^{i\lambda}))^{\top}(F(\lambda)+F_{\xi\eta}(\lambda)+F_{\eta\xi}(\lambda)+G(\lambda))\big]e^{-ij\lambda}d\lambda=\vec{0},\ j\in\mathbb{Z}_{-}\setminus S.$$

Hence, the function

$$\left[(A(e^{i\lambda}))^{\top}(F(\lambda)+F_{\xi\eta}(\lambda))-(h(e^{i\lambda}))^{\top}(F(\lambda)+F_{\xi\eta}(\lambda)+F_{\eta\xi}(\lambda)+G(\lambda))\right]$$

is of the form

$$(A(e^{i\lambda}))^{\top}(F(\lambda)+F_{\xi\eta}(\lambda))-(h(e^{i\lambda}))^{\top}(F(\lambda)+F_{\xi\eta}(\lambda)+F_{\eta\xi}(\lambda)+G(\lambda))=$$
$$=(C(e^{i\lambda}))^{\top},$$
$$C(e^{i\lambda})=\sum_{j\in U}\vec{c}(j)e^{ij\lambda},$$

where $U=S\cup\{0,1,\ldots\}$, and $\vec{c}(j)$, $j\in U$ are unknown coefficients to be determined.

From the last relation we deduce that the spectral characteristic of the optimal linear estimate $\hat{A}\vec{\xi}$ is of the form

$$(h(e^{i\lambda}))^{\top}=$$
$$=(A(e^{i\lambda}))^{\top}(F(\lambda)+F_{\xi\eta}(\lambda))(F_{\zeta}(\lambda))^{-1}-(C(e^{i\lambda}))^{\top}(F_{\zeta}(\lambda))^{-1},\quad(3.6)$$

where $F_{\zeta}(\lambda)=F(\lambda)+F_{\xi\eta}(\lambda)+F_{\eta\xi}(\lambda)+G(\lambda)$.

From the first condition, $\hat{A}\vec{\xi}\in H^{s}(\xi+\eta)$, which determines the optimal estimate of the functional $A\vec{\xi}$, it follows that

$$\frac{1}{2\pi}\int_{-\pi}^{\pi}h(e^{i\lambda})e^{-ij\lambda}d\lambda=\vec{0},\quad j\in U,$$

namely

$$\frac{1}{2\pi}\int_{-\pi}^{\pi}\left((A(e^{i\lambda}))^{\top}(F(\lambda)+F_{\xi\eta}(\lambda))(F_{\zeta}(\lambda))^{-1}-\right.$$
$$\left.-(C(e^{i\lambda}))^{\top}(F_{\zeta}(\lambda))^{-1}\right)e^{-ij\lambda}d\lambda=\vec{0},\ j\in U.$$

Disclose brackets and write the last equation in the form

$$\sum_{k=0}^{\infty}(\vec{a}(k))^{\top}\frac{1}{2\pi}\int_{-\pi}^{\pi}(F(\lambda)+F_{\xi\eta}(\lambda))(F_{\zeta}(\lambda))^{-1}e^{i(k-j)\lambda}d\lambda-$$
$$-\sum_{l\in U}(\vec{c}(l))^{\top}\frac{1}{2\pi}\int_{-\pi}^{\pi}(F_{\zeta}(\lambda))^{-1}e^{i(l-j)\lambda}d\lambda=\vec{0}. \quad (3.7)$$

To represent relation (3.7) in more appropriate form we introduce the Fourier coefficients of the functions

$$B(k-j)=\frac{1}{2\pi}\int_{-\pi}^{\pi}(F_{\zeta}(\lambda))^{-1}e^{-i(k-j)\lambda}d\lambda;$$
$$R(k-j)=\frac{1}{2\pi}\int_{-\pi}^{\pi}(F(\lambda)+F_{\xi\eta}(\lambda))(F_{\zeta}(\lambda))^{-1}e^{-i(k-j)\lambda}d\lambda; \quad (3.8)$$
$$Q(k-j)=$$
$$\frac{1}{2\pi}\int_{-\pi}^{\pi}\left(F(\lambda)(F_{\zeta}(\lambda))^{-1}G(\lambda)-F_{\xi\eta}(\lambda)(F_{\zeta}(\lambda))^{-1}F_{\eta\xi}(\lambda)\right)e^{-i(k-j)\lambda}d\lambda.$$

Denote by $\vec{\mathbf{a}}^\top = (0,0,\ldots,0,\vec{a}^\top)$ the vector which has first $T\cdot|S| = T\cdot\sum_{k=1}^{s}(N_k+1)$ zero components, and the last component $\vec{a}^\top = (\vec{a}(0)^\top, \vec{a}(1)^\top \ldots)$ is constructed from coefficients which define the functional $A\vec{\xi}$. Now we can represent relation (3.7) in the form

$$\mathbf{R}\vec{\mathbf{a}} = \mathbf{B}\vec{\mathbf{c}}, \tag{3.9}$$

where $\vec{\mathbf{c}}$ is the vector constructed of the unknown coefficients $\vec{c}(k), k\in U$. The linear operator $\mathbf{B}$ in the space ℓ_2 is defined by the matrix

$$B = \begin{pmatrix} B_{s,s} & B_{s,s-1} & \ldots & B_{s,1} & B_{s,n} \\ B_{s-1,s} & B_{s-1,s-1} & \ldots & B_{s-1,1} & B_{s-1,n} \\ \vdots & \vdots & \ddots & \vdots & \vdots \\ B_{1,s} & B_{1,s-1} & \ldots & B_{1,1} & B_{1,n} \\ B_{n,s} & B_{n,s-1} & \ldots & B_{n,1} & B_{n,n} \end{pmatrix},$$

where elements in the last column and in the last row are compound matrices constructed with the help of the block-matrices

$$B_{l,n}(k,j) = B(k-j),\ l = 1,2,\ldots,s;\ k = -M_l - N_l,\ldots,-M_l;\ j = 0,1,2,\ldots;$$
$$B_{n,m}(k,j) = B(k-j), m = 1,2,\ldots,s;\ k = 0,1,2,\ldots;\ j = -M_m - N_m,\ldots,-M_m;$$
$$B_{n,n}(k,j) = B(k-j),\ k,j = 0,1,2,\ldots,$$

and other elements of the matrix B are compound matrices constructed with the help of the block-matrices

$$B_{l,m}(j,k) = B(k-j),\ l,m = 1,2,\ldots,s;\ k = -M_l - N_l,\ldots,-M_l;$$
$$j = -M_m - N_m,\ldots,-M_m.$$

The linear operator $\mathbf{R}$ in the space ℓ_2 is defined by the corresponding matrix in the same manner.
The unknown coefficients $\vec{c}(k), k\in U$, which are determined by equation (3.9) can be calculated by the formula

$$\vec{c}(k) = (\mathbf{B}^{-1}\mathbf{R}\vec{\mathbf{a}})(k),$$

where $(\mathbf{B}^{-1}\mathbf{R}\vec{\mathbf{a}})(k)$ is the k-th component of the vector $\mathbf{B}^{-1}\mathbf{R}\vec{\mathbf{a}}$. We will suppose that the operator $\mathbf{B}$ is invertible (see the paper by H. Salehi [155]) for more details).

Hence the spectral characteristic $h(e^{i\lambda})$ of the estimate $\hat{A}\vec{\xi}$ can be calculated by the formula

$$(h(e^{i\lambda}))^{\top} = (A(e^{i\lambda}))^{\top}(F(\lambda)+F_{\xi\eta}(\lambda))(F_{\zeta}(\lambda))^{-1} - \\ - \left(\sum_{k\in T}(\mathbf{B}^{-1}\mathbf{R}\vec{\mathbf{a}})(k)e^{ik\lambda}\right)^{\top}(F_{\zeta}(\lambda))^{-1}. \quad (3.10)$$

The mean-square error of the estimate $\hat{A}\xi$ can be calculated by the formula (3.5) which can be represented in the form

$$\Delta(h;F,G,F_{\xi\eta},F_{\eta\xi}) = E\left|A\vec{\xi}-\hat{A}\vec{\xi}\right|^2 =$$

$$= \frac{1}{2\pi}\int_{-\pi}^{\pi}(a(\lambda))^{\top}F(\lambda)\overline{a(\lambda)}d\lambda + \frac{1}{2\pi}\int_{-\pi}^{\pi}(b(\lambda))^{\top}G(\lambda)\overline{b(\lambda}d\lambda -$$

$$-\frac{1}{2\pi}\int_{-\pi}^{\pi}(a(\lambda))^{\top}F_{\xi\eta}(\lambda)\overline{b(\lambda}d\lambda - \frac{1}{2\pi}\int_{-\pi}^{\pi}(b(\lambda))^{\top}F_{\eta\xi}(\lambda)\overline{a(\lambda)}d\lambda =$$

$$= \frac{1}{2\pi}\int_{-\pi}^{\pi}(A(e^{i\lambda}))^{\top}(F(\lambda)G(\lambda)-F_{\xi\eta}(\lambda)F_{\eta\xi}(\lambda))(F_{\zeta}(\lambda))^{-1}\overline{A(e^{i\lambda})}d\lambda$$

$$+\frac{1}{2\pi}\int_{-\pi}^{\pi}(C(e^{i\lambda}))^{\top}(F_{\zeta}(\lambda))^{-1}\overline{C(e^{i\lambda})}d\lambda =$$

$$= \langle\mathbf{R}\vec{\mathbf{a}},\mathbf{B}^{-1}\mathbf{R}\vec{\mathbf{a}}\rangle + \langle\mathbf{Q}\vec{\mathbf{a}},\vec{\mathbf{a}}\rangle, \quad (3.11)$$

where

$$(a(\lambda))^{\top} = (A(e^{i\lambda}))^{\top}(F_{\eta\xi}(\lambda)+G(\lambda))(F_{\zeta}(\lambda))^{-1} + (C(e^{i\lambda}))^{\top}(F_{\zeta}(\lambda))^{-1},$$

$$(b(\lambda))^{\top} = (A(e^{i\lambda}))^{\top}(F(\lambda)+F_{\xi\eta}(\lambda))(F_{\zeta}(\lambda))^{-1} - (C(e^{i\lambda}))^{\top}(F_{\zeta}(\lambda))^{-1},$$

and $\langle a,c\rangle = \sum_k a_k \overline{c}_k$ is the inner product in the space ℓ_2.

The linear operator $\mathbf{Q}$ in the space ℓ_2 is defined by the corresponding matrix in the same manner as the operator $\mathbf{B}$ is defined.

Thus we obtain the following theorem.

Theorem 3.1. *Let $\{\vec{\xi}(j), j \in \mathbb{Z}\}$ and $\{\vec{\eta}(j), j \in \mathbb{Z}\}$ be multidimensional stationary stochastic sequences with the spectral density matrices $F(\lambda), F_{\xi\eta}(\lambda), F_{\eta\xi}(\lambda), G(\lambda)$ and let the minimality condition (3.1) be satisfied. The spectral characteristic $h(e^{i\lambda})$ and the mean-square error $\Delta(h;F,G,F_{\xi\eta},F_{\eta\xi})$ of the optimal linear estimate of the functional $A\vec{\xi}$, based on observations of the sequence $\vec{\xi}(j)+\vec{\eta}(j)$, $j \in \mathbb{Z}_-\backslash S$, can be calculated by formulas (3.10), (3.11).*

Estimates Based on Observations of Uncorrelated Sequences

The corresponding results can be obtained for the uncorrelated sequences $\{\vec{\xi}(j), j \in \mathbb{Z}\}$ and $\{\vec{\eta}(j), j \in \mathbb{Z}\}$. In this case the spectral densities $F_{\xi\eta}(\lambda) = 0$, $F_{\eta\xi}(\lambda) = 0$ and we get the following corollary.

Corollary 3.1. *Let $\{\vec{\xi}(j), j \in \mathbb{Z}\}$ and $\{\vec{\eta}(j), j \in \mathbb{Z}\}$ be uncorrelated multidimensional stationary stochastic sequences with the spectral densities $F(\lambda)$ and $G(\lambda)$ satisfying the minimality condition*

$$\int_{-\pi}^{\pi} Tr(F(\lambda)+G(\lambda))^{-1} d\lambda < \infty. \tag{3.12}$$

The spectral characteristic $h(e^{i\lambda})$ and the mean-square error $\Delta(F,G)$ of the optimal linear estimate of the functional $A\vec{\xi}$, based on observations of the sequence $\vec{\xi}(j)+\vec{\eta}(j)$, $j \in \mathbb{Z}_-\backslash S$, can be calculated by the formulas

$$\begin{aligned}(h(e^{i\lambda}))^\top &= (A(e^{i\lambda}))^\top F(\lambda)(F(\lambda)+G(\lambda))^{-1} - \\ &- \left(\sum_{k\in U} (\mathbf{B}^{-1}\mathbf{R}\vec{\mathbf{a}})(k) e^{ik\lambda}\right)^\top (F(\lambda)+G(\lambda))^{-1},\end{aligned} \tag{3.13}$$

$$\Delta(h;F,G) = E\left|A\vec{\xi} - \hat{A}\vec{\xi}\right|^2 = \frac{1}{2\pi}\int_{-\pi}^{\pi} (r_G(\lambda))^\top F(\lambda)\overline{r_G(\lambda)} d\lambda +$$

$$+\frac{1}{2\pi}\int_{-\pi}^{\pi}(r_F(\lambda))^{\top}G(\lambda)\overline{r_F(\lambda)}d\lambda =$$

$$= \langle \mathbf{R}\vec{\mathbf{a}}, \mathbf{B}^{-1}\mathbf{R}\vec{\mathbf{a}}\rangle + \langle \mathbf{Q}\vec{\mathbf{a}}, \vec{\mathbf{a}}\rangle, \tag{3.14}$$

where

$$(r_F(\lambda))^{\top} =$$
$$= \left((A(e^{i\lambda}))^{\top}F(\lambda) - \left(\sum_{k\in U}(\mathbf{B}^{-1}\mathbf{R}\vec{\mathbf{a}})(k)e^{ik\lambda}\right)^{\top}\right)(F(\lambda)+G(\lambda))^{-1},$$

$$(r_G(\lambda))^{\top} =$$
$$= \left((A(e^{i\lambda}))^{\top}G(\lambda) + \left(\sum_{k\in U}(\mathbf{B}^{-1}\mathbf{R}\vec{\mathbf{a}})(k)e^{ik\lambda}\right)^{\top}\right)(F(\lambda)+G(\lambda))^{-1},$$

and $\mathbf{B}, \mathbf{R}, \mathbf{Q}$ *are linear operators in the space* ℓ_2 *determined by compound matrices constructed of the block-matrices* $B(k-j), R(k-j), Q(k-j)$ *respectively which are defined by the Fourier coefficients of the functions*

$$B(k-j) = \frac{1}{2\pi}\int_{-\pi}^{\pi}(F(\lambda)+G(\lambda))^{-1}e^{-i(k-j)\lambda}d\lambda;$$

$$R(k-j) = \frac{1}{2\pi}\int_{-\pi}^{\pi}F(\lambda)(F(\lambda)+G(\lambda))^{-1}e^{-i(k-j)\lambda}d\lambda; \tag{3.15}$$

$$Q(k-j) = \frac{1}{2\pi}\int_{-\pi}^{\pi}F(\lambda)(F(\lambda)+G(\lambda))^{-1}G(\lambda)e^{-i(k-j)\lambda}d\lambda.$$

Estimates Based on Observations without Noise

Consider, next, the estimation problem in the case where the stationary sequence $\{\vec{\xi}(j), j\in\mathbb{Z}\}$ is observed without noise. Since in this case $G(\lambda)=0$, the spectral characteristic of the estimate $\hat{A}\vec{\xi}$ is of the form

$$(h(e^{i\lambda}))^{\top} = (A(e^{i\lambda}))^{\top} - (C(e^{i\lambda}))^{\top}(F(\lambda))^{-1}, \tag{3.16}$$
$$C(e^{i\lambda}) = \sum_{j\in U} \vec{c}(j)e^{ij\lambda},$$

and the system of equations (3.9) can be represented in the form

$$\vec{\mathbf{a}} = \mathbf{B}\vec{\mathbf{c}}, \tag{3.17}$$

where $\mathbf{B}$ is the linear operator in the space ℓ_2 constructed with the help of the Fourier coefficients of the function $(F(\lambda))^{-1}$. It is of the similar form as operators defined before.

Hence, the unknown coefficients $\vec{c}(j), j \in U$, can be calculated by the formula

$$\vec{c}(j) = \left(\mathbf{B}^{-1}\vec{\mathbf{a}}\right)(j),$$

where $\left(\mathbf{B}^{-1}\vec{\mathbf{a}}\right)(j)$ is the j-th component of the vector $\mathbf{B}^{-1}\vec{\mathbf{a}}$, and the spectral characteristic of the estimate $\hat{A}\vec{\xi}$ is determined by the formula

$$(h(e^{i\lambda})^{\top} = \left(\sum_{j=0}^{\infty} \vec{a}(j)e^{ij\lambda}\right)^{\top} - \left(\sum_{j\in U} \left(\mathbf{B}^{-1}\vec{\mathbf{a}}\right)(j)e^{ij\lambda}\right)^{\top} (F(\lambda))^{-1}. \tag{3.18}$$

The mean-square error of the estimate $\hat{A}\vec{\xi}$ is determined by the formula

$$\Delta(h;F) = \langle \mathbf{B}^{-1}\vec{\mathbf{a}}, \vec{\mathbf{a}} \rangle. \tag{3.19}$$

Let us summarize the obtained results in the form of a corollary.

Corollary 3.2. *Let $\{\vec{\xi}(j), j \in \mathbb{Z}\}$ be a multidimensional stationary stochastic sequence with the spectral density $F(\lambda)$ satisfying the minimality condition*

$$\int_{-\pi}^{\pi} Tr\,(F(\lambda))^{-1} d\lambda < \infty. \tag{3.20}$$

The spectral characteristic $h(e^{i\lambda})$ and the mean-square error $\Delta(h,F)$ of the optimal linear estimate $\hat{A}\vec{\xi}$ of the functional $A\vec{\xi}$ based on observations of the sequence $\vec{\xi}(j)$ at points $j \in \mathbb{Z}_-\backslash S$, $S = \bigcup_{l=1}^{s}\{-M_l - N_l, \ldots, -M_l\}$, can be calculated by formulas (3.18), (3.19).

Estimation of the Functional $A_N\vec{\xi}$

Let $\{\vec{\xi}(j)\}$ and $\{\vec{\eta}(j)\}$ be uncorrelated stationary stochastic sequences. Consider the problem of mean-square optimal linear extrapolation of the functional

$$A_N\vec{\xi} = \sum_{j=0}^{N} \vec{a}(j)^{\top}\vec{\xi}(j)$$

based on observations of the sequence $\vec{\xi}(j)+\vec{\eta}(j)$ at points $j \in \mathbb{Z}_-\backslash S$, where $S = \bigcup_{l=1}^{s}\{-M_l-N_l, -M_l-N_l+1, \ldots, -M_l\}$. In order to find the spectral characteristic $h_N(e^{i\lambda})$ of the estimate

$$\hat{A}_N\vec{\xi} = \int_{-\pi}^{\pi}(h_N(e^{i\lambda}))^{\top}(Z_{\xi}(d\lambda)+Z_{\eta}(d\lambda))$$

and the mean-square error $\Delta(h_N;F,G)$ of the estimate of the functional $A_N\vec{\xi}$, we define the vector $\mathbf{a}_N^{\top} = (0,0,\ldots,0,\vec{a}_N^{\top})$ which has first $T\cdot|S| = T\cdot\sum_{k=1}^{s}(N_k+1)$ zero components and the last component is $\vec{a}_N^{\top} = (\vec{a}(0)^{\top},\vec{a}(1)^{\top},\ldots,\vec{a}(N)^{\top},0,0,\ldots)$.

Consider the linear operator $\mathbf{R}_N$ in the space ℓ_2 which is defined as follows: $\mathbf{R}_N(k,j)=\mathbf{R}(k,j)$, $j\leq N$, $\mathbf{R}_N(k,j)=0$, $j>N$.

Thus the spectral characteristic of the optimal estimate $\hat{A}_N\xi$ can be calculated by the formula

$$h_N(e^{i\lambda}))^{\top} = (A_N(e^{i\lambda}))^{\top}F(\lambda)(F(\lambda)+G(\lambda))^{-1} - \left(\sum_{k\in U}(\mathbf{B}^{-1}\mathbf{R}_N\mathbf{\vec{a}}_N)(k)e^{ik\lambda}\right)^{\top}(F(\lambda)+G(\lambda))^{-1}. \quad (3.21)$$

The mean-square error of the estimate $\hat{A}_N\vec{\xi}$ is determined by the formula

$$\Delta(h_N;F,G) = E\left|A_N\vec{\xi}-\hat{A}_N\vec{\xi}\right|^2 =$$

$$= \frac{1}{2\pi}\int_{-\pi}^{\pi}(r_G^N(\lambda))^{\top}F(\lambda)\overline{r_G^N(\lambda)}d\lambda + \frac{1}{2\pi}\int_{-\pi}^{\pi}(r_F^N(\lambda))^{\top}G(\lambda)\overline{r_F^N(\lambda)}d\lambda =$$

$$= \langle \mathbf{R}_N \vec{\mathbf{a}}_N \mathbf{B}^{-1} \mathbf{R}_N \vec{\mathbf{a}}_N \rangle + \langle \mathbf{Q}_N \vec{\mathbf{a}}_N, \vec{\mathbf{a}}_N \rangle, \tag{3.22}$$

where

$$(r_F^N(\lambda))^\top =$$

$$= \left((A_N(e^{i\lambda}))^\top F(\lambda) - \left(\sum_{k\in U} (\mathbf{B}^{-1}\mathbf{R}_N \vec{\mathbf{a}_N})(k) e^{ik\lambda} \right)^\top \right) (F(\lambda)+G(\lambda))^{-1},$$

$$(r_G^N(\lambda))^\top =$$

$$= \left((A_N(e^{i\lambda}))^\top G(\lambda) + \left(\sum_{k\in U} (\mathbf{B}^{-1}\mathbf{R}_N \vec{\mathbf{a}_N})(k) e^{ik\lambda} \right)^\top \right) (F(\lambda)+G(\lambda))^{-1},$$

and $\mathbf{Q}_N$ is the linear operator in the space ℓ_2, $\mathbf{Q}_N(k,j) = \mathbf{Q}(k,j)$, $k,j \leq N$, $\mathbf{Q}_N(k,j) = 0$ if $k > N$ or $j > N$. Note, that linear operators $\mathbf{B}$, $\mathbf{R}$, $\mathbf{Q}$ are defined in Corollary 3.1.

Corollary 3.3. *Let $\{\vec{\xi}(j), j \in \mathbb{Z}\}$ and $\{\vec{\eta}(j), j \in \mathbb{Z}\}$ be uncorrelated multidimensional stationary stochastic sequences with the spectral densities $F(\lambda)$ and $G(\lambda)$ satisfying the minimality condition (3.12). The spectral characteristic $h_N(e^{i\lambda})$ and the mean-square error $\Delta(h_N; F, G)$ of the optimal linear estimate of the functional $A_N\vec{\xi}$ based on observations of the sequence $\vec{\xi}(j) + \vec{\eta}(j)$, $j \in \mathbb{Z}_- \setminus S$ can be calculated by formulas (3.21), (3.22).*

Estimation of the Functional $A_N\vec{\xi}$ Based on Observations without Noise

In the case, where the sequence $\{\vec{\xi}(j), j \in \mathbb{Z}\}$ is observed without noise, we have the following corollary.

Corollary 3.4. *Let $\{\vec{\xi}(j), j \in \mathbb{Z}\}$ be a multidimensional stationary stochastic sequence with the spectral density $F(\lambda)$ satisfying the minimality condition (3.20). The spectral characteristic $h_N(e^{i\lambda})$ and the mean-square error $\Delta(h_N, F)$ of the optimal linear estimate $\hat{A}_N\vec{\xi}$ of the functional $A_N\vec{\xi}$ can be calculated by the formulas*

$$(h_N(e^{i\lambda}))^\top = \left(\sum_{j=0}^{N} \vec{a}(j) e^{ij\lambda} \right)^\top - \left(\sum_{j\in U} \left(\mathbf{B}^{-1}\vec{\mathbf{a}}_N\right)(j) e^{ij\lambda} \right)^\top (F(\lambda))^{-1}, \tag{3.23}$$

$$\Delta(h_N;F) = \langle \mathbf{B}^{-1}\vec{\mathbf{a}}_N, \vec{\mathbf{a}}_N \rangle. \tag{3.24}$$

The linear operator **B** *is defined in Corollary 3.1.*

To demonstrate the developed technique, we propose the following example.

Example 3.1. *Consider the problem of the optimal linear estimation of the functional*

$$A_1\vec{\xi} = \vec{a}(0)^\top \vec{\xi}(0) + \vec{a}(1)^\top \vec{\xi}(1)$$

constructed from unobserved values of a stationary sequence $\vec{\xi}(j) = \{\xi_k(j)\}_{k=1}^2$ *from observations of the sequence* $\vec{\xi}(j)$ *at points* $j \in \mathbb{Z}_- \backslash S$, *where* $S = \{-3,-2\}$, $\vec{a}(0) = (1,1)^\top$, $\vec{a}(1) = (1,1)^\top$. *Let* $\xi_1(n) = \xi(n)$ *be a stationary stochastic sequence with the spectral density* $f(\lambda)$, *and let* $\xi_2(n) = \xi(n) + \eta(n)$, *where* $\eta(n)$ *is an uncorrelated with* $\xi(n)$ *stationary stochastic sequence with the spectral density* $g(\lambda)$. *In this case the matrix of spectral densities is of the form*

$$F(\lambda) = \begin{pmatrix} f(\lambda) & f(\lambda) \\ f(\lambda) & f(\lambda)+g(\lambda) \end{pmatrix},$$

and the inverse matrix is as follows

$$(F(\lambda))^{-1} = \begin{pmatrix} \frac{1}{f(\lambda)} + \frac{1}{g(\lambda)} & \frac{-1}{g(\lambda)} \\ \frac{-1}{g(\lambda)} & \frac{1}{g(\lambda)} \end{pmatrix}.$$

Let

$$f(\lambda) = \frac{1}{\left|1-b_1e^{i\lambda}\right|^2}, \quad g(\lambda) = \frac{1}{\left|1-b_2e^{i\lambda}\right|^2}, \; |b_1|<1, |b_2|<1.$$

In this case the inverse matrix is of the form

$$(F(\lambda))^{-1} = \begin{pmatrix} \left|1-b_1e^{i\lambda}\right|^2 + \left|1-b_2e^{i\lambda}\right|^2 & -\left|1-b_2e^{i\lambda}\right|^2 \\ -\left|1-b_2e^{i\lambda}\right|^2 & \left|1-b_2e^{i\lambda}\right|^2 \end{pmatrix} =$$

$$= B(-1)e^{-i\lambda} + B(0) + B(1)e^{i\lambda},$$

where

$$B(0) = \begin{pmatrix} 2+b_1^2+b_2^2 & -1-b_2^2 \\ -1-b_2^2 & 1+b_2^2 \end{pmatrix},$$

$$B(1)=B(-1)=\begin{pmatrix} -b_1-b_2 & b_2 \\ b_2 & -b_2 \end{pmatrix},$$

are the Fourier coefficients of the function $(F(\lambda))^{-1}$.

According to the Corollary 3.4 the spectral characteristic of the optimal estimate $\hat{A}_1\vec{\xi}$ *of the functional* $A_1\vec{\xi}$ *is calculated by the formula*

$$(h_1(e^{i\lambda}))^\top = \left(\vec{a}(0)+\vec{a}(1)e^{i\lambda}\right)^\top -$$
$$-\left(\sum_{j\in U}\left(\mathbf{B}^{-1}\vec{\mathbf{a}}_1\right)(j)e^{ij\lambda}\right)^\top \left(B(-1)e^{-i\lambda}+B(0)+B(1)e^{i\lambda}\right),$$

where the vector $\vec{\mathbf{a}}_1^\top = (0,0,0,0,\vec{a}(0)^\top,\vec{a}(1)^\top,0,0,\ldots)$.

To find the unknown coefficients $\vec{c}(j) = \left(\mathbf{B}^{-1}\vec{\mathbf{a}}_1\right)(j), j \in U = S\cup\{0,1,2,\ldots\}$, *we use equation (3.17), where* $\vec{\mathbf{c}}^\top = (\vec{c}(-3)^\top,\vec{c}(-2)^\top,\vec{c}(0)^\top,\vec{c}(1)^\top,\vec{c}(2)^\top,\vec{c}(3)^\top,\ldots)$. *The operator* $\mathbf{B}$ *is defined by the matrix*

$$B=\begin{pmatrix}
B(0) & B(-1) & 0 & 0 & 0 & 0 & 0 & \ldots \\
B(1) & B(0) & 0 & 0 & 0 & 0 & 0 & \ldots \\
0 & 0 & B(0) & B(-1) & 0 & 0 & 0 & \ldots \\
0 & 0 & B(1) & B(0) & B(-1) & 0 & 0 & \ldots \\
0 & 0 & 0 & B(1) & B(0) & B(-1) & 0 & \ldots \\
0 & 0 & 0 & 0 & B(1) & B(0) & B(-1) & \ldots \\
0 & 0 & 0 & 0 & 0 & B(1) & B(0) & \ldots \\
\ldots
\end{pmatrix}.$$

We have to find the inverse matrix B^{-1} *which defines the inverse operator* $\mathbf{B}^{-1}$. *We first represent the matrix* B *in the form*

$$B=\begin{pmatrix} B_{00} & 0 \\ 0 & B_{11} \end{pmatrix},$$

where

$$B_{00}=\begin{pmatrix} B(0) & B(-1) \\ B(1) & B(0) \end{pmatrix},$$

$$B_{11} = \begin{pmatrix} B(0) & B(-1) & 0 & 0 & 0 & \dots \\ B(1) & B(0) & B(-1) & 0 & 0 & \dots \\ 0 & B(1) & B(0) & B(-1) & 0 & \dots \\ 0 & 0 & B(1) & B(0) & B(-1) & \dots \\ 0 & 0 & 0 & B(1) & B(0) & \dots \\ \dots \end{pmatrix}.$$

Making use of the indicated representation we conclude that the matrix B^{-1} *can be represented in the form*

$$B^{-1} = \begin{pmatrix} B_{00}^{-1} & 0 \\ 0 & B_{11}^{-1} \end{pmatrix},$$

where B_{00}^{-1}, B_{11}^{-1} *are inverse matrices to the matrices* B_{00}, B_{11} *respectively. The matrix* B_{00}^{-1} *can be found in the form*

$$B_{00}^{-1} = \begin{pmatrix} \frac{1+b_1^2}{A} & \frac{1+b_1^2}{A} & \frac{b_1}{A} & \frac{b_1}{A} \\ \frac{1+b_1^2}{A} & \frac{1+b_1^2}{A} + \frac{1+b_2^2}{B} & \frac{b_1}{A} & \frac{b_1}{A} + \frac{b_2}{B} \\ \frac{b_1}{A} & \frac{b_1}{A} & \frac{1+b_1^2}{A} & \frac{1+b_1^2}{A} \\ \frac{b_1}{A} & \frac{b_1}{A} + \frac{b_2}{B} & \frac{1+b_1^2}{A} & \frac{1+b_1^2}{A} + \frac{1+b_2^2}{B} \end{pmatrix},$$

where $A = 1 + b_1^2 + b_1^4$, $B = 1 + b_2^2 + b_2^4$. *In order to find the matrix* $(B_{11})^{-1}$ *we use the following method. The matrix* B_{11} *is constructed with the help of the Fourier coefficients of the function* $(F(\lambda))^{-1}$

$$B_{11}(k, j) = B(k - j), \quad k, j = 0, 1, 2, \dots.$$

The density $(F(\lambda))^{-1}$ *admits the factorization*

$$(F(\lambda))^{-1} = \sum_{p=-\infty}^{\infty} B(p)e^{ip\lambda} = \left(\sum_{j=0}^{\infty} \psi(j)e^{-ij\lambda}\right) \cdot \left(\sum_{j=0}^{\infty} \psi(j)e^{-ij\lambda}\right)^* =$$

$$= \left(\sum_{j=0}^{\infty} \theta(j)e^{-ij\lambda}\right)^* \cdot \left(\sum_{j=0}^{\infty} \theta(j)e^{-ij\lambda}\right),$$

where

$$\psi(0) = \begin{pmatrix} 1 & 1 \\ 0 & -1 \end{pmatrix}, \psi(1) = \begin{pmatrix} -b_1 & -b_2 \\ 0 & b_2 \end{pmatrix}, \psi(j) = 0, j > 1,$$

$$\theta(j) = \begin{pmatrix} b_1^j & b_1^j \\ 0 & -b_2^j \end{pmatrix}, j \geq 0.$$

Hence $B(p) = \sum_{k=0}^{\infty} \psi(k)(\psi(k+p))^*$, $p \geq 0$, *and* $B(-p) = (B(p))^*$, $p \geq 0$. *Then*

$$B(i-j) = \sum_{l=\max(i,j)}^{\infty} \psi(l-i)(\psi(l-j))^*.$$

Denote by Ψ *and* Θ *linear operators in the space* ℓ_2 *determined by matrices with elements* $\Psi(i,j) = \psi(j-i)$, $\Theta(i,j) = \theta(j-i)$, *for* $0 \leq i \leq j$, $\Psi(i,j) = 0$, $\Theta(i,j) = 0$, *for* $0 \leq j < i$. *Then elements of the matrix* B_{11} *can be represented in the form* $B_{11}(i,j) = (\Psi\Psi^*)(i,j)$. *Since* $\Psi\Theta = \Theta\Psi = I$, *elements of the matrix* B_{11}^{-1} *can be calculated by the formula*

$$B_{11}^{-1}(i,j) = (\Theta^*\Theta)(i,j) = \sum_{l=0}^{\min(i,j)} (\theta(i-l))^*\theta(j-l).$$

From equation (3.17) we can find the unknown coefficients $c(j), j \in U$,

$$\begin{aligned}
&\vec{c}(-3) = \vec{0},\\
&\vec{c}(-2) = \vec{0},\\
&\vec{c}(0) = B_{11}^{-1}(0,0)\vec{a}(0) + B_{11}^{-1}(0,1)\vec{a}(1),\\
&\vec{c}(1) = B_{11}^{-1}(1,0)\vec{a}(0) + B_{11}^{-1}(1,1)\vec{a}(1),\\
&\vec{c}(2) = B_{11}^{-1}(2,0)\vec{a}(0) + B_{11}^{-1}(2,1)\vec{a}(1),\\
&\dots\\
&\vec{c}(i) = B_{11}^{-1}(i,0)\vec{a}(0) + B_{11}^{-1}(i,1)\vec{a}(1), \quad i > 2.
\end{aligned}$$

Hence the spectral characteristic of the optimal estimate is calculated by the formula

$$(h_1(e^{i\lambda}))^{\top} = \left(\vec{a}(0) + \vec{a}(1)e^{i\lambda}\right)^{\top} - (\vec{c}(-3)e^{-i3\lambda} + \vec{c}(-2)e^{-i2\lambda} +$$

$$\vec{c}(0) + \vec{c}(1)e^{i\lambda} + \vec{c}(2)e^{i2\lambda} + \sum_{j>2} \vec{c}(j)e^{ij\lambda}) \left(B(-1)e^{-i\lambda} + B(0) + B(1)e^{i\lambda}\right) =$$

$$= -\vec{c}(0)^{\top}B(-1)e^{-i\lambda} - \vec{c}(1)^{\top}B(1)e^{i2\lambda} - \vec{c}(2)^{\top}B(0)e^{i2\lambda} - \vec{c}(2)^{\top}B(1)e^{i3\lambda} -$$
$$- \sum_{j>2} \vec{c}(j)^{\top} e^{ij\lambda} \left(B(-1)e^{-i\lambda} + B(0) + B(1)e^{i\lambda}\right).$$

Since coefficients $\vec{c}(j-1)^{\top}B(1) + \vec{c}(j)^{\top}B(0) + \vec{c}(j+1)^{\top}B(-1)$ for $j \geq 2$ are zero, the spectral characteristic of the estimate $\hat{A}_1\vec{\xi}$ is of the form

$$(h_1(e^{i\lambda}))^{\top} = -\vec{c}(0)^{\top}B(-1)e^{-i\lambda} = \left(b_2 + b_2^2 - 2(b_1 + b_1^2),\ -b_2 - b_2^2\right) e^{-i\lambda}.$$

The mean-square error of the estimate of the functional $A_1\vec{\xi}$ is calculated by the formula

$$\Delta(h_1; F) = \langle B^{-1}\vec{\mathbf{a}}_1, \vec{\mathbf{a}}_1 \rangle = 10 + 8b_1 + 4b_1^2 + 2b_2 + b_2^2.$$

1.2. Minimax Approach to Extrapolation Problem

Theorem 3.1 and its Corollaries 3.1 – 3.4 can be applied for finding solutions of the extrapolation problem for multidimensional stationary sequences with missing observations only in the case of spectral certainty, where the spectral densities $F(\lambda), F_{\xi\eta}(\lambda), F_{\eta\xi}(\lambda), G(\lambda)$ are exactly known. If the complete information about the spectral densities is impossible while a class of admissible spectral density matrices D is given, the minimax robust method of extrapolation is reasonable. According to this method, we find an estimate which minimizes the maximal values of the mean square errors of estimates for all spectral density matrices from a given class D of admissible spectral densities simultaneously.

For description of the minimax method of estimation we introduce the following definitions (see papers by M. P. Moklyachuk [55, 56], and M. P. Moklyachuk and O. Yu. Masyutka [90] - [96] for more details).

Definition 3.1. *For a given class of spectral density matrices D the spectral densities $(F^0(\lambda), F^0_{\xi\eta}(\lambda), F^0_{\eta\xi}(\lambda), G^0(\lambda)) \in D$ are called the least favourable in the class D for the optimal linear extrapolation of the functional $A\vec{\xi}$ if the following relation is true*

$$\Delta\left(h\left(F^0, F^0_{\xi\eta}, F^0_{\eta\xi}, G^0\right); F^0, F^0_{\xi\eta}, F^0_{\eta\xi}, G^0\right) =$$
$$= \max_{(F, F_{\xi\eta}, F_{\eta\xi}, G) \in D} \Delta\left(h\left(F, F_{\xi\eta}, F_{\eta\xi}, G\right); F, F_{\xi\eta}, F_{\eta\xi}, G\right).$$

Definition 3.2. *For a given class of spectral density matrices D the spectral characteristic $h^0(e^{i\lambda})$ of the optimal linear estimate of the functional $A\vec{\xi}$ is called minimax robust if there are satisfied conditions*

$$h^0(e^{i\lambda}) \in H_D = \bigcap_{(F,F_{\xi\eta},F_{\eta\xi},G)\in D} L_2^s(F+G),$$

$$\min_{h\in H_D} \max_{(F,F_{\xi\eta},F_{\eta\xi},G)\in D} \Delta\left(h; F, F_{\xi\eta}, F_{\eta\xi}, G\right) =$$

$$= \max_{(F,F_{\xi\eta},F_{\eta\xi},G)\in D} \Delta\left(h^0; F, F_{\xi\eta}, F_{\eta\xi}, G\right).$$

From the introduced definitions and derived formulas we can obtain the following statement.

Lemma 3.1. *The spectral densities $(F^0(\lambda), F^0_{\xi\eta}(\lambda), F^0_{\eta\xi}(\lambda), G^0(\lambda)) \in D$, satisfying the minimality condition (3.1), are the least favourable in the class D for the optimal linear extrapolation of the functional $A\vec{\xi}$ if the Fourier coefficients* (3.8) *of the functions*

$$(F^0_\zeta(\lambda))^{-1}, \quad (F^0(\lambda)+F^0_{\xi\eta}(\lambda))(F^0_\zeta(\lambda))^{-1},$$

$$F^0(\lambda)(F^0_\zeta(\lambda))^{-1}G^0(\lambda) - F^0_{\xi\eta}(\lambda)(F^0_\zeta(\lambda))^{-1}F^0_{\eta\xi}(\lambda)$$

define operators $\mathbf{B}^0, \mathbf{R}^0, \mathbf{Q}^0$ which determine a solution of the constrained optimization problem

$$\max_{(F,F_{\xi\eta},F_{\eta\xi},G)\in D} (\langle \mathbf{R}\vec{\mathbf{a}}, \mathbf{B}^{-1}\mathbf{R}\vec{\mathbf{a}}\rangle + \langle \mathbf{Q}\vec{\mathbf{a}}, \vec{\mathbf{a}}\rangle) =$$

$$= \langle \mathbf{R}^0\vec{\mathbf{a}}, (\mathbf{B}^0)^{-1}\mathbf{R}^0\vec{\mathbf{a}}\rangle + \langle \mathbf{Q}^0\vec{\mathbf{a}}, \vec{\mathbf{a}}\rangle. \tag{3.25}$$

The minimax spectral characteristic $h^0(e^{i\lambda}) = h(F^0, F^0_{\xi\eta}, F^0_{\eta\xi}, G^0)$ is calculated by the formula (3.10) if $h(F^0, F^0_{\xi\eta}, F^0_{\eta\xi}, G^0) \in H_D$.

In the case of uncorrelated stationary sequences the corresponding definitions and lemmas are as follows.

Definition 3.3. *For a given class of spectral densities $D = D_F \times D_G$ the spectral densities $F^0(\lambda) \in D_F$, $G^0(\lambda) \in D_G$ are called the least favourable in the class D for the optimal linear extrapolation of the functional $A\vec{\xi}$ based on observations of the uncorrelated sequences if the following relation is true*

$$\Delta\left(F^0, G^0\right) = \Delta\left(h\left(F^0, G^0\right); F^0, G^0\right) = \max_{(F,G) \in D_F \times D_G} \Delta\left(h\left(F, G\right); F, G\right).$$

Definition 3.4. *For a given class of spectral densities $D = D_F \times D_G$ the spectral characteristic $h^0(e^{i\lambda})$ of the optimal linear estimate of the functional $A\vec{\xi}$ based on observations of the uncorrelated sequences is called minimax robust if there are satisfied conditions*

$$h^0(e^{i\lambda}) \in H_D = \bigcap_{(F,G) \in D_F \times D_G} L_2^s(F+G),$$

$$\min_{h \in H_D} \max_{(F,G) \in D} \Delta\left(h; F, G\right) = \max_{(F,G) \in D} \Delta\left(h^0; F, G\right).$$

Lemma 3.2. *The spectral densities $F^0(\lambda) \in D_F$, $G^0(\lambda) \in D_G$ satisfying the minimality condition (3.12) are the least favourable in the class $D = D_F \times D_G$ for the optimal linear extrapolation of the functional $A\vec{\xi}$ based on observations of the uncorrelated sequences if the Fourier coefficients* (3.15) *of the functions*

$$(F^0(\lambda) + G^0(\lambda))^{-1}, F^0(\lambda)(F^0(\lambda) + G^0(\lambda))^{-1}, F^0(\lambda)(F^0(\lambda) + G^0(\lambda))^{-1} G^0(\lambda)$$

define operators $\mathbf{B}^0, \mathbf{R}^0, \mathbf{Q}^0$ which determine a solution of the constrained optimization problem

$$\max_{(F,G) \in D_F \times D_G} \left(\langle \mathbf{R}\vec{\mathbf{a}}, \mathbf{B}^{-1}\mathbf{R}\vec{\mathbf{a}} \rangle + \langle \mathbf{Q}\vec{\mathbf{a}}, \vec{\mathbf{a}} \rangle\right) =$$

$$= \langle \mathbf{R}^0\vec{\mathbf{a}}, (\mathbf{B}^0)^{-1}\mathbf{R}^0\vec{\mathbf{a}} \rangle + \langle \mathbf{Q}^0\vec{\mathbf{a}}, \vec{\mathbf{a}} \rangle. \tag{3.26}$$

The minimax spectral characteristic $h^0 = h(F^0, G^0)$ is calculated by the formula (3.13) if $h(F^0, G^0) \in H_D$.

In the case of observations of the sequence without noise we obtain the following corollary.

Corollary 3.5. *Let the spectral density $F^0(\lambda) \in D_F$ satisfy the minimality condition (3.20). The spectral density $F^0(\lambda) \in D_F$ is the least favourable in the class D_F for the optimal linear extrapolation of the functional $A\vec{\xi}$ if the Fourier coefficients of the function $(F^0(\lambda))^{-1}$ define the operator $\mathbf{B}^0$ which determines a solution of the optimization problem*

$$\max_{F \in D_F} \langle \mathbf{B}^{-1}\vec{\mathbf{a}}, \vec{\mathbf{a}} \rangle = \langle (\mathbf{B}^0)^{-1}\vec{\mathbf{a}}, \vec{\mathbf{a}} \rangle. \tag{3.27}$$

The minimax spectral characteristic $h^0 = h(F^0)$ is calculated by the formula (3.18) if $h(F^0) \in H_{D_F}$.

The least favourable spectral densities $F^0(\lambda)$, $G^0(\lambda)$ and the minimax spectral characteristic $h^0 = h(F^0, G^0)$ form a saddle point of the function $\Delta(h; F, G)$ on the set $H_D \times D$. The saddle point inequalities

$$\Delta\left(h; F^0, G^0\right) \geq \Delta\left(h^0; F^0, G^0\right) \geq \Delta\left(h^0; F, G\right)$$

$$\forall h \in H_D, \forall F \in D_F, \forall G \in D_G$$

are true if $h^0 = h(F^0, G^0)$ and $h(F^0, G^0) \in H_D$, where (F^0, G^0) is a solution of the constrained optimization problem

$$\sup_{(F,G) \in D_F \times D_G} \Delta\left(h(F^0, G^0); F, G\right) = \Delta\left(h(F^0, G^0); F^0, G^0\right), \tag{3.28}$$

$$\Delta\left(h(F^0, G^0); F, G\right) =$$
$$= \frac{1}{2\pi} \int_{-\pi}^{\pi} (r_G^0(\lambda))^{\top} F(\lambda) \overline{r_G^0(\lambda)} d\lambda + \frac{1}{2\pi} \int_{-\pi}^{\pi} (r_F^0(\lambda))^{\top} G(\lambda) \overline{r_F^0(\lambda)} d\lambda,$$

$$(r_F^0(\lambda))^{\top} =$$
$$= \left((A(e^{i\lambda}))^{\top} F^0(\lambda) - \left(\sum_{k \in U} ((\mathbf{B}^0)^{-1}\mathbf{R}^0\vec{\mathbf{a}})(k) e^{ik\lambda} \right)^{\top} \right) (F^0(\lambda) + G^0(\lambda))^{-1},$$

$$(r_G^0(\lambda))^\top =$$

$$= \left((A(e^{i\lambda}))^\top G^0(\lambda) + \left(\sum_{k\in U} ((\mathbf{B}^0)^{-1}\mathbf{R}^0\vec{\mathbf{a}})(k)e^{ik\lambda} \right)^\top \right) (F^0(\lambda)+G^0(\lambda))^{-1}.$$

The constrained optimization problem (3.28) is equivalent to the unconstrained optimization problem (see B. N. Pshenichnyj [156]):

$$\Delta_D(F,G) = -\Delta(h(F^0,G^0);F,G) + \delta((F,G)\,|D_F \times D_G) \to \inf, \qquad (3.29)$$

where $\delta((F,G)\,|D_F \times D_G)$ is the indicator function of the set $D = D_F \times D_G$.

A solution of the problem (3.29) is determined by the condition $0 \in \partial\Delta_D(F^0,G^0)$, where $\partial\Delta_D(F^0,G^0)$ is the subdifferential of the convex functional $\Delta_D(F,G)$ at point (F^0,G^0). This is the necessary and sufficient condition under which the pair (F^0,G^0) belongs to the set of minima of the convex functional $\Delta(h(F^0,G^0);F,G)$. This condition makes it possible to find the least favourable spectral densities in some special classes of spectral densities D (see books by A. D. Ioffe and V. M. Tihomirov [161], B. N. Pshenichnyj [156], T. Rockafellar [158]).

Note, that the form of the functional $\Delta\left(h^0;F,G\right)$ is convenient for application of the Lagrange method of indefinite multipliers for finding solutions of the problem (3.29). Making use of the method of Lagrange multipliers and the form of subdifferentials of the indicator functions we describe relations that determine the least favourable spectral densities in some special classes of spectral densities (see books by M. P. Moklyachuk [55, 163], M. P. Moklyachuk and O. Yu. Masyutka [96] for additional details).

1.3. Least Favourable Spectral Densities in the Class $D = D_0 \times D_V^U$

Consider the problem of extrapolation of the functional $A\vec{\xi}$ based on observations of the uncorrelated sequences in the case where the spectral densities of the sequences belong to the classes $D_0^k \times D_V^{U\,k}, k = 1,2,3,4$, where

$$D_0^1 = \left\{ F(\lambda) \left| \frac{1}{2\pi}\int_{-\pi}^{\pi} \operatorname{Tr} F(\lambda)d\lambda = p \right. \right\},$$

$$D_V^{U\,1} = \left\{ G(\lambda) \middle| \operatorname{Tr} V(\lambda) \leq \operatorname{Tr} G(\lambda) \leq \operatorname{Tr} U(\lambda), \frac{1}{2\pi}\int_{-\pi}^{\pi} \operatorname{Tr} G(\lambda) d\lambda = q \right\},$$

$$D_0^2 = \left\{ F(\lambda) \middle| \frac{1}{2\pi}\int_{-\pi}^{\pi} f_{kk}(\lambda) d\lambda = p_k, k = \overline{1,T} \right\},$$

$$D_V^{U\,2} = \left\{ G(\lambda) \middle| v_{kk}(\lambda) \leq g_{kk}(\lambda) \leq u_{kk}(\lambda), \right.$$
$$\left. \frac{1}{2\pi}\int_{-\pi}^{\pi} g_{kk}(\lambda) d\lambda = q_k, k = \overline{1,T} \right\},$$

$$D_0^3 = \left\{ F(\lambda) \middle| \frac{1}{2\pi}\int_{-\pi}^{\pi} \langle B_1, F(\lambda)\rangle \, d\lambda = p \right\},$$

$$D_V^{U\,3} = \left\{ G(\lambda) \middle| \langle B_2, V(\lambda)\rangle \leq \langle B_2, G(\lambda)\rangle \leq \langle B_2, U(\lambda)\rangle, \right.$$
$$\left. \frac{1}{2\pi}\int_{-\pi}^{\pi} \langle B_2, G(\lambda)\rangle \, d\lambda = q \right\},$$

$$D_0^4 = \left\{ F(\lambda) \middle| \frac{1}{2\pi}\int_{-\pi}^{\pi} F(\lambda) d\lambda = P \right\},$$

$$D_V^{U\,4} = \left\{ G(\lambda) \middle| V(\lambda) \leq G(\lambda) \leq U(\lambda), \frac{1}{2\pi}\int_{-\pi}^{\pi} G(\lambda) d\lambda = Q \right\}.$$

Here the spectral densities $V(\lambda), U(\lambda)$ are known and fixed, $p, q, p_k, q_k, k = \overline{1,T}$ are given numbers, P, Q, B_1, B_2 are given positive definite Hermitian matrices.

From the condition $0 \in \partial\Delta_D(F^0, G^0)$ we find the following equations which determine the least favourable spectral densities for these given sets of admissible spectral densities.

For the first pair $D_0^1 \times D_V^{U\,1}$ we have equations

$$(r_G^0(\lambda))^*(r_G^0(\lambda))^\top = \alpha^2(F^0(\lambda) + G^0(\lambda))^2, \tag{3.30}$$

$$(r_F^0(\lambda))^*(r_F^0(\lambda))^\top = (\beta^2 + \gamma_1(\lambda) + \gamma_2(\lambda))(F^0(\lambda) + G^0(\lambda))^2, \tag{3.31}$$

where α^2, β^2 are Lagrange multipliers, $\gamma_1(\lambda) \leq 0$ and $\gamma_1(\lambda) = 0$ if $\operatorname{Tr} G^0(\lambda) > \operatorname{Tr} V(\lambda)$, $\gamma_2(\lambda) \geq 0$ and $\gamma_2(\lambda) = 0$ if $\operatorname{Tr} G^0(\lambda) < \operatorname{Tr} U(\lambda)$.

For the second pair $D_0^2 \times D_V^{U2}$ we have equations

$$(r_G^0(\lambda))^*(r_G^0(\lambda))^\top = (F^0(\lambda) + G^0(\lambda)) \left\{\alpha_k^2 \delta_{kl}\right\}_{k,l=1}^T (F^0(\lambda) + G^0(\lambda)), \quad (3.32)$$

$$(r_F^0(\lambda))^*(r_F^0(\lambda))^\top =$$
$$= (F^0(\lambda) + G^0(\lambda)) \left\{(\beta_k^2 + \gamma_{1k}(\lambda) + \gamma_{2k}(\lambda))\delta_{kl}\right\}_{k,l=1}^T (F^0(\lambda) + G^0(\lambda)), \quad (3.33)$$

where α_k^2, β_k^2 are Lagrange multipliers, δ_{kl} are Kronecker symbols, $\gamma_{1k}(\lambda) \le 0$ and $\gamma_{1k}(\lambda) = 0$ if $g_{kk}^0(\lambda) > v_{kk}(\lambda)$, $\gamma_{2k}(\lambda) \ge 0$ and $\gamma_{2k}(\lambda) = 0$ if $g_{kk}^0(\lambda) < u_{kk}(\lambda)$.

For the third pair $D_0^3 \times D_V^{U3}$ we have equations

$$(r_G^0(\lambda))^*(r_G^0(\lambda))^\top = \alpha^2 (F^0(\lambda) + G^0(\lambda)) B_1^\top (F^0(\lambda) + G^0(\lambda)), \quad (3.34)$$

$$(r_F^0(\lambda))^*(r_F^0(\lambda))^\top =$$
$$= (\beta^2 + \gamma_1'(\lambda) + \gamma_2'(\lambda))(F^0(\lambda) + G^0(\lambda)) B_2^\top (F^0(\lambda) + G^0(\lambda)), \quad (3.35)$$

where α^2, β^2 are Lagrange multipliers, $\gamma_1'(\lambda) \le 0$ and $\gamma_1'(\lambda) = 0$ if $\langle B_2, G^0(\lambda)\rangle > \langle B_2, V(\lambda)\rangle$, $\gamma_2'(\lambda) \ge 0$ and $\gamma_2'(\lambda) = 0$ if $\langle B_2, G^0(\lambda)\rangle < \langle B_2, U(\lambda)\rangle$.

For the fourth pair $D_0^4 \times D_V^{U4}$ we have equations

$$(r_G^0(\lambda))^*(r_G^0(\lambda))^\top = (F^0(\lambda) + G^0(\lambda)) \vec{\alpha} \cdot \vec{\alpha}^* (F^0(\lambda) + G^0(\lambda)), \quad (3.36)$$

$$(r_F^0(\lambda))^*(r_F^0(\lambda))^\top =$$
$$= (F^0(\lambda) + G^0(\lambda))(\vec{\beta} \cdot \vec{\beta}^* + \Gamma_1(\lambda) + \Gamma_2(\lambda))(F^0(\lambda) + G^0(\lambda)) \quad (3.37)$$

where $\vec{\alpha}, \vec{\beta}$ are Lagrange multipliers, $\Gamma_1(\lambda) \le 0$ and $\Gamma_1(\lambda) = 0$ if $G^0(\lambda) > V(\lambda)$, $\Gamma_2(\lambda) \ge 0$ and $\Gamma_2(\lambda) = 0$ if $G^0(\lambda) < U(\lambda)$.

The following theorem and corollaries are true.

Theorem 3.2. *The least favourable spectral densities $F^0(\lambda)$, $G^0(\lambda)$ in the classes $D_0^k \times D_V^{Uk}, k = 1,2,3,4$ for the optimal linear extrapolation of the functional $A\vec{\xi}$ are determined by relations (3.30) – (3.37), the minimality condition (3.12), the constrained optimization problem (3.26), and restrictions on densities from the corresponding classes $D_0^k \times D_V^{Uk}, k = 1,2,3,4$. The minimax-robust spectral characteristic of the optimal estimate of the functional $A\vec{\xi}$ is determined by formula (3.13).*

Corollary 3.6. *The least favourable spectral densities $F^0(\lambda)$ in the classes D_0^k, $k=1,2,3,4$, for the optimal linear extrapolation of the functional $A\vec{\xi}$ based on observations of the sequence $\vec{\xi}(j)$ at points $j \in \mathbb{Z}_-\backslash S$, $S=\bigcup_{l=1}^{s}\{-M_l-N_l,\dots,-M_l\}$, are determined by the following equations, respectively,*

$$((C^0(\lambda))^\top)^* \cdot (C^0(\lambda))^\top = \alpha^2 (F^0(\lambda))^2, \tag{3.38}$$

$$((C^0(\lambda))^\top)^* \cdot (C^0(\lambda))^\top = F^0(\lambda)\left\{\alpha_k^2 \delta_{kl}\right\}_{k,l=1}^T F^0(\lambda), \tag{3.39}$$

$$((C^0(\lambda))^\top)^* \cdot (C^0(\lambda))^\top = \alpha^2 F^0(\lambda) B_1^\top F^0(\lambda), \tag{3.40}$$

$$((C^0(\lambda))^\top)^* \cdot (C^0(\lambda))^\top = F^0(\lambda)\vec{\alpha}\cdot\vec{\alpha}^* F^0(\lambda), \tag{3.41}$$

the minimality condition (3.20), the constrained optimization problem (3.27), and restrictions on densities from the corresponding classes D_0^k, $k=1,2,3,4$. The minimax spectral characteristic of the optimal estimate of the functional $A\vec{\xi}$ is determined by formula (3.18).

Corollary 3.7. *The least favourable spectral densities $F^0(\lambda)$ in the classes $D_V^{U\,k}$, $k=1,2,3,4$, for the optimal linear extrapolation of the functional $A\vec{\xi}$ based on observations of the sequence $\vec{\xi}(j)$ at points $j \in \mathbb{Z}_-\backslash S$, $S=\bigcup_{l=1}^{s}\{-M_l-N_l,\dots,-M_l\}$, are determined by the following equations, respectively,*

$$((C^0(\lambda))^\top)^* \cdot (C^0(\lambda))^\top = (\beta^2+\gamma_1(\lambda)+\gamma_2(\lambda))(F^0(\lambda))^2, \tag{3.42}$$

$$((C^0(\lambda))^\top)^* \cdot (C^0(\lambda))^\top = F^0(\lambda)\left\{(\beta_k^2+\gamma_{1k}(\lambda)+\gamma_{2k}(\lambda))\delta_{kl}\right\}_{k,l=1}^T F^0(\lambda), \tag{3.43}$$

$$((C^0(\lambda))^\top)^* \cdot (C^0(\lambda))^\top = (\beta^2+\gamma_1'(\lambda)+\gamma_2'(\lambda))F^0(\lambda)B_2^\top F^0(\lambda), \tag{3.44}$$

$$((C^0(\lambda))^\top)^* \cdot (C^0(\lambda))^\top = F^0(\lambda)(\vec{\beta}\cdot\vec{\beta}^*+\Gamma_1(\lambda)+\Gamma_2(\lambda))F^0(\lambda), \tag{3.45}$$

the minimality condition (3.20), the constrained optimization problem (3.27), and restrictions on densities from the corresponding classes $D_V^{U\,k}$, $k=1,2,3,4$. The minimax spectral characteristic of the optimal estimate of the functional $A_s\vec{\xi}$ is determined by formula (3.18).

1.4. Least Favourable Spectral Densities in the Class $D = D_\varepsilon \times D_{1\delta}$

Consider the problem of minimax extrapolation of the functional $A\vec{\xi}$ based on observations of the uncorrelated sequences in the case where the spectral densities of the sequences belong to the classes $D = D_\varepsilon^k \times D_{1\delta}^k, k = 1,2,3,4,$

$$D_\varepsilon^1 = \left\{ F(\lambda) \middle| \operatorname{Tr} F(\lambda) = (1-\varepsilon) \operatorname{Tr} F_1(\lambda) + \varepsilon \operatorname{Tr} W(\lambda), \right. \\ \left. \frac{1}{2\pi} \int_{-\pi}^{\pi} \operatorname{Tr} F(\lambda) d\lambda = p \right\};$$

$$D_{1\delta}^1 = \left\{ G(\lambda) \middle| \frac{1}{2\pi} \int_{-\pi}^{\pi} |\operatorname{Tr}(G(\lambda) - G_1(\lambda))| d\lambda \leq \delta \right\};$$

$$D_\varepsilon^2 = \left\{ F(\lambda) \middle| f_{kk}(\lambda) = (1-\varepsilon) f_{kk}^1(\lambda) + \varepsilon w_{kk}(\lambda), \right. \\ \left. \frac{1}{2\pi} \int_{-\pi}^{\pi} f_{kk}(\lambda) d\lambda = p_k, k = \overline{1,T} \right\};$$

$$D_{1\delta}^2 = \left\{ G(\lambda) \middle| \frac{1}{2\pi} \int_{-\pi}^{\pi} \left| g_{kk}(\lambda) - g_{kk}^1(\lambda) \right| d\lambda \leq \delta_k, k = \overline{1,T} \right\};$$

$$D_\varepsilon^3 = \left\{ F(\lambda) \middle| \langle B_1, F(\lambda) \rangle = (1-\varepsilon) \langle B_1, F_1(\lambda) \rangle + \varepsilon \langle B_1, W(\lambda) \rangle, \right. \\ \left. \frac{1}{2\pi} \int_{-\pi}^{\pi} \langle B_1, F(\lambda) \rangle d\lambda = p \right\};$$

$$D_{1\delta}^3 = \left\{ G(\lambda) \middle| \frac{1}{2\pi} \int_{-\pi}^{\pi} |\langle B_2, G(\lambda) - G_1(\lambda) \rangle| d\lambda \leq \delta \right\};$$

$$D_\varepsilon^4 = \left\{ F(\lambda) \middle| F(\lambda) = (1-\varepsilon) F_1(\lambda) + \varepsilon W(\lambda), \frac{1}{2\pi} \int_{-\pi}^{\pi} F(\lambda) d\lambda = P \right\},$$

$$D_{1\delta}^4 = \left\{ G(\lambda) \middle| \frac{1}{2\pi} \int_{-\pi}^{\pi} \left| g_{ij}(\lambda) - g_{ij}^1(\lambda) \right| d\lambda \leq \delta_i^j, i,j = \overline{1,T} \right\}.$$

Here $F_1(\lambda), G_1(\lambda)$ are known and fixed spectral density matrices, $W(\lambda)$ is an unknown spectral density matrix, $p, \delta, \delta_k, p_k, k=\overline{1,T}$, $\delta_i^j, i,j=\overline{1,T}$, are given numbers, P is a given positive definite Hermitian matrices.

From the condition $0 \in \partial\Delta_D(F^0, G^0)$ we find the following equations which determine the least favourable spectral densities for these given sets of admissible spectral densities.

For the first pair $D_\varepsilon^1 \times D_{1\delta}^1$ we have equations

$$(r_G^0(\lambda))^*(r_G^0(\lambda))^\top = (\alpha^2+\gamma_1(\lambda))(F^0(\lambda)+G^0(\lambda))^2, \tag{3.46}$$

$$(r_F^0(\lambda))^*(r_F^0(\lambda))^\top = \beta^2\gamma_2(\lambda)(F^0(\lambda)+G^0(\lambda))^2, \tag{3.47}$$

$$\frac{1}{2\pi}\int_{-\pi}^{\pi}\left|\mathrm{Tr}\,(G^0(\lambda)-G_1(\lambda))\right|d\lambda=\delta, \tag{3.48}$$

where α^2, β^2 are Lagrange multipliers, $\gamma_1(\lambda) \leq 0$ and $\gamma_1(\lambda)=0$ if $\mathrm{Tr}\,F^0(\lambda) > (1-\varepsilon)\mathrm{Tr}\,F_1(\lambda)$, $|\gamma_2(\lambda)| \leq 1$ and

$$\gamma_2(\lambda)=\mathrm{sign}\,(\mathrm{Tr}\,(G^0(\lambda)-G_1(\lambda))):\ \mathrm{Tr}\,(G^0(\lambda)-G_1(\lambda)) \neq 0.$$

For the second pair $D_\varepsilon^2 \times D_{1\delta}^2$ we have equations

$$(r_G^0(\lambda))^*(r_G^0(\lambda))^\top =$$
$$= (F^0(\lambda)+G^0(\lambda))\left\{(\alpha_k^2+\gamma_k^1(\lambda))\delta_{kl}\right\}_{k,l=1}^T(F^0(\lambda)+G^0(\lambda)), \tag{3.49}$$

$$(r_F^0(\lambda))^*(r_F^0(\lambda))^\top =$$
$$= (F^0(\lambda)+G^0(\lambda))\left\{\beta_k^2\gamma_k^2(\lambda)\delta_{kl}\right\}_{k,l=1}^T(F^0(\lambda)+G^0(\lambda)), \tag{3.50}$$

$$\frac{1}{2\pi}\int_{-\pi}^{\pi}\left|g_{kk}^0(\lambda)-g_{kk}^1(\lambda)\right|d\lambda=\delta_k, \tag{3.51}$$

where α_k^2, β_k^2 are Lagrange multipliers, $\gamma_k^1(\lambda) \leq 0$ and $\gamma_k^1(\lambda)=0$ if $f_{kk}^0(\lambda) > (1-\varepsilon)f_{kk}^1(\lambda)$, $\left|\gamma_k^2(\lambda)\right| \leq 1$ and

$$\gamma_k^2(\lambda)=\mathrm{sign}\,(g_{kk}^0(\lambda)-g_{kk}^1(\lambda)):\ g_{kk}^0(\lambda)-g_{kk}^1(\lambda) \neq 0,\ k=\overline{1,T}.$$

For the third pair $D_\varepsilon^3 \times D_{1\delta}^3$ we have equations

$$(r_G^0(\lambda))^*(r_G^0(\lambda))^\top =$$
$$= (\alpha^2 + \gamma_1'(\lambda))(F^0(\lambda) + G^0(\lambda))B_1^\top(F^0(\lambda) + G^0(\lambda)), \quad (3.52)$$

$$(r_F^0(\lambda))^*(r_F^0(\lambda))^\top = \beta^2\gamma_2'(\lambda)(F^0(\lambda) + G^0(\lambda))B_2^\top(F^0(\lambda) + G^0(\lambda)), \quad (3.53)$$

$$\frac{1}{2\pi}\int_{-\pi}^{\pi} \left|\left\langle B_2, G^0(\lambda) - G_1(\lambda)\right\rangle\right| d\lambda = \delta, \quad (3.54)$$

where α^2, β^2 are Lagrange multipliers, $\gamma_1'(\lambda) \leq 0$ and $\gamma_1'(\lambda) = 0$ if $\langle B_1, F^0(\lambda)\rangle > (1-\varepsilon)\langle B_1, F_1(\lambda)\rangle$, $|\gamma_2'(\lambda)| \leq 1$ and

$$\gamma_2'(\lambda) = \text{sign}\left\langle B_2, G^0(\lambda) - G_1(\lambda)\right\rangle : \left\langle B_2, G^0(\lambda) - G_1(\lambda)\right\rangle \neq 0.$$

For the fourth pair $D_\varepsilon^4 \times D_{1\delta}^4$ we have equations

$$(r_G^0(\lambda))^*(r_G^0(\lambda))^\top =$$
$$= (F^0(\lambda) + G^0(\lambda))(\vec{\alpha}\cdot\vec{\alpha}^* + \Gamma(\lambda))(F^0(\lambda) + G^0(\lambda)), \quad (3.55)$$

$$(r_F^0(\lambda))^*(r_F^0(\lambda))^\top =$$
$$= (F^0(\lambda) + G^0(\lambda))\left\{\beta_{ij}\gamma_{ij}(\lambda)\right\}_{i,j=1}^{T}(F^0(\lambda) + G^0(\lambda)), \quad (3.56)$$

$$\frac{1}{2\pi}\int_{-\pi}^{\pi} \left|g_{ij}^0(\lambda) - g_{ij}^1(\lambda)\right| d\lambda = \delta_i^j, \quad (3.57)$$

where $\vec{\alpha}, \beta_{ij}$ are Lagrange multipliers, $\Gamma(\lambda) \leq 0$ and $\Gamma(\lambda) = 0$ if $F^0(\lambda) > (1-\varepsilon)F_1(\lambda)$, $\left|\gamma_{ij}(\lambda)\right| \leq 1$ and

$$\gamma_{ij}(\lambda) = \frac{g_{ij}^0(\lambda) - g_{ij}^1(\lambda)}{\left|g_{ij}^0(\lambda) - g_{ij}^1(\lambda)\right|} : g_{ij}^0(\lambda) - g_{ij}^1(\lambda) \neq 0,\ i, j = \overline{1, T}.$$

The following theorem and corollaries are true.

Theorem 3.3. *The least favourable spectral densities $F^0(\lambda)$, $G^0(\lambda)$ in the classes $D = D_\varepsilon^k \times D_{1\delta}^k, k = 1,2,3,4$, for the optimal linear extrapolation of the functional $A\vec{\xi}$ are determined by relations (3.46) – (3.57), the minimality condition (3.12), the constrained optimization problem (3.26), and restrictions on densities from the corresponding classes $D = D_\varepsilon^k \times D_{1\delta}^k, k = 1,2,3,4$. The minimax-robust spectral characteristic of the optimal estimate of the functional $A\vec{\xi}$ is determined by formula (3.13).*

Corollary 3.8. *The least favourable spectral densities $F^0(\lambda)$ in the classes D_ε^k, $k = 1,2,3,4$, for the optimal linear extrapolation of the functional $A\vec{\xi}$ based on observations of the sequence $\vec{\xi}(j)$ at points $j \in \mathbb{Z}_-\backslash S$, $S = \bigcup_{l=1}^{s}\{-M_l - N_l, \ldots, -M_l\}$, are determined by the following equations, respectively,*

$$((C^0(\lambda))^\top)^* \cdot (C^0(\lambda))^\top = (\alpha^2 + \gamma(\lambda))(F^0(\lambda))^2, \tag{3.58}$$

$$((C^0(\lambda))^\top)^* \cdot (C^0(\lambda))^\top = F^0(\lambda)\left\{(\alpha_k^2 + \gamma_k(\lambda))\delta_{kl}\right\}_{k,l=1}^{T} F^0(\lambda), \tag{3.59}$$

$$((C^0(\lambda))^\top)^* \cdot (C^0(\lambda))^\top = (\alpha^2 + \gamma'(\lambda))F^0(\lambda)B_1^\top F^0(\lambda), \tag{3.60}$$

$$((C^0(\lambda))^\top)^* \cdot (C^0(\lambda))^\top = F^0(\lambda)(\vec{\alpha}\cdot\vec{\alpha}^* + \Gamma(\lambda))F^0(\lambda), \tag{3.61}$$

the minimality condition (3.20), the constrained optimization problem (3.27), and restrictions on densities from the corresponding classes D_ε^k, $k = 1,2,3,4$. The minimax spectral characteristic of the optimal estimate of the functional $A\vec{\xi}$ is determined by formula (3.18).

Corollary 3.9. *The least favourable spectral densities $F^0(\lambda)$ in the classes $D_{1\delta}^k$, $k = 1,2,3,4$, for the optimal linear extrapolation of the functional $A\vec{\xi}$ based on observations of the sequence $\vec{\xi}(j)$ at points $j \in \mathbb{Z}_-\backslash S$, where $S = \bigcup_{l=1}^{s}\{-M_l - N_l, \ldots, -M_l\}$, are determined by the following equations, respectively,*

$$((C^0(\lambda))^\top)^* \cdot (C^0(\lambda))^\top = \beta^2\gamma_2(\lambda)(F^0(\lambda))^2, \tag{3.62}$$

$$((C^0(\lambda))^\top)^* \cdot (C^0(\lambda))^\top = F^0(\lambda)\left\{\beta_k^2\gamma_k^2(\lambda)\delta_{kl}\right\}_{k,l=1}^{T} F^0(\lambda), \tag{3.63}$$

$$((C^0(\lambda))^\top)^* \cdot (C^0(\lambda))^\top = \beta^2\gamma_2'(\lambda)F^0(\lambda)B_2^\top F^0(\lambda), \tag{3.64}$$

$$((C^0(\lambda))^\top)^* \cdot (C^0(\lambda))^\top = F^0(\lambda)\left\{\beta_{ij}(\lambda)\gamma_{ij}(\lambda)\right\}_{i,j=1}^T F^0(\lambda), \tag{3.65}$$

the minimality condition (3.20), the constrained optimization problem (3.27), and the following restrictions on densities from the corresponding classes $D_{1\delta}^k$, $k = 1,2,3,4$, *respectively,*

$$\frac{1}{2\pi}\int_{-\pi}^{\pi}\left|\mathrm{Tr}\,(F^0(\lambda) - G_1(\lambda))\right| d\lambda = \delta, \tag{3.66}$$

$$\frac{1}{2\pi}\int_{-\pi}^{\pi}\left|f_{kk}^0(\lambda) - g_{kk}^1(\lambda)\right| d\lambda = \delta_k, \tag{3.67}$$

$$\frac{1}{2\pi}\int_{-\pi}^{\pi}\left|\langle B_2, F^0(\lambda) - G_1(\lambda)\rangle\right| d\lambda = \delta, \tag{3.68}$$

$$\frac{1}{2\pi}\int_{-\pi}^{\pi}\left|f_{ij}^0(\lambda) - g_{ij}^1(\lambda)\right| d\lambda = \delta_i^j. \tag{3.69}$$

The minimax spectral characteristic of the optimal estimate of the functional $A\vec{\xi}$ *is determined by formula (3.18).*

2. Interpolation Problem for Multidimensional Stationary Stochastic Sequences with Missing Observations

In this section we consider the problem of mean-square optimal estimation of the linear functional

$$A_s\vec{\xi} = \sum_{l=0}^{s-1}\sum_{j=M_l}^{M_l+N_{l+1}} \vec{a}(j)^\top \vec{\xi}(j),$$

$$M_l = \sum_{k=0}^{l}(N_k + K_k),\, N_0 = K_0 = 0$$

constructed from unobserved values of a multidimensional stationary stochastic sequence $\vec{\xi}(j) = \{\xi_k(j)\}_{k=1}^T$, $j \in \mathbb{Z}$. Estimates are based on observations of the sequence at points $j \in \mathbb{Z}\backslash S$, where $S = \bigcup_{l=0}^{s-1}\{M_l, M_l+1, \ldots, M_l+N_{l+1}\}$. The problem is investigated in the case of spectral certainty, where the spectral density of the sequence is exactly known, and in the case of spectral uncertainty, where the spectral density of the sequence is unknown while a class of admissible spectral densities is given.

2.1. Hilbert Space Projection Method of Interpolation

Let $\vec{\xi}(j) = \{\xi_k(j)\}_{k=1}^{T}$, $j \in \mathbb{Z}$, be a multidimensional (wide sense) stationary stochastic sequence withzero mean values, $E\vec{\xi}(j) = \vec{0}$, and with the correlation function $R(n) = E\vec{\xi}(j+n)(\vec{\xi}(j))^* = \left\{E\xi_k(j+n)\overline{\xi_l(j)}\right\}_{k,l=1}^{T}$ which admit the spectral representation

$$R(n) = \frac{1}{2\pi}\int_{-\pi}^{\pi} e^{ik\lambda}F(\lambda)d\lambda,$$

where $F(\lambda) = \{f_{kl}(\lambda)\}_{k,l=1}^{T}$ is the spectral density matrix of the sequence $\vec{\xi}(j)$ that satisfies the minimality condition

$$\int_{-\pi}^{\pi} \mathrm{Tr}\left[(F(\lambda))^{-1}\right] d\lambda < \infty. \tag{3.70}$$

This is the necessary and sufficient condition for the error-free interpolation of the unobserved sequence values to be impossible.

The stationary stochastic sequence $\vec{\xi}(j)$, $j \in \mathbb{Z}$, admits the spectral representation

$$\vec{\xi}(j) = \int_{-\pi}^{\pi} e^{ij\lambda}Z(d\lambda), \tag{3.71}$$

where $Z(\Delta)$ is the vector–valued orthogonal stochastic measure of the sequence such that

$$EZ(\Delta_1)(Z(\Delta_2))^* = \frac{1}{2\pi}\int_{\Delta_1\cap\Delta_2} F(\lambda)d\lambda.$$

Consider the problem of mean-square optimal estimation of the linear functional

$$A_s\vec{\xi} = \sum_{l=0}^{s-1}\sum_{j=M_l}^{M_l+N_{l+1}} \vec{a}(j)^{\top}\vec{\xi}(j),$$

$$M_l = \sum_{k=0}^{l} (N_k + K_k),\, N_0 = K_0 = 0,$$

based on observations of the sequence $\vec{\xi}(j)$ at points $j \in \mathbb{Z} \backslash S$, where $S = \bigcup_{l=0}^{s-1} \{M_l, M_l+1, \dots, M_l+N_{l+1}\}$.

It follows from the spectral representation (3.71) of the sequence $\vec{\xi}(j)$ that we can represent the functional $A_s\vec{\xi}$ in the form

$$A_s\vec{\xi} = \int_{-\pi}^{\pi} (A_s(e^{i\lambda}))^{\top} Z(d\lambda), \tag{3.72}$$

where

$$A_s(e^{i\lambda}) = \sum_{l=0}^{s-1} \sum_{j=M_l}^{M_l+N_{l+1}} \vec{a}(j) e^{ij\lambda}.$$

We will consider values $\xi_k(j), k = 1, \dots, T, j \in \mathbb{Z}$, of the sequence $\vec{\xi}(j)$ as elements of the Hilbert space $H = L_2(\Omega, \mathcal{F}, P)$ generated by random variables ξ with zero mathematical expectations, $E\xi = 0$, finite variations, $E|\xi|^2 < \infty$, and the inner product $(\xi, \eta) = E(\xi\overline{\eta})$.

Denote by $H^s(\xi)$ the subspace of the Hilbert space $H = L_2(\Omega, \mathcal{F}, P)$ generated by elements $\{\xi_k(j) : j \in \mathbb{Z} \backslash S, k = 1, \dots, T\}$.
Denote by $L_2(F)$ the Hilbert space of vector-valued functions $\vec{a}(\lambda) = \{a_k(\lambda)\}_{k=1}^{T}$ such that

$$\int_{-\pi}^{\pi} \vec{a}(\lambda)^{\top} F(\lambda) \overline{\vec{a}(\lambda)} d\lambda = \int_{-\pi}^{\pi} \sum_{k,l=1}^{T} a_k(\lambda) \overline{a_l(\lambda)} f_{kl}(\lambda)\, d\lambda < \infty.$$

Denote by $L_2^s(F)$ the subspace of $L_2(F)$ generated by functions of the form

$$e^{in\lambda}\delta_k,\ \delta_k = \{\delta_{kl}\}_{l=1}^{T},\ k = 1, \dots, T,\ n \in Z \backslash S.$$

The mean square optimal linear estimate $\hat{A}_s\vec{\xi}$ of the functional $A_s\vec{\xi}$ is an element of the space $H^s(\xi)$. It can be represented in the form

$$\hat{A}_s\vec{\xi} = \int_{-\pi}^{\pi} (h(e^{i\lambda}))^{\top} Z(d\lambda), \tag{3.73}$$

where $h(e^{i\lambda}) = \left\{h_k(e^{i\lambda})\right\}_{k=1}^{T} \in L_2^s(F)$ is the spectral characteristic of the estimate $\hat{A}_s\vec{\xi}$.

The mean square error $\Delta(h;F)$ of the estimate $\hat{A}_s\vec{\xi}$ is given by the formula

$$\Delta(h;F) = E\left|A_s\vec{\xi} - \hat{A}_s\vec{\xi}\right|^2 =$$
$$= \frac{1}{2\pi}\int_{-\pi}^{\pi}(A_s(e^{i\lambda}) - h(e^{i\lambda}))^{\top}F(\lambda)\overline{(A_s(e^{i\lambda}) - h(e^{i\lambda}))}d\lambda.$$

The Hilbert space projection method proposed by Kolmogorov (see selected works of A. N. Kolmogorov [1]) makes it possible to find the spectral characteristic $h(e^{i\lambda})$ and mean square error $\Delta(h;F)$ of the optimal linear estimate of the functional $A_s\vec{\xi}$ in the case where the spectral density $F(\lambda)$ of the sequence is exactly known and the minimality condition (3.70) is satisfied. The spectral characteristic is determined by the following two conditions:

$$1)h(e^{i\lambda}) \in L_2^s(F),$$
$$2)A_s(e^{i\lambda}) - h(e^{i\lambda}) \perp L_2^s(F).$$

It follows from the second condition that for any $\eta \in H^s(\xi)$ the following equations should be satisfied

$$\left(A_s\vec{\xi} - \hat{A}_s\vec{\xi}, \eta\right) = E\left[(A_s\vec{\xi} - \hat{A}_s\vec{\xi})\overline{\eta}\right] = 0.$$

The last relation is equivalent to equations

$$E\left[(A_s\vec{\xi} - \hat{A}_s\vec{\xi})\overline{\xi_k(j)}\right] = 0, \;\; j \in \mathbb{Z}\backslash S, \; k = \overline{1,T}.$$

By using representations (3.72), (3.73) and definition of the inner product in the space H, we get equations

$$\int_{-\pi}^{\pi}\left(A_s(e^{i\lambda}) - h(e^{i\lambda})\right)^{\top}F(\lambda)e^{-ik\lambda}d\lambda = 0, \; k \in \mathbb{Z}\backslash S.$$

It follows from these equations that the function $(A_s(e^{i\lambda}) - h(e^{i\lambda}))^{\top}F(\lambda)$ is of the form

$$(A_s(e^{i\lambda}) - h(e^{i\lambda}))^{\top}F(\lambda) = (C_s(e^{i\lambda}))^{\top},$$

$$C_s(e^{i\lambda}) = \sum_{l=0}^{s-1} \sum_{j=M_l}^{M_l+N_{l+1}} \vec{c}(j)e^{ij\lambda},$$

where $\vec{c}(j), j \in S$ are unknown coefficicnts that we have to find.
From the last relation we deduce that the spectral characteristic $h(e^{i\lambda})$ of the optimal linear estimate of the functional $A_s\vec{\xi}$ is of the form

$$h(e^{i\lambda}) = A_s(e^{i\lambda}) - (F^{-1}(\lambda))^\top C_s(e^{i\lambda}). \tag{3.74}$$

To find equations for calculating the unknown coefficients $\vec{c}(j), j \in S$, we use the decomposition of the function $(F^{-1}(\lambda))^\top$ into the Fourier series

$$(F^{-1}(\lambda))^\top = \sum_{m=-\infty}^{\infty} B(m)e^{im\lambda}, \tag{3.75}$$

where $B(m)$ are the Fourier coefficients of the function $(F^{-1}(\lambda))^\top$.
Inserting (3.75) into (3.74), we obtain the following representation of the spectral characteristic

$$h(e^{i\lambda}) = \left(\sum_{l=0}^{s-1} \sum_{j=M_l}^{M_l+N_{l+1}} \vec{a}(j)e^{ij\lambda} \right) - \\ - \left(\sum_{m=-\infty}^{\infty} B(m)e^{im\lambda} \right) \left(\sum_{l=0}^{s-1} \sum_{j=M_l}^{M_l+N_{l+1}} \vec{c}(j)e^{ij\lambda} \right). \tag{3.76}$$

It follows from the first condition, $h(e^{i\lambda}) \in L_2^s(F)$, that the Fourier coefficients of the function $h(e^{i\lambda}) = \{h_k(e^{i\lambda})\}_{k=1}^T$ are equal to zero for $j \in S$, namely

$$\int_{-\pi}^{\pi} h_k(e^{i\lambda})e^{-ij\lambda}d\lambda = 0, \; j \in S, \; k = 1, \ldots, T.$$

Using the last relations and (3.76), we get the following system of equations that determine the unknown coefficients $\vec{c}(j), j \in S$,

$$\begin{aligned}
\vec{a}(M_{k-1}) - \sum_{l=0}^{s-1} \sum_{j=M_l}^{M_l+N_{l+1}} B(M_{k-1}-j)\vec{c}(j) = 0;\\
\vec{a}(M_{k-1}+1) - \sum_{l=0}^{s-1} \sum_{j=M_l}^{M_l+N_{l+1}} B(M_{k-1}+1-j)\vec{c}(j) = 0;\\
\dots\\
\vec{a}(M_{k-1}+N_k) - \sum_{l=0}^{s-1} \sum_{j=M_l}^{M_l+N_{l+1}} B(M_{k-1}+N_k-j)\vec{c}(j) = 0,
\end{aligned} \tag{3.77}$$

where $k = 1, \dots, s$.
Denote by $\vec{\mathbf{a}}_s^\top$ the following vector $\vec{\mathbf{a}}_s^\top = (\vec{a}_1^\top, \vec{a}_2^\top, \dots, \vec{a}_s^\top), \vec{a}_k^\top = (\vec{a}(M_{k-1})^\top, \dots, \vec{a}(M_{k-1}+N_k)^\top)$, $k = 1, \dots, s$, $q = N_1 + N_2 + \dots + N_s + s$.
Let B_s be the $q \times q$ matrix

$$B_s = \begin{pmatrix} B_{11} & B_{12} & \dots & B_{1s} \\ B_{21} & B_{22} & \dots & B_{2s} \\ \vdots & \vdots & \ddots & \vdots \\ B_{s1} & B_{s2} & \dots & B_{ss} \end{pmatrix},$$

where B_{mn} are $(N_m+1) \times (N_n+1)$ compound matrices constructed from the block-matrices of dimension $T \times T$ that are Fourier coefficients of the function $(F^{-1}(\lambda))^\top$:

$$B_{mn}(k,j) = \frac{1}{2\pi} \int_{-\pi}^{\pi} (F^{-1}(\lambda))^\top e^{-i(k-j)\lambda} d\lambda = B(k-j),$$

$$\begin{aligned}
&k = M_{m-1}, \dots, M_{m-1}+N_m,\\
&j = M_{n-1}, \dots, M_{n-1}+N_n,\\
&m, n = 1, \dots, s.
\end{aligned}$$

Making use of the introduced notations, we can write equations (3.77) in the form

$$\vec{\mathbf{a}}_s = B_s \vec{\mathbf{c}}_s, \tag{3.78}$$

where $\vec{\mathbf{c}}_s^\top = (\vec{c}_1^\top, \vec{c}_2^\top, \ldots, \vec{c}_s^\top)$, $\vec{c}_k^\top = (\vec{c}(M_{k-1})^\top, \ldots, \vec{c}(M_{k-1}+N_k)^\top)$, $k = 1,\ldots,s$, is a vector, constructed from the unknown coefficients $\vec{c}(j), j \in S$. Since the matrix B_s is invertible (see paper by H. Salehi [155] for more details), we get the formula

$$\vec{\mathbf{c}}_s = B_s^{-1}\vec{\mathbf{a}}_s. \tag{3.79}$$

Hence, the unknown coefficients $\vec{c}(j), j \in S$, are calculated by the formula

$$\vec{c}(j) = \left(B_s^{-1}\vec{\mathbf{a}}_s\right)(j),$$

where $\left(B_s^{-1}\vec{\mathbf{a}}_s\right)(j)$ is the j-th component of the vector $B_s^{-1}\vec{\mathbf{a}}_s$. The formula for calculating the spectral characteristic of the estimate $\hat{A}_s\xi$ is of the form

$$h(e^{i\lambda}) = \left(\sum_{l=0}^{s-1}\sum_{j=M_l}^{M_l+N_{l+1}} \vec{a}(j)e^{ij\lambda}\right) - \\ - \left(\sum_{m=-\infty}^{\infty} B(m)e^{im\lambda}\right)\left(\sum_{l=0}^{s-1}\sum_{j=M_l}^{M_l+N_{l+1}} \left(B_s^{-1}\vec{\mathbf{a}}_s\right)(j)e^{ij\lambda}\right). \tag{3.80}$$

The mean square error of the estimate of the function can be calculated by the formula

$$\Delta(h;F) = \frac{1}{2\pi}\int_{-\pi}^{\pi} (C_s(e^{i\lambda}))^\top F^{-1}(\lambda)\overline{C_s(e^{i\lambda})}d\lambda = \\ = \frac{1}{2\pi}\int_{-\pi}^{\pi} \left(\sum_{l=0}^{s-1}\sum_{k=M_l}^{M_l+N_{l+1}} \vec{c}(k)e^{ik\lambda}\right)^\top \left(\sum_{m=-\infty}^{\infty} B(m)e^{im\lambda}\right) \times \\ \times \left(\sum_{l=0}^{s-1}\sum_{j=M_l}^{M_l+N_{l+1}} \overline{\vec{c}(j)}e^{-ij\lambda}\right) d\lambda = \\ = \langle \vec{\mathbf{c}}_s, B_s\vec{\mathbf{c}}_s\rangle = \left\langle B_s^{-1}\vec{\mathbf{a}}_s, \vec{\mathbf{a}}_s\right\rangle, \tag{3.81}$$

where $\langle\cdot,\cdot\rangle$is the inner product in the space ℓ_2.

Let us summarize our results and present them in the form of a theorem.

Theorem 3.4. *Let $\vec{\xi}(j) = \{\xi_k(j)\}_{k=1}^T$ be a multidimensional stationary stochastic sequence with the spectral density $F(\lambda)$ that satisfies the minimality condition (3.70). The spectral characteristic $h(e^{i\lambda})$ and mean square error $\Delta(h,F)$*

of the optimal linear estimate $\hat{A}_s\vec{\xi}$ of the functional $A_s\vec{\xi}$ based on observations of the sequence $\vec{\xi}(j)$ at points $j \in \mathbb{Z}\backslash S$, $S = \bigcup_{l=0}^{s-1}\{M_l, \dots, M_l + N_{l+1}\}$, can be calculated by formulas (3.80), (3.81).

Example 3.2. *Consider the problem of linear interpolation of the functional $A_2\vec{\xi} = \vec{a}(0)^\top\vec{\xi}(0) + \vec{a}(n)^\top\vec{\xi}(n)$, $n > 1$, based on observations of the sequence at points $j \in \mathbb{Z}\backslash S$, $S = \{0\} \cup \{n\}$. In this case the spectral characteristic (3.76) of the estimate $\hat{A}_2\vec{\xi}$ is calculated by the formula*

$$h(e^{i\lambda}) = (\vec{a}(0) + \vec{a}(n)e^{in\lambda}) - (F^{-1}(\lambda))^\top \cdot (\vec{c}(0) + \vec{c}(n)e^{in\lambda}), \tag{3.82}$$

where $F(\lambda)$ is a spectral density, the function $(F^{-1}(\lambda))^\top$ admits the decomposition $(F^{-1}(\lambda))^\top = \sum_{m=-\infty}^{\infty} B(m)e^{i\lambda m}$, and coefficients $\vec{c}(0), \vec{c}(n)$ are determined by the system of equations

$$\vec{a}(0) = B(0)\vec{c}(0) + B(-n)\vec{c}(n),$$
$$\vec{a}(n) = B(n)\vec{c}(0) + B(0)\vec{c}(n).$$

The matrix B_2 is of the form

$$B_2 = \begin{pmatrix} B_{11} & B_{12} \\ B_{21} & B_{22} \end{pmatrix},$$

$$B_{11} = B(0), B_{12} = B(-n), B_{21} = B(n), B_{22} = B(0).$$

Let $\vec{\mathbf{a}}_2^\top = (\vec{a}_1^\top, \vec{a}_2^\top)$, where $\vec{a}_1 = \vec{a}(0)$, $\vec{a}_2 = \vec{a}(n)$ and let $\vec{\mathbf{c}}_2^\top = (\vec{c}_1^\top, \vec{c}_2^\top)$, where $\vec{c}_1 = \vec{c}(0)$, $\vec{c}_2 = \vec{c}(n)$. In this case equations (3.78) and (3.79) are of the form

$$\vec{\mathbf{a}}_2 = B_2\vec{\mathbf{c}}_2,$$

$$\vec{\mathbf{c}}_2 = B_2^{-1}\vec{\mathbf{a}}_2.$$

Consider this problem in the case of two-dimensional stationary sequence $\vec{\xi}(n) = \{\xi_k(n)\}_{k=1}^2$, where $\xi_1(n) = \xi(n)$ is a stationary stochastic sequence with the spectral density $f(\lambda)$, and $\xi_2(n) = \xi(n) + \eta(n)$, where $\eta(n)$ is uncorrelated with $\xi(n)$ stationary stochastic sequence with the spectral density $g(\lambda)$. The matrix of spectral densities is of the form

$$F(\lambda) = \begin{pmatrix} f(\lambda) & f(\lambda) \\ f(\lambda) & f(\lambda) + g(\lambda) \end{pmatrix}.$$

Its determinant equals

$$D = |F(\lambda)| = f(\lambda)\,g(\lambda),$$

and the inverse matrix is as follows

$$F(\lambda)^{-1} = \begin{pmatrix} \frac{1}{f(\lambda)} + \frac{1}{g(\lambda)} & \frac{-1}{g(\lambda)} \\ \frac{-1}{g(\lambda)} & \frac{1}{g(\lambda)} \end{pmatrix}.$$

Let the spectral densities $f(\lambda)$ and $g(\lambda)$ be of the form

$$f(\lambda) = \left|1 - b_1 e^{i\lambda}\right|^2, \quad g(\lambda) = \left|1 - b_2 e^{i\lambda}\right|^2, \ |b_1| < 1, |b_2| < 1.$$

In this case the matrix

$$B_2 = \begin{pmatrix} 2+b_1^2+b_2^2 & -1-b_2^2 & 0 & 0 \\ -1-b_2^2 & 1+b_2^2 & 0 & 0 \\ 0 & 0 & 2+b_1^2+b_2^2 & -1-b_2^2 \\ 0 & 0 & -1-b_2^2 & 1+b_2^2 \end{pmatrix}.$$

Its determinant equals

$$D = (1+b_1^2)^2(1+b_2^2)^2.$$

The inverse matrix B_2^{-1} is equal to

$$\begin{pmatrix} \frac{1}{1+b_1^2} & \frac{1}{1+b_1^2} & 0 & 0 \\ \frac{1}{1+b_1^2} & \frac{2+b_1^2+b_2^2}{(1+b_1^2)(1+b_2^2)} & 0 & 0 \\ 0 & 0 & \frac{1}{1+b_1^2} & \frac{1}{1+b_1^2} \\ 0 & 0 & \frac{1}{1+b_1^2} & \frac{2+b_1^2+b_2^2}{(1+b_1^2)(1+b_2^2)} \end{pmatrix}.$$

The vector $\vec{\mathbf{c}}_2$ is as follows

$$\vec{\mathbf{c}}_2 = \left(\frac{\alpha+\beta}{1+b_1^2}, \frac{\alpha}{1+b_1^2} + \frac{(2+b_1^2+b_2^2)\beta}{(1+b_1^2)(1+b_2^2)}, \frac{\gamma+\delta}{1+b_1^2}, \right.$$
$$\left. \frac{\gamma}{1+b_1^2} + \frac{(2+b_1^2+b_2^2)\delta}{(1+b_1^2)(1+b_2^2)}\right)^{\top},$$

where $\vec{a}(0) = (\alpha, \beta)^\top$, $\vec{a}(n) = (\gamma, \delta)^\top$.
Thus, the spectral characteristic of the optimal estimate of the functional $A_2\vec{\xi}$ *is calculated by the formula*

$$(h(e^{i\lambda}))^\top = (h_1(e^{i\lambda}), h_2(e^{i\lambda}))^\top,$$

where

$$h_1(e^{i\lambda}) = \left(\frac{(\alpha+\beta)b_1}{1+b_1^2} - \frac{\beta b_2}{1+b_2^2}\right)(e^{i\lambda}+e^{-i\lambda}) + \\ + \left(\frac{(\gamma+\delta)b_1}{1+b_1^2} - \frac{\delta b_2}{1+b_2^2}\right)(e^{i(n+1)\lambda}+e^{i(n-1)\lambda}),$$

$$h_2(e^{i\lambda}) = \frac{\beta b_2}{1+b_2^2}(e^{i\lambda}+e^{-i\lambda}) + \frac{\delta b_2}{1+b_2^2}(e^{i(n+1)\lambda}+e^{i(n-1)\lambda}).$$

The mean square error of the optimal estimate of the functional is of the form

$$\Delta(F) = \frac{(\alpha+\beta)^2+(\gamma+\delta)^2}{2\pi(1+b_1^2)} + \frac{\beta^2+\delta^2}{2\pi(1+b_2^2)}.$$

Example 3.3. *Let all conditions of the previous example are satisfied. Consider the problem of linear interpolation of the functional*

$$A_2\vec{\xi} = \vec{a}(0)^\top\vec{\xi}(0) + \vec{a}(1)^\top\vec{\xi}(1) + \vec{a}(n)^\top\vec{\xi}(n),$$

where $\vec{a}(0) = (\alpha_1, \beta_1)^\top$, $\vec{a}(1) = (\alpha_2, \beta_2)^\top$, $\vec{a}(n) = (\gamma, \delta)^\top$, $n > 2$. *In this case the matrix* B_2 *is of the form*

$$B_2 = \begin{pmatrix} B_{11} & B_{12} \\ B_{21} & B_{22} \end{pmatrix},$$

where

$$B_{11} = \begin{pmatrix} B(0) & B(-1) \\ B(1) & B(0) \end{pmatrix}, B_{12} = \begin{pmatrix} B(-n) \\ B(-n+1) \end{pmatrix},$$

$$B_{12} = \begin{pmatrix} B(n) & B(n-1) \end{pmatrix}, B_{22} = \begin{pmatrix} B(0) \end{pmatrix},$$

$$B(0) = \begin{pmatrix} 2+b_1^2+b_2^2 & -1-b_2^2 \\ -1-b_2^2 & 1+b_2^2 \end{pmatrix},$$

$$B(1)=B(-1)=\begin{pmatrix} -b_1-b_2 & b_2 \\ b_2 & -b_2 \end{pmatrix},$$

$$B(-n)=B(-n+1)=B(n)=B(n-1)=\begin{pmatrix} 0 & 0 \\ 0 & 0 \end{pmatrix}.$$

The inverse matrix B_2^{-1} *is equal to*

$$\begin{pmatrix}
\frac{1+b_1^2}{A} & \frac{1+b_1^2}{A} & \frac{b_1}{A} & \frac{b_1}{A} & 0 & 0 \\
\frac{1+b_1^2}{A} & \frac{1+b_1^2}{A}+\frac{1+b_2^2}{B} & \frac{b_1}{A} & \frac{b_1}{A}+\frac{b_2}{B} & 0 & 0 \\
\frac{b_1}{A} & \frac{b_1}{A} & \frac{1+b_1^2}{A} & \frac{1+b_1^2}{A} & 0 & 0 \\
\frac{b_1}{A} & \frac{b_1}{A}+\frac{b_2}{B} & \frac{1+b_1^2}{A} & \frac{1+b_1^2}{A}+\frac{1+b_2^2}{B} & 0 & 0 \\
0 & 0 & 0 & 0 & \frac{1}{1+b_1^2} & \frac{1}{1+b_1^2} \\
0 & 0 & 0 & 0 & \frac{1}{1+b_1^2} & \frac{1}{1+b_1^2}+\frac{1}{1+b_2^2}
\end{pmatrix},$$

where $A=1+b_1^2+b_1^4$, $B=1+b_2^2+b_2^4$. *In this case we have* $\vec{\mathbf{c}}_2^\top=(\vec{c}(0)^\top,\vec{c}(1)^\top,\vec{c}(n)^\top)$, *where* $\vec{c}(0)=(c_1(0),c_2(0))^\top,\vec{c}(1)=(c_1(1),c_2(1))^\top$,

$$c_1(0)=\frac{(1+b_1^2)(\alpha_1+\beta_1)+b_1(\alpha_2+\beta_2)}{A},$$

$$c_2(0)=\frac{(1+b_1^2)(\alpha_1+\beta_1)+b_1(\alpha_2+\beta_2)}{A}+\frac{(1+b_2^2)\beta_1+b_2\beta_2}{B},$$

$$c_1(1)=\frac{b_1(\alpha_1+\beta_1)+(1+b_1^2)(\alpha_2+\beta_2)}{A},$$

$$c_2(1)=\frac{b_1(\alpha_1+\beta_1)+(1+b_1^2)(\alpha_2+\beta_2)}{A}+\frac{b_2\beta_1+(1+b_2^2)\beta_2}{B},$$

$$\vec{c}(n)=\left(\frac{\gamma+\delta}{1+b_1^2},\frac{\gamma}{1+b_1^2}+\frac{\delta}{(1+b_1^2)}+\frac{\delta}{(1+b_2^2)}\right)^\top.$$

Thus, the spectral characteristic of the optimal estimate of the functional $A_2\vec{\xi}$ *is calculated by the formula*

$$(h(e^{i\lambda}))^\top=(h_1(e^{i\lambda}),h_2(e^{i\lambda}))^\top,$$

where

$$h_1(e^{i\lambda}) = \\ = \left(b_1\frac{(1+b_1^2)(\alpha_1+\beta_1)+b_1(\alpha_2+\beta_2)}{1+b_1^2+b_1^4} - b_2\frac{(1+b_2^2)\beta_1+b_2\beta_2}{1+b_2^2+b_2^4}\right)e^{-i\lambda}+ \\ +\left(b_1\frac{b_1(\alpha_1+\beta_1)+(1+b_1^2)(\alpha_2+\beta_2)}{1+b_1^2+b_1^4} - b_2\frac{b_2\beta_1+(1+b_2^2)\beta_2}{1+b_2^2+b_2^4}\right)e^{2i\lambda}+ \\ +\left(\frac{(\gamma+\delta)b_1}{1+b_1^2} - \frac{\delta b_2}{1+b_2^2}\right)(e^{i(n+1)\lambda}+e^{i(n-1)\lambda}),$$

$$h_2(e^{i\lambda}) = b_2\frac{(1+b_2^2)\beta_1+b_2\beta_2}{1+b_2^2+b_2^4}e^{-i\lambda}+ \\ +b_2\frac{b_2\beta_1+(1+b_2^2)\beta_2}{1+b_2^2+b_2^4}e^{2i\lambda}+\frac{\delta b_2}{1+b_2^2}(e^{i(n+1)\lambda}+e^{i(n-1)\lambda}).$$

The mean square error of the optimal estimate of the functional is of the form

$$\Delta(F) = \\ = \left(\frac{((\alpha_1+\beta_1)^2+(\alpha_2+\beta_2)^2)(1+b_1^2)+2(\alpha_1+\beta_1)(\alpha_2+\beta_2)b_1}{1+b_1^2+b_1^4}\right)+ \\ +\left(\frac{(1+b_2^2)(\beta_1^2+\beta_2^2)+2\beta_1\beta_2 b_2}{1+b_2^2+b_2^4}\right)+ \\ +\frac{1}{2\pi}\left(\frac{(\gamma+\delta)^2}{(1+b_1^2)}+\frac{\delta^2}{(1+b_2^2)}\right).$$

2.2. Minimax Robust Method of Interpolation

The traditional methods of estimation of the functional $A_s\vec{\xi}$ constructed from unobserved values of a stationary stochastic sequence $\vec{\xi}(j)$ can be applied in the case where the spectral density $F(\lambda)$ of the considered stochastic sequence $\vec{\xi}(j)$ is exactly known. In practice, however, we do not have complete information on the spectral density of the sequence. For this reason, we apply the minimax robust method of estimation of the functional $A_s\vec{\xi}$, that is we find an estimate that minimizes the maximal values of mean square errors for all spectral densities from a given class of admissible spectral densities D.

For description of the minimax method of estimation of the optimal linear interpolation of the functional we propose the following definitions.

Definition 3.5. *For a given class of spectral densities D the spectral density $F_0(\lambda) \in D$ is called the least favourable in D for the optimal linear interpolation of the functional $A_s\vec{\xi}$ if the following relation is true*

$$\Delta(F_0) = \Delta(h(F_0);F_0) = \max_{F\in D} \Delta(h(F);F).$$

Definition 3.6. *For a given class of spectral densities D the spectral characteristic $h^0(e^{i\lambda})$ of the optimal linear interpolation of the functional $A_s\vec{\xi}$ is called minimax robust if the following relations are true*

$$h^0(e^{i\lambda}) \in H_D = \bigcap_{F\in D} L_2^s(F),$$

$$\min_{h\in H_D} \max_{F\in D} \Delta(h;F) = \sup_{F\in D} \Delta\left(h^0;F\right).$$

It follows from the introduced definitions and the obtained formulas that the following statement is true.

Lemma 3.3. *The spectral density $F_0(\lambda) \in D$ is the least favourable in the class of admissible spectral densities D for the optimal linear interpolation of the functional $A_s\vec{\xi}$ if the Fourier coefficients of the function $F_0^{-1}(\lambda)$ define the matrix B_s^0 that determines a solution of the optimization problem*

$$\max_{F\in D} \left\langle B_s^{-1}\vec{\mathbf{a}}_s, \vec{\mathbf{a}}_s\right\rangle = \left\langle (B_s^0)^{-1}\vec{\mathbf{a}}_s, \vec{\mathbf{a}}_s\right\rangle. \tag{3.83}$$

The minimax spectral characteristic $h^0 = h(F_0)$ can be calculated by the formula (3.80) if $h(F_0) \in H_D$.

The least favourable spectral density F_0 and the minimax spectral characteristic h^0 form a saddle point of the function $\Delta(h;F)$ on the set $H_D \times D$. The saddle point inequalities

$$\Delta(h;F_0) \geq \Delta\left(h^0;F_0\right) \geq \Delta\left(h^0;F\right) \quad \forall F \in D, \forall h \in H_D$$

are true if $h^0 = h(F_0)$ and $h(F_0) \in H_D$, where F_0 is a solution to the constrained optimization problem

$$\tilde{\Delta}(F) = -\Delta\left(h^0;F\right) \to \inf,\ F(\lambda) \in D, \tag{3.84}$$

$$\Delta\left(h^0;F\right) = \frac{1}{2\pi}\int_{-\pi}^{\pi}(C_s^0(e^{i\lambda}))^\top[F^0(\lambda)]^{-1}F(\lambda)[F^0(\lambda)]^{-1}\overline{C_s^0(e^{i\lambda})}d\lambda,$$

$$C_s^0(e^{i\lambda}) = \sum_{l=0}^{s-1}\sum_{j=M_l}^{M_l+N_{l+1}}\left((B_s^0)^{-1}\vec{\mathbf{a}}_s\right)(j)e^{ij\lambda}.$$

The constrained optimization problem (3.84) is equivalent to the unconstrained optimization problem

$$\Delta_D(F) = \tilde{\Delta}(F) + \delta(F\,|D) \to \inf,$$

where $\delta(F\,|D)$ is the indicator function of the set D. A solution F_0 of this problem is characterized by the condition $0 \in \partial\Delta_D(F_0)$, where $\partial\Delta_D(F_0)$ is the subdifferential of the convex functional $\Delta_D(F)$ at point F_0. This condition makes it possible to find the least favourable spectral densities in some special classes of spectral densities D (see books by A. D. Ioffe and V. M. Tihomirov [161], B. N. Pshenichnyj [156], T. Rockafellar [158]).

Note, that the form of the functional $\Delta\left(h^0;F\right)$ is convenient for application of the Lagrange method of indefinite multipliers for finding a solution of the problem (3.84). Making use of the method of Lagrange multipliers and the form of subdifferentials of the indicator functions we describe relations that determine the least favourable spectral densities in some special classes of spectral densities (see books by M. P. Moklyachuk [55, 163], M. P. Moklyachuk and O. Yu. Masyutka [96] for additional details).

2.3. Least Favourable Spectral Densities in the Class D_0^-

Consider the problem of the optimal interpolation of the functional $A_s\vec{\xi}$ in the case where the spectral density is from the class

$$D_0^- = \left\{F(\lambda)\left|\frac{1}{2\pi}\int_{-\pi}^{\pi}(F^{-1}(\lambda))^\top d\lambda = P\right.\right\},$$

where $P = \left\{p_{ij}\right\}_{i,j=1}^{T}$ is a given matrix. To find solutions of the constrained optimization problem (3.84) we use the method of Lagrange multipliers. Withthe help of this method we get the equation

$$(F^0(\lambda)^\top)^{-1}C_s^0(e^{i\lambda})(C_s^0(e^{i\lambda}))^*(F^0(\lambda)^\top)^{-1} = (F^0(\lambda)^\top)^{-1}\vec{\alpha}\cdot\vec{\alpha}^*(F^0(\lambda)^\top)^{-1},$$

where $\vec{\alpha} = \{\alpha_k\}_{k=1}^{T}$ is a vector composed of the Lagrange multipliers. From this relation we find that the Fourier coefficients of the matrix function $(F^0(\lambda)^\top)^{-1}$ satisfy the equation

$$\left(\sum_{l=0}^{s-1}\sum_{k=M_l}^{M_l+N_{l+1}} \vec{c}(k)e^{ik\lambda}\right)\cdot\left(\sum_{l=0}^{s-1}\sum_{k=M_l}^{M_l+N_{l+1}} \vec{c}(k)e^{ik\lambda}\right)^* = \vec{\alpha}\cdot\vec{\alpha}^*, \tag{3.85}$$

where $\vec{c}(k), k \in S$, are components of the vector $\vec{\mathbf{c}}_s$ that satisfies the equation $B_s^0\vec{\mathbf{c}}_s = \vec{\mathbf{a}}_s$, the matrix B_s^0 consists of matrices $B_{mn}^0(k,j)$, each of which is determined by the Fourier coefficients of the function $(F^0(\lambda)^\top)^{-1}$

$$B_{mn}^0(k,j) = \frac{1}{2\pi}\int_{-\pi}^{\pi}(F^0(\lambda)^\top)^{-1}e^{-i(k-j)\lambda}d\lambda = B^0(k-j),$$

$$k = M_{m-1},\ldots,M_{m-1}+N_m,$$
$$j = M_{n-1},\ldots,M_{n-1}+N_n,$$
$$m,n = 1,\ldots,s.$$

The Fourier coefficients $B(k) = B(-k), k \in S$, satisfy both equation (3.85) and equation $B_s^0\vec{\mathbf{c}}_s = \vec{\mathbf{a}}_s$. These coefficients can be found from the equation $B_s^0\vec{\alpha}_s = \vec{\mathbf{a}}_s$, where $\vec{\alpha}_s = (\vec{\alpha}, 0, \ldots, 0)$. The last relation can be presented in the form of the system of equations

$$B(k)\vec{\alpha} = \vec{a}(k), k \in S.$$

From the first equation of the system (for $k = 0$) we find the unknown value $\vec{\alpha} = B^{-1}(0)\vec{a}(0)$. It follows from the extremum condition (3.83) and the restriction on the spectral densities from the class D_0^- that the Fourier coefficient

$$B(0) = \frac{1}{2\pi}\int_{-\pi}^{\pi}(F^0(\lambda)^\top)^{-1}d\lambda = P.$$

Thus,

$$B(k) = \vec{a}(k)\vec{a}(0)^{-1}P,$$

where $\vec{a}(0)^{-1}$ is a vector that satisfies the equation

$$\vec{a}(0)^{-1} \cdot \vec{a}(0) = 1.$$

Let

$$B(k) = B(-k) = \begin{cases} \vec{a}(k)\vec{a}(0)^{-1}P & \text{if } k \in S; \\ 0 & \text{if } k \in \{0, \dots, M_{s-1} + N_s\} \setminus S. \end{cases}$$

Let the vector sequences $\vec{a}(k), k \in S$ be such that the matrix function

$$(F^0(\lambda)^\top)^{-1} = \sum_{k=-(M_{s-1}+N_s)}^{M_{s-1}+N_s} B(k)e^{ik\lambda}.$$

is positive definite and has the determinant which does not equal zero identically. In this case the matrix function can be represented in the form (see the book by M. G. Krein and A. A. Nudelman [157] for more details)

$$(F^0(\lambda))^{-1} = \left(\sum_{k=0}^{M_{s-1}+N_s} A_k e^{-ik\lambda}\right) \cdot \left(\sum_{k=0}^{M_{s-1}+N_s} A_k e^{-ik\lambda}\right)^*,$$

where $A_k = 0, k \in \{0, \dots, M_{s-1} + N_s\} \setminus S$. Hence, $F_0(\lambda)$ is the spectral density of the autoregressive stochastic sequence of order $M_{s-1} + N_s$ generated by the equation

$$\sum_{k=0}^{M_{s-1}+N_s} A_k \vec{\xi}(n-k) = \sum_{l=0}^{s-1} \sum_{k=M_l}^{M_l+N_{l+1}} A_k \vec{\xi}(n-k) = \vec{\varepsilon}_n, \tag{3.86}$$

where $\vec{\varepsilon}(n)$ is a vector-valued white noise stochastic sequence.

The minimax spectral characteristic $h(F_0)$ of the optimal linear estimate of the functional $A_s\vec{\xi}$ can be calculated by the formula (3.74), where

$$C_s(e^{i\lambda}) = \sum_{l=0}^{s-1} \sum_{k=M_l}^{M_l+N_{l+1}} \vec{c}(k)e^{ik\lambda} = \vec{\alpha} = P^{-1}\vec{a}(0),$$

namely

$$h(F_0) = \sum_{l=0}^{s-1} \sum_{k=M_l}^{M_l+N_{l+1}} \vec{a}(k)e^{ik\lambda} - \left(\sum_{k=-(M_{s-1}+N_s)}^{M_{s-1}+N_s} B(k)e^{ik\lambda} \right) P^{-1}\vec{a}(0)$$
$$= \sum_{k=1}^{N_1} \vec{a}(k)e^{-ik\lambda} + \sum_{l=1}^{s-1} \sum_{k=M_l}^{M_l+N_{l+1}} \vec{a}(k)e^{-ik\lambda}. \tag{3.87}$$

Summing up our reasoning we come to the conclusion that the following theorem is true.

Theorem 3.5. *Suppose that the sequence $\vec{a}(k), k \in S$, that determine the functional $A_s\vec{\xi}$, is such that the matrix function*

$$\sum_{k=-(M_{s-1}+N_s)}^{M_{s-1}+N_s} B(k)e^{ik\lambda}, \quad B(k) = B(-k) = \vec{a}(k)\vec{a}(0)^{-1}P,\ k \in S,$$

is positive definite and has the determinant which does not equal to zero identically. The least favourable in the class D_0^- spectral density for the optimal linear interpolation of the functional $A_s\vec{\xi}$ is the spectral density of the autoregressive sequence (3.86) with the Fourier coefficients $B(k)$. The minimax spectral characteristics $h(F_0)$ is given by formula (3.87).

2.4. Least Favourable Spectral Densities in the Class D_W

Consider the problem of the optimal interpolation of the functional $A_s\vec{\xi}$ in the case where the spectral density of the sequence is from the set of spectral densities with restrictions on the moments of the function $(F^{-1}(\lambda))^\top$. Let

$$D_W = \left\{ F(\lambda) \left| \frac{1}{2\pi} \int_{-\pi}^{\pi} (F^{-1}(\lambda))^\top \cos(w\lambda)d\lambda = B(w)\ , w = 0, 1, \dots, W \right. \right\},$$

where the sequence of matrices $B(w) = B(-w)$, $w = 0, \dots, W$, are such that the matrix function $\sum_{w=-W}^{W} B(w)e^{iw\lambda}$ is positive definite and has the determinant which does not equal zero identically (see the book by M. G. Krein and A. A. Nudelman [157] for more details). To find solutions of the constrained optimization problem (3.84) for the set D_W of admissible spectral densities we use the method of Lagrange multipliers and get the equation

$$\left(\sum_{l=0}^{s-1}\sum_{k=M_l}^{M_l+N_{l+1}} \vec{c}(k)e^{ik\lambda}\right)\cdot\left(\sum_{l=0}^{s-1}\sum_{k=M_l}^{M_l+N_{l+1}} \vec{c}(k)e^{ik\lambda}\right)^{*} =$$
$$= \left(\sum_{w=0}^{W} \vec{\alpha}(w)e^{iw\lambda}\right)\cdot\left(\sum_{w=0}^{W} \vec{\alpha}(w)e^{iw\lambda}\right)^{*}, \quad (3.88)$$

where $\vec{\alpha}(w), w=0,1,\dots,W$ are Lagrange multipliers and $\vec{c}(k), k=0,\dots,W$ are solutions of the equation

$$B_s^0\vec{\mathbf{c}}_s = \vec{\mathbf{a}}_s.$$

Consider two cases: $W \geq M_{s-1}+N_s$ and $W < M_{s-1}+N_s$.
Let $W \geq M_{s-1}+N_s$. In this case the given Fourier coefficients $B(w)$ define the matrix B_s^0 and the optimization problem (3.83) is degenerate. Let $\vec{\alpha}(M_{s-1}+N_s+1)=\dots=\vec{\alpha}(W)=\vec{0}$ and $\vec{\alpha}(j)=\vec{0}, j\notin S$. Components $\vec{\alpha}(j), j\in S$, of the vector $\vec{\alpha}_s$ can be found from the equation $B_s^0\vec{\alpha}_s=\vec{\mathbf{a}}_s$. Hence, the relation (3.88) is true. Thus the least favourable is every density $F(\lambda)\in D_W$ and the density of the autoregressive stochastic sequence

$$F^0(\lambda)) = \left(\sum_{w=-W}^{W} B(|w|)^{\top}e^{iw\lambda}\right)^{-1} = \left(\sum_{k=0}^{W} A_k e^{ik\lambda}\right)\left(\sum_{k=0}^{W} A_k e^{ik\lambda}\right)^{*} \quad (3.89)$$

is least favourable, too.
Let $W < M_{s-1}+N_s$. In this case the matrix B_s is determined by the known $B(w), w\in S\cap\{0,\dots,W\}$, and the unknown $B(w), w\in S\setminus\{0,\dots,W\}$, Fourier coefficients of the function $(F^{-1}(\lambda))^{\top}$. The unknown coefficients $\vec{\alpha}(k)$, $k\in S\cap\{0,\dots,W\}$, and $B(w), w\in S\setminus\{0,\dots,W\}$, can be found from the equation $B_s\vec{\alpha}_s^0=\vec{\mathbf{a}}_s$, with $\vec{\alpha}_s^0=(\vec{\alpha}(0),\dots,\vec{\alpha}(W_1),0,\dots,0)$, where W_1 is determined from the relation $\{0,\dots,W_1\}=\{0,\dots,W\}\cap S$.

The last equation can be presented as the system of equations

$$B(0)\vec{\alpha}(0)+B(1)\vec{\alpha}(1)+\ldots+B(-W_1)\vec{\alpha}(W_1)=\vec{a}(0);$$
$$B(1)\vec{\alpha}(0)+B(0)\vec{\alpha}(0)+\ldots+B(-W_1+1)\vec{\alpha}(W_1)=\vec{a}(1);$$
$$\ldots$$
$$B(W_1)\vec{\alpha}(0)+B(W_1-1)\vec{\alpha}(0)+\ldots+B(0)\vec{\alpha}(W_1)=\vec{a}(W_1);$$
$$\ldots$$
$$B(M_{s-1}+N_s)\vec{\alpha}(0)+B(1)\vec{\alpha}(1)+\ldots+B(0)\vec{\alpha}(W_1)=\vec{a}(M_{s-1}+N_s).$$

From the first W_1 equations we can find the unknown coefficients $\vec{\alpha}(k)$ and from the next equations we find the Fourier coefficients $B(w), w\in S\setminus\{0,\ldots,W\}$.
If the sequence $B(w)=B(-w), w\in S$, constructed from the sequence $B(w), w\in S\cap\{0,\ldots,W\}$ and the calculated coefficients $B(w), w\in S\setminus\{0,\ldots,W\}$, determines a positive definite function with the determinant which does not equal zero identically, then the least favourable spectral density $F^0(\lambda)$ is determined by the Fourier coefficients $B(w), w\in S$ of the function $(F^0(\lambda)^\top)^{-1}$

$$F^0(\lambda)=\left(\sum_{k=0}^{M_{s-1}+N_s}(B(k))^\top e^{ik\lambda}+(B(-k))^\top e^{-ik\lambda}\right)^{-1}=$$
$$=\left(\sum_{k=0}^{M_{s-1}+N_s}A_k e^{ik\lambda}\right)\left(\sum_{k=0}^{M_{s-1}+N_s}A_k e^{ik\lambda}\right)^*. \quad (3.90)$$

Let us summarize our results and present them in the form of a theorem.

Theorem 3.6. *The least favourable spectral density in the class D_W for the optimal linear estimate of the functional $A_s\vec{\xi}$ in the case where $W\geq M_{s-1}+N_s$ is the spectral density (3.89) of the autoregressive stochastic sequence of order W determined by coefficients $B(w), w=0,1,\ldots,W$. In the case where $W<M_{s-1}+N_s$ and solutions $B(w)=B(-w), w\in S\setminus\{0,\ldots,W\}$, of the equation $B_s\vec{\alpha}_s^0=\mathbf{\vec{a}}_s$ together with coefficients $B(w)=B(-w), w\in S\cap\{0,\ldots,W\}$, form a positive definite function with the determinant which does not equal zero identically, the least favourable spectral density in D_W is the density (3.90) of the autoregressive stochastic sequence of order $M_{s-1}+N_s$. The minimax characteristic of the estimate is calculated by formula (3.80).*

2.5. Least Favourable Spectral Densities in the Class D_V^U

Consider the problem of the optimal interpolation of the functional $A_s\vec{\xi}$ in the case where the spectral density of the sequence is from the set of spectral densities

$$D_V^U = \left\{ F(\lambda) \left| 0 \leq V(\lambda) \leq F(\lambda) \leq U(\lambda), \frac{1}{2\pi}\int_{-\pi}^{\pi} (F^{-1}(\lambda))^\top d\lambda = P \right. \right\},$$

where $V(\lambda), U(\lambda)$ are given spectral density matrices, P is a given positive definite matrix. To find solutions of the constrained optimization problem (3.84) for the set D_V^U of admissible spectral densities we use the condition $0 \in \partial\Delta_D(F^0)$. It follows from the condition $0 \in \partial\Delta_D(F^0)$ for $D = D_V^U$ that the Fourier coefficients of the function $(F^0(\lambda)^\top)^{-1}$ satisfy both equation

$$B_s^0 \vec{\mathbf{c}}_s = \vec{\mathbf{a}}_s$$

and the equation

$$\left(\sum_{l=0}^{s-1} \sum_{k=M_l}^{M_l+N_{l+1}} \left(\left(B_s^0\right)^{-1} \vec{\mathbf{a}}_s \right)(k) e^{ik\lambda} \right) \times \\ \times \left(\sum_{l=0}^{s-1} \sum_{k=M_l}^{M_l+N_{l+1}} \left(\left(B_s^0\right)^{-1} \vec{\mathbf{a}}_s \right)(k) e^{ik\lambda} \right)^* = \\ = \Gamma_1(\lambda) + \Gamma_2(\lambda) + \vec{\alpha} \cdot \vec{\alpha}^*, \quad (3.91)$$

where $\vec{\alpha}$ is a vector of Lagrange multipliers, $\Gamma_1(\lambda) \geq 0$ and $\Gamma_1(\lambda) = 0$ if $F^0(\lambda) \geq V(\lambda)$; $\Gamma_2(\lambda) \leq 0$ and $\Gamma_2(\lambda) = 0$ if $F_0(\lambda) \leq U(\lambda)$. Therefore, in the case where $V(\lambda) \leq F^0(\lambda) \leq U(\lambda)$, the function $(F^0(\lambda))^{-1}$ is of the form

$$(F^0(\lambda))^{-1} = \sum_{k=0}^{M_{s-1}+N_s} (B(k))^\top e^{ik\lambda} + (B(-k))^\top e^{-ik\lambda}) = \\ = \left(\sum_{k=0}^{M_{s-1}+N_s} A_k e^{ik\lambda} \right) \left(\sum_{k=0}^{M_{s-1}+N_s} A_k e^{ik\lambda} \right)^*,$$

where $B(k) = B(-k) = \vec{a}(k)\vec{a}(0)^{-1}P$. The least favourable in the class D_V^U is the spectral density of the autoregressive stochastic sequence of order $M_{s-1}+N_s$ if the following inequality is true

$$V(\lambda) \leq \left(\sum_{k=0}^{M_{s-1}+N_s} (B(k))^\top e^{ik\lambda} + (B(-k))^\top e^{-ik\lambda} \right)^{-1} \leq U(\lambda). \qquad (3.92)$$

The following theorem is true.

Theorem 3.7. *If the coefficients $B(k) = B(-k) = \vec{a}(k)\vec{a}(0)^{-1}P$, $k \in S$, satisfy the inequality (3.92) and form a positive definite function with the determinant which does not equal zero identically, then the least favourable in the class D_V^U spectral density for the optimal linear estimate of the functional $A_s\vec{\xi}$ is density (3.86) of the autoregressive stochastic sequence of order $M_{s-1}+N_s$. The minimax characteristic $h(F^0)$ of the estimate can be calculated by the formula (3.87). If the inequality (3.92) is not satisfied, then the least favourable spectral density in D_V^U is determined by relation (3.91) and the constrained optimization problem (3.83). The minimax characteristic of the estimate is calculated by formula (3.80).*

3. Filtering of Multidimensional Stationary Stochastic Sequences with Missing Observations

In this section, we propose results of the investigation of the problem of mean-square optimal linear estimation of the functional

$$A\vec{\xi} = \sum_{j \in Z^S} \vec{a}(j)^\top \vec{\xi}(-j)$$

constructed from unobserved values of a multidimensional stationary stochastic sequence $\{\vec{\xi}(j), j \in \mathbb{Z}\}$. Estimates are based on observations of the sequence $\{\vec{\xi}(j) + \vec{\eta}(j)\}$ at points $j \in \mathbb{Z}_- \backslash S$, where $\{\vec{\eta}(j), j \in \mathbb{Z}\}$ is uncorrelated with $\{\vec{\xi}(j), j \in \mathbb{Z}\}$ multidimensional stationary sequence,

$$S = \bigcup_{l=1}^{s} \{-(M_l+N_l), \ldots, -M_l\}, \quad Z^S = \{1,2,\ldots\} \backslash S^+,$$

$$S^{+}=\bigcup_{l=1}^{s}\{M_l,\ldots,M_l+N_l\},\, M_0=0, N_0=0.$$

The problem is investigated in the case where both spectral densities of the sequences $\{\vec{\xi}(j), j\in\mathbb{Z}\}$ and $\{\vec{\eta}(j), j\in\mathbb{Z}\}$ are known. In this case we derive formulas for calculating the spectral characteristic and the mean-square error of the optimal estimate of the functional using the method of orthogonal projections in the Hilbert space of random variables with finite second moments. In the case of spectral uncertainty, where the spectral densities of the sequences are not exactly known, while a set of admissible spectral densities is given, the minimax method of estimation is applied. Making use of this method, formulas that determine the least favourable spectral densities and the minimax-robust spectral characteristics of the optimal estimate of the functional are proposed for some specific classes of admissible spectral densities.

3.1. Hilbert Space Projection Method of Filtering

Consider uncorrelated multidimensional stationary stochastic sequences $\vec{\xi}(j)=\{\xi_k(j)\}_{k=1}^{T}$, $j\in\mathbb{Z}$, and $\vec{\eta}(j)=\{\eta_k(j)\}_{k=1}^{T}$, $j\in\mathbb{Z}$, with absolutely continuous spectral functions and the correlation functions of the form

$$R_{\xi}(n)=E\vec{\xi}(j+n)(\vec{\xi}(j))^{*}=\frac{1}{2\pi}\int_{-\pi}^{\pi}e^{in\lambda}F(\lambda)d\lambda,$$

$$R_{\eta}(n)=E\vec{\eta}(j+n)(\vec{\eta}(j))^{*}=\frac{1}{2\pi}\int_{-\pi}^{\pi}e^{in\lambda}G(\lambda)d\lambda,$$

where $F(\lambda)=\{f_{kl}(\lambda)\}_{k,l=1}^{T}$, $G(\lambda)=\{g_{kl}(\lambda)\}_{k,l=1}^{T}$ are the spectral density matrices of the sequences $\{\vec{\xi}(j), j\in\mathbb{Z}\}$ and $\{\vec{\eta}(j), j\in\mathbb{Z}\}$ respectively.
We will suppose that the spectral densities $F(\lambda)$ and $G(\lambda)$ satisfy the minimality condition

$$\int_{-\pi}^{\pi}\mathrm{Tr}\,(F(\lambda)+G(\lambda))^{-1}d\lambda<\infty. \tag{3.93}$$

This is the necessary and sufficient condition for the error-free filtering of the unobserved values of the sequence to be impossible (see, for example, Yu. A. Rozanov [10]).
The stationary stochastic sequences $\{\vec{\xi}(j)\}$ and $\{\vec{\eta}(j)\}$ admit the spectral representations

$$\xi(j)=\int\limits_{-\pi}^{\pi} e^{ij\lambda}Z_{\xi}(d\lambda), \qquad \eta(j)=\int\limits_{-\pi}^{\pi} e^{ij\lambda}Z_{\eta}(d\lambda), \tag{3.94}$$

where $Z_{\xi}(d\lambda)$ and $Z_{\eta}(d\lambda)$ are the orthogonal stochastic measures defined on $[-\pi,\pi)$ satisfying relations

$$EZ_{\xi}(\Delta_1)(Z_{\xi}(\Delta_2))^* = \frac{1}{2\pi}\int_{\Delta_1\cap\Delta_2} F(\lambda)d\lambda,$$

$$EZ_{\eta}(\Delta_1)(Z_{\eta}(\Delta_2))^* = \frac{1}{2\pi}\int_{\Delta_1\cap\Delta_2} G(\lambda)d\lambda.$$

Suppose that we have observations of the sequence $\{\vec{\xi}(j)+\vec{\eta}(j)\}$ at points $j\in \mathbb{Z}_-\backslash S$, where $S=\bigcup\limits_{l=1}^{s}\{-(M_l+N_l),\dots,-M_l\}$. The problem is to find the mean-square optimal linear estimate of the functional

$$A\vec{\xi}=\sum_{j\in Z^S}\vec{a}(j)^{\top}\vec{\xi}(-j)$$

constructed from unobserved values of the sequence $\{\vec{\xi}(j)\}$, $Z^S = \{1,2,\dots\}\backslash S^+$, $S^+ = \bigcup\limits_{l=1}^{s}\{M_l,\dots,M_l+N_l\}$.

Suppose that the coefficients $\{\vec{a}(j), j=0,1,\dots\}$, defining the functional $A\vec{\xi}$, satisfy the condition

$$\sum_{j\in Z^S}\sum_{k=1}^{T}|a_k(j)|<\infty. \tag{3.95}$$

This condition ensures that the functional $A\vec{\xi}$ has a finite second moment.

It follows from the spectral representation of the sequence $\{\vec{\xi}(j)\}$ that the functional $A\vec{\xi}$ can be represented in the form

$$A\vec{\xi} = \int_{-\pi}^{\pi} (A(e^{i\lambda}))^{\top} Z_{\xi}(d\lambda), \quad A(e^{i\lambda}) = \sum_{j \in Z^S} \vec{a}(j) e^{-ij\lambda}.$$

Consider values $\xi_k(j), k = 1, \ldots, T; j \in \mathbb{Z}$ and $\eta_k(j), k = 1, \ldots, T; j \in \mathbb{Z}$ of the stationary stochastic sequences $\{\vec{\xi}(j)\}$ and $\{\vec{\eta}(j)\}$ as elements of the Hilbert space $H = L_2(\Omega, \mathcal{F}, P)$ generated by random variables ξ with zero mathematical expectations, $E\xi = 0$, finite variations, $E|\xi|^2 < \infty$, and the inner product $(\xi, \eta) = E\xi\overline{\eta}$.

Denote by $H^s(\xi + \eta)$ the closed linear subspace generated by elements $\{\xi_k(j) + \eta_k(j) : j \in \mathbb{Z}_- \backslash S, k = \overline{1, T}\}$ in the Hilbert space $H = L_2(\Omega, \mathcal{F}, P)$.
Denote by $L_2(F + G)$ the Hilbert space of vector-valued functions $\vec{a}(\lambda) = \{a_k(\lambda)\}_{k=1}^{T}$ such that

$$\int_{-\pi}^{\pi} \vec{a}(\lambda)^{\top} (F(\lambda) + G(\lambda)) \overline{\vec{a}(\lambda)} d\lambda < \infty.$$

Denote by $L_2^s(F + G)$ the subspace of $L_2(F + G)$ generated by functions of the form

$$e^{in\lambda} \delta_k, \; \delta_k = \{\delta_{kl}\}_{l=1}^{T}, \; k = 1, \ldots, T, \; n \in Z_- \backslash S.$$

The mean square optimal linear estimate $\hat{A}\vec{\xi}$ of the functional $A\vec{\xi}$ based on observations of the sequence $\{\vec{\xi}(j) + \vec{\eta}(j)\}$ at points $j \in \mathbb{Z}_- \backslash S$ can be represented in the form

$$\hat{A}\vec{\xi} = \int_{-\pi}^{\pi} (h(e^{i\lambda}))^{\top} (Z_{\xi}(d\lambda) + Z_{\eta}(d\lambda)), \tag{3.96}$$

where $h(e^{i\lambda}) = \left\{h_k(e^{i\lambda})\right\}_{k=1}^{T} \in L_2^s(F + G)$ is the spectral characteristic of the estimate.

The mean square error $\Delta(h;F,G)$ of the estimate $\hat{A}\vec{\xi}$ is given by the formula

$$\Delta(h;F,G) = E\left|A\vec{\xi}-\hat{A}\vec{\xi}\right|^2$$

$$= \frac{1}{2\pi}\int_{-\pi}^{\pi}\left(A(e^{i\lambda})-h(e^{i\lambda})\right)^{\top}F(\lambda)\overline{\left(A(e^{i\lambda})-h(e^{i\lambda})\right)}d\lambda$$

$$+\frac{1}{2\pi}\int_{-\pi}^{\pi}\left(h(e^{i\lambda})\right)^{\top}G(\lambda)\overline{\left(h(e^{i\lambda})\right)}d\lambda.$$

The Hilbert space projection method proposed by A. N. Kolmogorov [1] makes it possible to find the spectral characteristic $h(e^{i\lambda})$ and mean square error $\Delta(h;F,G)$ of the optimal linear estimate of the functional $A\vec{\xi}$ in the case where the spectral densities $F(\lambda)$ and $G(\lambda)$ of the sequences are exactly known and the minimality condition (3.93) is satisfied. According to this method the optimal estimate of the functional $A\vec{\xi}$ is a projection of the element $A\vec{\xi}$ of the space H on the subspace $H^s(\xi+\eta)$. The projection is determined by the following two conditions:

$$1)\hat{A}\vec{\xi}\in H^s(\xi+\eta),$$
$$2)A\vec{\xi}-\hat{A}\vec{\xi}\perp H^s(\xi+\eta).$$

It follows from the second condition that the spectral characteristic $h(e^{i\lambda})$ for any $j\in\mathbb{Z}_-\backslash S$ satisfies the equation

$$\frac{1}{2\pi}\int_{-\pi}^{\pi}\left(A(e^{i\lambda})-h(e^{i\lambda})\right)^{\top}F(\lambda)e^{-ij\lambda}d\lambda-\frac{1}{2\pi}\int_{-\pi}^{\pi}(h(e^{-i\lambda}))^{\top}G(\lambda)e^{ij\lambda}d\lambda=\vec{0}.$$

The last relation is equivalent to equations

$$\frac{1}{2\pi}\int_{-\pi}^{\pi}\left[(A(e^{i\lambda}))^{\top}F(\lambda)-(h(e^{i\lambda}))^{\top}(F(\lambda)+G(\lambda))\right]e^{-ij\lambda}d\lambda=\vec{0},\ j\in\mathbb{Z}_-\backslash S.$$

Hence the function $\left[(A(e^{i\lambda}))^{\top}F(\lambda)-(h(e^{i\lambda}))^{\top}(F(\lambda)+G(\lambda))\right]$ is of the form

$$(A(e^{i\lambda}))^{\top}F(\lambda)-(h(e^{i\lambda}))^{\top}(F(\lambda)+G(\lambda))=(C(e^{i\lambda}))^{\top},$$

where

$$C(e^{i\lambda}) = \sum_{j\in S}\vec{c}(j)e^{ij\lambda} + \sum_{j=0}^{\infty}\vec{c}(j)e^{ij\lambda}.$$

Here $\vec{c}(j)$, $j \in U = S \cup \{0,1,2,\ldots\}$ are unknown coefficients that we have to find.

From the last relation we deduce that the spectral characteristic of the optimal linear estimate $\hat{A}\vec{\xi}$ is of the form

$$\begin{aligned}(h(e^{i\lambda}))^{\top} &= \\ &= (A(e^{i\lambda}))^{\top}F(\lambda)(F(\lambda)+G(\lambda))^{-1} - (C(e^{i\lambda}))^{\top}(F(\lambda)+G(\lambda))^{-1}.\end{aligned} \tag{3.97}$$

It follows from the first condition, $\hat{A}\vec{\xi} \in H^s(\xi+\eta)$, which determine the optimal linear estimate of the functional $A\vec{\xi}$, that the Fourier coefficients of the function $h(e^{i\lambda})$ are equal to zero for $k \in U$, namely

$$\frac{1}{2\pi}\int_{-\pi}^{\pi}\Big((A(e^{i\lambda}))^{\top}F(\lambda)(F(\lambda)+G(\lambda))^{-1} \\ - (C(e^{i\lambda}))^{\top}(F(\lambda)+G(\lambda))^{-1}\Big)e^{-ik\lambda}d\lambda = \vec{0},\ k \in U.$$

We will use the last equality to find equations which determine the unknown coefficients $\vec{c}(j), j \in U$. After disclosing the brackets we get the relation

$$\begin{aligned}&\sum_{j\in Z^S}\vec{a}(j)^{\top}\frac{1}{2\pi}\int_{-\pi}^{\pi}F(\lambda)(F(\lambda)+G(\lambda))^{-1}e^{-i(k+j)\lambda}d\lambda- \\ &-\sum_{j\in S}\vec{c}(j)^{\top}\frac{1}{2\pi}\int_{-\pi}^{\pi}(F(\lambda)+G(\lambda))^{-1}e^{-i(k-j)\lambda}d\lambda- \\ &-\sum_{j=0}^{\infty}\vec{c}(j)^{\top}\frac{1}{2\pi}\int_{-\pi}^{\pi}(F(\lambda)+G(\lambda))^{-1}e^{-i(k-j)\lambda}d\lambda = 0,\ k \in U.\end{aligned} \tag{3.98}$$

To represent relations (3.98) in more appropriate form we introduce the Fourier coefficients

$$B(k,j) = \frac{1}{2\pi}\int_{-\pi}^{\pi}(F(\lambda)+G(\lambda))^{-1}e^{-i(k-j)\lambda}d\lambda, \tag{3.99}$$

$$R(k,j) = \frac{1}{2\pi}\int_{-\pi}^{\pi}F(\lambda)(F(\lambda)+G(\lambda))^{-1}e^{-i(k+j)\lambda}d\lambda, \tag{3.100}$$

$$Q(k,j) = \frac{1}{2\pi}\int_{-\pi}^{\pi}F(\lambda)(F(\lambda)+G(\lambda))^{-1}G(\lambda)e^{-i(k-j)\lambda}d\lambda. \tag{3.101}$$

of the functions

$$(F(\lambda)+G(\lambda))^{-1}, \quad F(\lambda)(F(\lambda)+G(\lambda))^{-1}, \quad F(\lambda)(F(\lambda)+G(\lambda))^{-1}G(\lambda).$$

Making use of the introduced notations, we can verify that relation (3.98) is equivalent to the following system of equations:

$$\sum_{j\in Z^S} R(k,j)\vec{a}(j) = \sum_{j\in S} B(k,j)\vec{c}(j) + \sum_{j=0}^{\infty} B(k,j)\vec{c}(j), \qquad k\in U.$$

Let us introduce the next notations: $\vec{a}(j)=\vec{0},\ j\in S, \vec{a}(0)=\vec{0}, \vec{a}(j)=\vec{0},\ j\in S^{+}$. Making use of the introduced notations, we can write the last equations in the form

$$\sum_{j\in U} R(k,j)\vec{a}(j) = \sum_{j\in S} B(k,j)\vec{c}(j) + \sum_{j=0}^{\infty} B(k,j)\vec{c}(j), \qquad k\in U.$$

These equations can be written in the form

$$\mathbf{R}\vec{\mathbf{a}} = \mathbf{B}\vec{\mathbf{c}}, \tag{3.102}$$

where $\vec{\mathbf{c}}$ is a vector constructed of the unknown coefficients $\vec{c}(j), j\in U$, the vector $\vec{\mathbf{a}}$ has the same with the vector $\vec{\mathbf{c}}$ dimension. It is of the form

$$\vec{\mathbf{a}}^{\top} = (\vec{0}_0^{\top}, \vec{a}_1^{\top}, \vec{0}_1^{\top}, \vec{a}_2^{\top}, \vec{0}_2^{\top}, \ldots \vec{a}_i^{\top}, \vec{0}_i^{\top}, \ldots, \vec{a}_s^{\top}, \vec{0}_s^{\top}, \vec{a}_{s+1}^{\top}),$$

where $\vec{0}_0$ is a vector which consists of $(|S|+1)T$ zeros, where $|S| = \sum_{k=1}^{s}(N_k+1)$ is the number of missed values, vectors $\vec{0}_i$, $i = 1,2,\dots,s$, consist of $(N_i+1)T$ zeros, vectors

$$\vec{a}_1^\top = (\vec{a}(1)^\top, \dots, \vec{a}(M_1-1)^\top),$$
$$\vec{a}_i^\top = (\vec{a}(M_{i-1}+N_{i-1}+1)^\top, \dots, \vec{a}(M_i-1)^\top), \quad i = 2,\dots,s,$$
$$\vec{a}_{s+1}^\top = (\vec{a}(M_s+N_s+1)^\top, \vec{a}(M_s+N_s+2)^\top, \dots),$$

are constructed from the coefficients that determine the functional $A\vec{\xi}$.
The operator $\mathbf{B}$ in equation (3.102) is a linear operator in the space ℓ_2 defined by the matrix

$$B = \begin{pmatrix} B_{s,s} & B_{s,s-1} & \dots & B_{s,1} & B_{s,n} \\ B_{s-1,s} & B_{s-1,s-1} & \dots & B_{s-1,1} & B_{s-1,n} \\ \vdots & \vdots & \ddots & \vdots & \vdots \\ B_{1,s} & B_{1,s-1} & \dots & B_{1,1} & B_{1,n} \\ B_{n,s} & B_{n,s-1} & \dots & B_{n,1} & B_{n,n} \end{pmatrix},$$

where elements in the last column and the last row are compound matrices constructed from the block-matrices

$$B_{l,n}(k,j) = B(k,j),\ l = 1,2,\dots,s,\ k = -M_l-N_l,\dots,-M_l,$$
$$j = 0,1,2,\dots, B_{n,m}(k,j) = B(k,j),\ m = 1,2,\dots,s,\ k = 0,1,2,\dots,$$
$$j = -M_m-N_m,\dots,-M_m, B_{n,n}(k,j) = B(k,j),\ k,j = 0,1,2,\dots,$$

and other elements of the matrix B are compound matrices with elements of the form

$$B_{l,m}(j,k) = B(k,j),\ l,m = 1,2,\dots,s,$$
$$k = -M_l-N_l,\dots,-M_l, j = -M_m-N_m,\dots,-M_m.$$

The linear operator $\mathbf{R}$ in the space ℓ_2 is defined by the corresponding matrix in the same manner.

The unknown coefficients $\vec{c}(k), k \in U$, defined by equations (3.102), can be calculated by the formula

$$\vec{c}(k) = (\mathbf{B}^{-1}\mathbf{R}\vec{\mathbf{a}})(k),$$

where $(\mathbf{B}^{-1}\mathbf{R}\vec{\mathbf{a}})(k)$ is the k-th component of the vector $\mathbf{B}^{-1}\mathbf{R}\vec{\mathbf{a}}$. (see paper by H. Salehi [155] for more details).

The formula for calculating the spectral characteristic $h(e^{i\lambda})$ of the optimal estimate $\hat{A}\vec{\xi}$ of the functional is of the form

$$(h(e^{i\lambda}))^{\top} = (A(e^{i\lambda}))^{\top} F(\lambda)(F(\lambda)+G(\lambda))^{-1} - \\ -\left(\sum_{k\in U}(\mathbf{B}^{-1}\mathbf{R}\vec{\mathbf{a}})(k)e^{ik\lambda}\right)^{\top}(F(\lambda)+G(\lambda))^{-1}. \tag{3.103}$$

The mean square error of the estimate $\hat{A}\vec{\xi}$ can be calculated by the formula

$$\Delta(F,G) = E\left|A\vec{\xi}-\hat{A}\vec{\xi}\right|^2 = \frac{1}{2\pi}\int_{-\pi}^{\pi}(r_G(\lambda))^{\top}F(\lambda)\overline{r_G(\lambda)}d\lambda +$$

$$+\frac{1}{2\pi}\int_{-\pi}^{\pi}(r_F(\lambda))^{\top}G(\lambda)\overline{r_F(\lambda)}d\lambda = \langle\mathbf{R}\vec{\mathbf{a}},\mathbf{B}^{-1}\mathbf{R}\vec{\mathbf{a}}\rangle + \langle\mathbf{Q}\vec{\mathbf{a}},\vec{\mathbf{a}}\rangle, \tag{3.104}$$

$$(r_F(\lambda))^{\top} = \\ = \left((A(e^{i\lambda}))^{\top}F(\lambda) - \left(\sum_{k\in U}(\mathbf{B}^{-1}\mathbf{R}\vec{\mathbf{a}})(k)e^{ik\lambda}\right)^{\top}\right)(F(\lambda)+G(\lambda))^{-1},$$

$$(r_G(\lambda))^{\top} = \\ = \left((A(e^{i\lambda}))^{\top}G(\lambda) + \left(\sum_{k\in U}(\mathbf{B}^{-1}\mathbf{R}\vec{\mathbf{a}})(k)e^{ik\lambda}\right)^{\top}\right)(F(\lambda)+G(\lambda))^{-1},$$

where $\mathbf{Q}$ is a linear operator in the space ℓ_2 defined by the matrix with coefficients $Q(k,j), k,j\in U$ in the same way as the operator $\mathbf{B}$ is defined.

Let us summarize the obtained results and present them in the form of a theorem.

Theorem 3.8. *Let $\{\vec{\xi}(j)\}$ and $\{\vec{\eta}(j)\}$ be uncorrelated multidimensional stationary stochastic sequences with the spectral densities $F(\lambda)$ and $G(\lambda)$ satisfying the minimality condition (3.93). The spectral characteristic $h(e^{i\lambda})$ and the mean square error $\Delta(F,G)$ of the optimal linear estimate of the functional $A\vec{\xi}$, based on observations of the sequence $\{\vec{\xi}(j)+\vec{\eta}(j)\}$ at points $j\in\mathbb{Z}_-\backslash S$, can be calculated by formulas (3.103), (3.104).*

Estimates Based on Observations Missed in One Interval

Consider the problem of mean-square optimal linear estimation of the functional $A\vec{\xi}$, based on observations of the sequence $\{\vec{\xi}(j)+\vec{\eta}(j)\}$ at points $j\in\mathbb{Z}_-\backslash S$, where $S=\{-(M+N),\ldots,-M\}$, $Z^S=\{1,2,\ldots\}\backslash S^+$, $S^+=\{M,\ldots,M+N\}$.

From Theorem 3.8 the following corollary can be derived in this particular case.

Corollary 3.10. *Let $\{\vec{\xi}(j)\}$ and $\{\vec{\eta}(j)\}$ be uncorrelated multidimensional stationary sequences with the spectral densities $F(\lambda)$ and $G(\lambda)$ satisfying the minimality condition (3.93). The spectral characteristic $h(e^{i\lambda})$ and mean square error $\Delta(F,G)$ of the optimal linear estimate of the functional $A\vec{\xi}$ based on observations of the sequence $\{\vec{\xi}(j)+\vec{\eta}(j)\}$ at points $j\in\mathbb{Z}_-\backslash S$, $S=\{-(M+N),\ldots,-M\}$, can be calculated by formulas (3.103), (3.104), where* **B**, **R**, **Q** *are linear operators in the space ℓ_2 defined by compound matrices constructed from coefficients $B(k,j)$, $R(k,j)$, $Q(k,j)$, $k,j\in U$, $(U=S\cup\{0,1,2,\ldots\})$. For example, the matrix B is of the form*

$$B=\begin{pmatrix} B_{s,s} & B_{s,n} \\ B_{n,s} & B_{n,n} \end{pmatrix},$$

where its components are matrices constructed from the block-matrices

$$\begin{aligned}
&B_{s,n}(k,j)=B(k,j), && k=-M-N,\ldots,-M, && j=0,1,2,\ldots,\\
&B_{n,s}(k,j)=B(k,j), && k=0,1,2,\ldots, && j=-M-N,\ldots,-M,\\
&B_{n,n}(k,j)=B(k,j), && k,j=0,1,2,\ldots, && \\
&B_{s,s}(k,j)=B(k,j), && k=-M-N,\ldots,-M, && j=-M-N,\ldots,-M.
\end{aligned}$$

Estimates Based on Observations with One Missed Observation

Consider the problem of mean-square optimal linear estimation of the functional $A\vec{\xi}$ based on observations of the sequence $\{\vec{\xi}(j)+\vec{\eta}(j)\}$ at points $j \in \mathbb{Z}_{-}\setminus\{-s\}$, $Z^S = \{1,2,\ldots\}\setminus\{s\}$.

It follows from Theorem 3.8 that the following corollary is true.

Corollary 3.11. *Let $\vec{\xi}(j)$ and $\vec{\eta}(j)$ be uncorrelated multidimensional stationary stochastic sequences with the spectral densities $F(\lambda)$ and $G(\lambda)$ satisfying the minimality condition (3.93). The spectral characteristic $h(e^{i\lambda})$ and mean square error $\Delta(F,G)$ of the optimal linear estimate of the functional $A\vec{\xi}$ based on observations of the sequence $\vec{\xi}(j)+\vec{\eta}(j)$, $j \in \mathbb{Z}_{-}\setminus\{-s\}$ can be calculated by formulas (3.103), (3.104), where* **B**, **R**, **Q** *are linear operators in the space ℓ_2 defined by compound matrices constructed from coefficients $B(k,j)$, $R(k,j)$, $Q(k,j)$, $k,j \in U$, $(U = S \cup \{0,1,2,\ldots\})$,*

$$B = \begin{pmatrix} B(-s,-s) & B_{-s,n} \\ B_{n,-s} & B_{n,n} \end{pmatrix},$$

where elements in the last column and the last row are matrices with the elements

$$\begin{aligned} B_{-s,n}(k,j) &= B(k,j), \quad k=-s, \quad j=0,1,2,\ldots, \\ B_{n,-s}(k,j) &= B(k,j), \quad k=0,1,2,\ldots, \quad j=-s, \\ B_{n,n}(k,j) &= B(k,j), \quad k,j=0,1,2,\ldots. \end{aligned}$$

Estimation of the Functional $A_N\vec{\xi}$

Consider, next, the problem of mean-square optimal linear estimation of the functional

$$A_N\vec{\xi} = \sum_{j\in Z^S\cap\{0,\ldots,N\}} \vec{a}(j)^\top \vec{\xi}(-j),$$

based on observations of the sequence $\vec{\xi}(j)+\vec{\eta}(j)$ at points $j \in \mathbb{Z}_{-}\setminus S$, where S is defined in the introduction. Each linear estimate of the functional $A_N\vec{\xi}$ has the representation

$$\hat{A}_N\vec{\xi} = \int_{-\pi}^{\pi} (h_N(e^{i\lambda})^\top (Z_\xi(d\lambda)+Z_\eta(d\lambda)).$$

Define the vector $\vec{\mathbf{a}}_N$ as follows: elements with indices from the set $U \cap (S \cup \{0, \dots, N\})$ coincide with the elements of the vector $\vec{\mathbf{a}}$ with the same indices, and elements with indices from the set $U \backslash (S \cup \{0, \dots, N\})$ are zeros. **B**, **R**, **Q** are linear operators in the space ℓ_2 defined in Theorem 3.8.

The spectral characteristic $h_N(e^{i\lambda})$ and mean square error $\Delta(h_N; F, G)$ of the optimal linear estimate of the functional $A_N \vec{\xi}$ can be calculated by formulas

$$(h_N(e^{i\lambda}))^\top = (A_N(e^{i\lambda}))^\top F(\lambda)(F(\lambda)+G(\lambda))^{-1} -$$
$$- \left(\sum_{k \in U} (\mathbf{B}^{-1}\mathbf{R}\vec{\mathbf{a}}_N)(k) e^{ik\lambda} \right)^\top (F(\lambda)+G(\lambda))^{-1}, \quad (3.105)$$

$$\Delta(h_N; F, G) = \langle \mathbf{R}\vec{\mathbf{a}}_N, \mathbf{B}^{-1}\mathbf{R}\vec{\mathbf{a}}_N \rangle + \langle \mathbf{Q}\vec{\mathbf{a}}_N, \vec{\mathbf{a}}_N \rangle, \quad (3.106)$$

where

$$A_N(e^{i\lambda}) = \sum_{j \in Z^S \cap \{0, \dots, N\}} \vec{a}(j) e^{-ij\lambda}.$$

The following corollary is true.

Corollary 3.12. *Let $\vec{\xi}(j)$ and $\vec{\eta}(j)$ be uncorrelated multidimensional stationary stochastic sequences with the spectral densities $F(\lambda)$ and $G(\lambda)$ satisfying the minimality condition (3.93). The spectral characteristic $h_N(e^{i\lambda})$ and mean square error $\Delta(h_N; F, G)$ of the optimal linear estimate of the functional $A_N \vec{\xi}$ based on observation of the sequence $\{\vec{\xi}(j) + \vec{\eta}(j)\}$ at points $j \in \mathbb{Z}_- \backslash S$ can be calculated by formulas (3.105), (3.106).*

3.2. Minimax Robust Method of Filtering

The formulas obtained in the previous section may be applied to filtering the functional $A\vec{\xi}$ only in the case where the spectral densities of the sequences are precisely known. In the case where complete information on the spectral densities is impossible while a class of admissible densities is given, it is reasonable to apply the minimax robust method of filtering. According to this method, we find an estimate which minimizes the maximal values of mean square errors for all spectral densities from a given class D of admissible spectral densities simultaneously.

For a description of the minimax method of estimation we propose the following definitions.

Definition 3.7. *For a given class of spectral densities $D = D_F \times D_G$ the spectral densities $F^0(\lambda) \in D_F$, $G^0(\lambda) \in D_G$ are called the least favourable in the class D for the optimal linear filtering of the functional $A\vec{\xi}$ if the following relation is true*

$$\Delta\left(F^0, G^0\right) = \Delta\left(h\left(F^0, G^0\right); F^0, G^0\right) = \max_{(F,G)\in D_F\times D_G} \Delta\left(h\left(F,G\right); F,G\right).$$

Definition 3.8. *For a given class of spectral densities $D = D_F \times D_G$ the spectral characteristic $h^0(e^{i\lambda})$ of the optimal linear estimate of the functional $A\vec{\xi}$ is called minimax robust if there are satisfied conditions*

$$h^0(e^{i\lambda}) \in H_D = \bigcap_{(F,G)\in D_F\times D_G} L_2^s(F+G),$$

$$\min_{h\in H_D} \max_{(F,G)\in D} \Delta\left(h; F,G\right) = \max_{(F,G)\in D} \Delta\left(h^0; F,G\right).$$

From the introduced definitions and formulas derived above we can obtain the following statement.

Lemma 3.4. *The spectral densities $F^0(\lambda) \in D_F$, $G^0(\lambda) \in D_G$ satisfying the minimality condition (3.93) are the least favourable in the class $D = D_F \times D_G$ for the optimal linear filtering of the functional $A\vec{\xi}$ if the operators B^0, R^0, Q^0 defined by the Fourier coefficients of the functions*

$$(F^0(\lambda)+G^0(\lambda))^{-1},\ F^0(\lambda)(F^0(\lambda)+G^0(\lambda))^{-1},\ F^0(\lambda)(F^0(\lambda)+G^0(\lambda))^{-1}G^0(\lambda)$$

determine a solution of the constrain optimization problem

$$\begin{aligned}\max_{(F,G)\in D_F\times D_G} &\langle \mathbf{R}\vec{\mathbf{a}}, \mathbf{B}^{-1}\mathbf{R}\vec{\mathbf{a}}\rangle + \langle \mathbf{Q}\vec{\mathbf{a}}, \vec{\mathbf{a}}\rangle = \\ &= \langle \mathbf{R}^0\vec{\mathbf{a}}, (\mathbf{B}^0)^{-1}\mathbf{R}^0\vec{\mathbf{a}}\rangle + \langle \mathbf{Q}^0\vec{\mathbf{a}}, \vec{\mathbf{a}}\rangle.\end{aligned} \tag{3.107}$$

The minimax spectral characteristic $h^0 = h(F^0, G^0)$ is determined by the formula (3.103) if $h(F^0, G^0) \in H_D$.

The least favourable spectral densities $F^0(\lambda)$, $G^0(\lambda)$ and the minimax spectral characteristic $h^0 = h(F^0, G^0)$ form a saddle point of the function $\Delta(h; F, G)$ on the set $H_D \times D$. The saddle point inequalities

$$\Delta\left(h; F^0, G^0\right) \geq \Delta\left(h^0; F^0, G^0\right) \geq \Delta\left(h^0; F, G\right)$$

$$\forall h \in H_D, \forall F \in D_F, \forall G \in D_G$$

are true if $h^0 = h(F^0, G^0)$, $h(F^0, G^0) \in H_D$, and (F^0, G^0) is a solution of the constrained optimization problem

$$\sup_{(F,G) \in D_F \times D_G} \Delta\left(h(F^0, G^0); F, G\right) = \Delta\left(h(F^0, G^0); F^0, G^0\right), \tag{3.108}$$

where the functional $\Delta\left(h(F^0, G^0); F, G\right)$ is calculated by the formula

$$\Delta\left(h(F^0, G^0); F, G\right) =$$
$$= \frac{1}{2\pi} \int_{-\pi}^{\pi} (r_G^0(\lambda))^\top F(\lambda) \overline{r_G^0(\lambda)} d\lambda + \frac{1}{2\pi} \int_{-\pi}^{\pi} (r_F^0(\lambda))^\top G(\lambda) \overline{r_F^0(\lambda)} d\lambda,$$

$$(r_F^0(\lambda))^\top =$$
$$= \left((A(e^{i\lambda}))^\top F^0(\lambda) - \left(\sum_{k \in U} ((\mathbf{B}^0)^{-1} \mathbf{R}^0 \vec{\mathbf{a}})(k) e^{ik\lambda} \right)^\top \right) (F^0(\lambda) + G^0(\lambda))^{-1},$$

$$(r_G^0(\lambda))^\top =$$
$$= \left((A(e^{i\lambda}))^\top G^0(\lambda) + \left(\sum_{k \in U} ((\mathbf{B}^0)^{-1} \mathbf{R}^0 \vec{\mathbf{a}})(k) e^{ik\lambda} \right)^\top \right) (F^0(\lambda) + G^0(\lambda))^{-1}.$$

The constrained optimization problem (3.108) is equivalent to the unconstrained optimization problem (see, for example, B. N. Pshenichnyj [156])

$$\Delta_D(F, G) = -\Delta(h(F^0, G^0); F, G) + \delta((F, G) \,|D_F \times D_G) \to \inf, \tag{3.109}$$

where $\delta((F,G)|D_F \times D_G)$ is the indicator function of the set $D = D_F \times D_G$. A solution of the problem (3.109) is characterized by the condition $0 \in \partial\Delta_D(F^0,G^0)$, where $\partial\Delta_D(F^0,G^0)$ is the subdifferential of the convex functional $\Delta_D(F,G)$ at point (F^0,G^0). This condition makes it possible to find the least favourable spectral densities in some special classes of spectral densities D (see books by A. D. Ioffe and V. M. Tihomirov [161], B. N. Pshenichnyj [156], T. Rockafellar [158]).

Note, that the form of the functional $\Delta\left(h^0;F,G\right)$ is convenient for application of the Lagrange method of indefinite multipliers for finding a solution of the problem (3.109). Making use of the method of Lagrange multipliers and the form of subdifferentials of the indicator functions we describe relations that determine the least favourable spectral densities in some special classes of spectral densities (see books by M. P. Moklyachuk [55, 163], M. P. Moklyachuk and O. Yu. Masyutka [96] for additional details).

3.3. Least Favourable Spectral Densities in the Class $D = D_0 \times D_{2\delta}$

Consider the problem of filtering of the functional $A\vec{\xi}$ in the case where the spectral densities $F(\lambda)$, $G(\lambda)$ belong to the classes of admissible spectral densities $D_0^k \times D_{2\delta}^k, k = 1,2,3,4$, where

$$D_0^1 = \left\{F(\lambda)\left|\frac{1}{2\pi}\int_{-\pi}^{\pi}\operatorname{Tr}F(\lambda)d\lambda = p\right.\right\},$$

$$D_{2\delta}^1 = \left\{G(\lambda)\left|\frac{1}{2\pi}\int_{-\pi}^{\pi}|\operatorname{Tr}(G(\lambda)-G_1(\lambda))|^2\,d\lambda \leq \delta\right.\right\};$$

$$D_0^2 = \left\{F(\lambda)\left|\frac{1}{2\pi}\int_{-\pi}^{\pi}f_{kk}(\lambda)d\lambda = p_k, k=\overline{1,T}\right.\right\},$$

$$D_{2\delta}^2 = \left\{G(\lambda)\left|\frac{1}{2\pi}\int_{-\pi}^{\pi}\left|g_{kk}(\lambda)-g_{kk}^1(\lambda)\right|^2\,d\lambda \leq \delta_k, k=\overline{1,T}\right.\right\};$$

$$D_0^3 = \left\{F(\lambda)\left|\frac{1}{2\pi}\int_{-\pi}^{\pi}\langle B_1, F(\lambda)\rangle\,d\lambda = p\right.\right\},$$

$$D_{2\delta}^3 = \left\{G(\lambda)\left|\frac{1}{2\pi}\int_{-\pi}^{\pi}|\langle B_2, G(\lambda)-G_1(\lambda)\rangle|^2\,d\lambda \leq \delta\right.\right\};$$

$$D_0^4 = \left\{ F(\lambda) \left| \frac{1}{2\pi} \int_{-\pi}^{\pi} F(\lambda) d\lambda = P \right. \right\}.$$

$$D_{2\delta}^4 = \left\{ G(\lambda) \left| \frac{1}{2\pi} \int_{-\pi}^{\pi} \left| g_{ij}(\lambda) - g_{ij}^1(\lambda) \right|^2 d\lambda \le \delta_i^j, i, j = \overline{1,T} \right. \right\},$$

Here the spectral density matrix $G_1(\lambda)$ is known and fixed, $p, \delta, p_k, \delta_k, k = \overline{1,T}$, $\delta_i^j, i, j = \overline{1,T}$, are given numbers, P, B_1, B_2 are given positive definite Hermitian matrices. From the condition $0 \in \partial\Delta_D(F^0, G^0)$ we find the following equations which determine the least favourable spectral densities for these given sets of admissible spectral densities.

For the first pair $D_0^1 \times D_{2\delta}^1$ we have equations

$$(r_G^0(\lambda))^* (r_G^0(\lambda))^\top = \alpha^2 (F^0(\lambda) + G^0(\lambda))^2, \tag{3.110}$$

$$(r_F^0(\lambda))^* (r_F^0(\lambda))^\top = \beta^2 \mathrm{Tr}\,(G^0(\lambda) - G_1(\lambda))(F^0(\lambda) + G^0(\lambda))^2, \tag{3.111}$$

$$\frac{1}{2\pi} \int_{-\pi}^{\pi} |\mathrm{Tr}\,(G(\lambda) - G_1(\lambda))|^2 d\lambda = \delta, \tag{3.112}$$

where α^2, β^2 are Lagrange multipliers.

For the second pair $D_0^2 \times D_{2\delta}^2$ we have equations

$$(r_G^0(\lambda))^* (r_G^0(\lambda))^\top =$$
$$= (F^0(\lambda) + G^0(\lambda)) \left\{ \alpha_k^2 \delta_{kl} \right\}_{k,l=1}^T (F^0(\lambda) + G^0(\lambda)), \tag{3.113}$$

$$(r_F^0(\lambda))^* (r_F^0(\lambda))^\top =$$
$$= (F^0(\lambda) + G^0(\lambda)) \left\{ \beta_k^2 (g_{kk}^0(\lambda) - g_{kk}^1(\lambda)) \delta_{kl} \right\}_{k,l=1}^T (F^0(\lambda) + G^0(\lambda)), \tag{3.114}$$

$$\frac{1}{2\pi} \int_{-\pi}^{\pi} \left| g_{kk}(\lambda) - g_{kk}^1(\lambda) \right|^2 d\lambda = \delta_k, \; k = \overline{1,T}, \tag{3.115}$$

where α_k^2, β_k^2 are Lagrange multipliers, δ_{kl} are Kronecker symbols.

For the third pair $D_0^3 \times D_{2\delta}^3$ we have equations

$$(r_G^0(\lambda))^* (r_G^0(\lambda))^\top = \alpha^2 (F^0(\lambda) + G^0(\lambda)) B_1^\top (F^0(\lambda) + G^0(\lambda)), \tag{3.116}$$

$$(r_F^0(\lambda))^* (r_F^0(\lambda))^\top = \beta^2 \left\langle B_2, G^0(\lambda) - G_1(\lambda) \right\rangle (F^0(\lambda) + G^0(\lambda))^2, \tag{3.117}$$

$$\frac{1}{2\pi}\int_{-\pi}^{\pi}|\langle B_2, G(\lambda)-G_1(\lambda)\rangle|^2\,d\lambda=\delta, \tag{3.118}$$

where α^2, β^2 are Lagrange multipliers.

For the fourth pair $D_0^4 \times D_{2\delta}^4$ we have equations

$$(r_G^0(\lambda))^*(r_G^0(\lambda))^\top=(F^0(\lambda)+G^0(\lambda))\vec{\alpha}\cdot\vec{\alpha}^*(F^0(\lambda)+G^0(\lambda)), \tag{3.119}$$

$$(r_F^0(\lambda))^*(r_F^0(\lambda))^\top=$$
$$=(F^0(\lambda)+G^0(\lambda))\left\{\beta_{ij}(g_{ij}^0(\lambda)-g_{ij}^1(\lambda))\right\}_{i,j=1}^T(F^0(\lambda)+G^0(\lambda)), \tag{3.120}$$

$$\frac{1}{2\pi}\int_{-\pi}^{\pi}\left|g_{ij}(\lambda)-g_{ij}^1(\lambda)\right|^2 d\lambda=\delta_i^j,\ i,j=\overline{1,T}, \tag{3.121}$$

where $\vec{\alpha}, \beta_{ij}$ are Lagrange multipliers.

The following theorem and corollaries are true.

Theorem 3.9. *The least favourable spectral densities $F^0(\lambda)$, $G^0(\lambda)$ in the classes $D_0^k \times D_{2\delta}^k, k=1,2,3,4,$ for the optimal linear filtering of the functional $A\vec{\xi}$ are determined by relations (3.110) —- (3.121), the minimality condition (3.93), the constrained optimization problem (3.107), and restrictions on densities from the corresponding classes $D_0^k \times D_{2\delta}^k, k=1,2,3,4$. The minimax-robust spectral characteristic of the optimal estimate of the functional $A\vec{\xi}$ is determined by the formula (3.103).*

Corollary 3.13. *Assume that the spectral density $G(\lambda)$ is known. Let the function $F^0(\lambda)+G(\lambda)$ satisfy the minimality condition (3.93). The spectral density $F^0(\lambda)$ is the least favourable in the classes D_0^k, $k=\overline{1,4}$, for the optimal linear filtering of the functional $A\vec{\xi}$ if it satisfies relations (3.110), (3.113), (3.116), (3.119), respectively, and the pair $(F^0(\lambda), G(\lambda))$ determines a solution of the optimization problem (3.107). The minimax-robust spectral characteristic of the optimal estimate of the functional $A\vec{\xi}$ is determined by formula (3.103).*

Corollary 3.14. *Assume that the spectral density $F(\lambda)$ is known. Let the function $F(\lambda)+G^0(\lambda)$ satisfy the minimality condition (3.93). The spectral density $G^0(\lambda)$ is the least favourable in the classes $D_{2\delta}^k$, $k=\overline{1,4}$, for the optimal linear filtering of the functional $A\vec{\xi}$ if it satisfies relations (3.111) – (3.112),*

(3.114) – (3.115), (3.117) – (3.118), (3.120) – (3.121), respectively, and the pair $(F(\lambda), G^0(\lambda))$ determines a solution of the optimization problem (3.107). The minimax-robust spectral characteristic of the optimal estimate of the functional $A\vec{\xi}$ is determined by formula (3.103).

3.4. Least Favourable Spectral Densities in the Class $D = D_{1\delta} \times D_V^U$

Consider the problem of filtering of the functional $A\vec{\xi}$ in the case where spectral densities $F(\lambda)$, $G(\lambda)$ belong to the classes of admissible spectral densities $D_{1\delta}^k \times D_V^{U\,k}$, $k = 1, 2, 3, 4$,, where

$$D_{1\delta}^1 = \left\{ F(\lambda) \left| \frac{1}{2\pi} \int_{-\pi}^{\pi} |\mathrm{Tr}(F(\lambda) - F_1(\lambda))| d\lambda \leq \delta \right. \right\},$$

$$D_V^{U\,1} = \left\{ G(\lambda) \left| \mathrm{Tr}\, V(\lambda) \leq \mathrm{Tr}\, G(\lambda) \leq \mathrm{Tr}\, U(\lambda), \frac{1}{2\pi} \int_{-\pi}^{\pi} \mathrm{Tr}\, G(\lambda) d\lambda = q \right. \right\},$$

$$D_{1\delta}^2 = \left\{ F(\lambda) \left| \frac{1}{2\pi} \int_{-\pi}^{\pi} \left| f_{kk}(\lambda) - f_{kk}^1(\lambda) \right| d\lambda \leq \delta_k, k = \overline{1,T} \right. \right\},$$

$$D_V^{U\,2} = \left\{ G(\lambda) \left| v_{kk}(\lambda) \leq g_{kk}(\lambda) \leq u_{kk}(\lambda), \right. \right.$$
$$\left. \frac{1}{2\pi} \int_{-\pi}^{\pi} g_{kk}(\lambda) d\lambda = q_k, k = \overline{1,T} \right\},$$

$$D_{1\delta}^3 = \left\{ F(\lambda) \left| \frac{1}{2\pi} \int_{-\pi}^{\pi} |\langle B_1, F(\lambda) - F_1(\lambda) \rangle| d\lambda \leq \delta \right. \right\},$$

$$D_V^{U\,3} = \left\{ G(\lambda) \left| \langle B_2, V(\lambda) \rangle \leq \langle B_2, G(\lambda) \rangle \leq \langle B_2, U(\lambda) \rangle, \right. \right.$$
$$\left. \frac{1}{2\pi} \int_{-\pi}^{\pi} \langle B_2, G(\lambda) \rangle d\lambda = q \right\},$$

$$D_{1\delta}^4 = \left\{ F(\lambda) \left| \frac{1}{2\pi} \int_{-\pi}^{\pi} \left| f_{ij}(\lambda) - f_{ij}^1(\lambda) \right| d\lambda \leq \delta_i^j, i, j = \overline{1,T} \right. \right\},$$

$$D_V^{U\,4} = \left\{ G(\lambda) \middle| V(\lambda) \le G(\lambda) \le U(\lambda), \frac{1}{2\pi}\int_{-\pi}^{\pi} G(\lambda)d\lambda = Q \right\}.$$

Here the spectral densities $F_1(\lambda), V(\lambda), U(\lambda)$ are known and fixed, $\delta, q, \delta_k, q_k, k = \overline{1,T}$, $\delta_i^j, i, j = \overline{1,T}$, are given numbers, Q, B_1, B_2 are given positive definite Hermitian matrices.

From the condition $0 \in \partial\Delta_D(F^0, G^0)$ we find the following equations which determine the least favourable spectral densities for these given sets of admissible spectral densities.

For the first pair $D_{1\delta}^1 \times D_V^{U\,1}$ we have equations

$$(r_G^0(\lambda))^*(r_G^0(\lambda))^\top = \alpha^2\gamma(\lambda)(F^0(\lambda)+G^0(\lambda))^2, \tag{3.122}$$

$$\frac{1}{2\pi}\int_{-\pi}^{\pi} \left|\mathrm{Tr}\,(F^0(\lambda)-F_1(\lambda))\right| d\lambda = \delta, \tag{3.123}$$

$$(r_F^0(\lambda))^*(r_F^0(\lambda))^\top = (\beta^2+\gamma_1(\lambda)+\gamma_2(\lambda))(F^0(\lambda)+G^0(\lambda))^2, \tag{3.124}$$

where α^2, β^2 are Lagrange multipliers, $|\gamma(\lambda)| \le 1$ and

$$\gamma(\lambda) = \mathrm{sign}\,(\mathrm{Tr}\,(F^0(\lambda)-F_1(\lambda))) : \ \mathrm{Tr}\,(F^0(\lambda)-F_1(\lambda)) \neq 0,$$

$\gamma_1(\lambda) \le 0$ and $\gamma_1(\lambda) = 0$ if $\mathrm{Tr}\,G^0(\lambda) > \mathrm{Tr}\,V(\lambda)$, $\gamma_2(\lambda) \ge 0$ and $\gamma_2(\lambda) = 0$ if $\mathrm{Tr}\,G^0(\lambda) < \mathrm{Tr}\,U(\lambda)$.

For the second pair $D_{1\delta}^2 \times D_V^{U\,2}$ we have equations

$$(r_G^0(\lambda))^*(r_G^0(\lambda))^\top = \\ = (F^0(\lambda)+G^0(\lambda))\left\{\alpha_k^2\gamma_k(\lambda)\delta_{kl}\right\}_{k,l=1}^{T}(F^0(\lambda)+G^0(\lambda)), \tag{3.125}$$

$$\frac{1}{2\pi}\int_{-\pi}^{\pi} \left|f_{kk}^0(\lambda)-f_{kk}^1(\lambda)\right| d\lambda = \delta_k, \ k = \overline{1,T}, \tag{3.126}$$

$$(r_F^0(\lambda))^*(r_F^0(\lambda))^\top = \\ = (F^0(\lambda)+G^0(\lambda))\left\{(\beta_k^2+\gamma_{1k}(\lambda)+\gamma_{2k}(\lambda))\delta_{kl}\right\}_{k,l=1}^{T}(F^0(\lambda)+G^0(\lambda)), \tag{3.127}$$

where α_k^2, β_k^2 are Lagrange multipliers, $|\gamma_k(\lambda)| \leq 1$ and

$$\gamma_k(\lambda) = \text{sign}\,(f_{kk}^0(\lambda) - f_{kk}^1(\lambda)) : \; f_{kk}^0(\lambda) - f_{kk}^1(\lambda) \neq 0, \; k = \overline{1,T},$$

$\gamma_{1k}(\lambda) \leq 0$ and $\gamma_{1k}(\lambda) = 0$ if $g_{kk}^0(\lambda) > v_{kk}(\lambda)$, $\gamma_{2k}(\lambda) \geq 0$ and $\gamma_{2k}(\lambda) = 0$ if $g_{kk}^0(\lambda) < u_{kk}(\lambda)$.

For the third pair $D_{1\delta}^3 \times D_V^{U\,3}$ we have equations

$$(r_G^0(\lambda))^*(r_G^0(\lambda))^\top = \alpha^2\gamma'(\lambda)(F^0(\lambda)+G^0(\lambda))B_1^\top(F^0(\lambda)+G^0(\lambda)), \quad (3.128)$$

$$\frac{1}{2\pi}\int_{-\pi}^{\pi}\left|\left\langle B_1, F^0(\lambda) - F_1(\lambda)\right\rangle\right| d\lambda = \delta, \quad (3.129)$$

$$(r_F^0(\lambda))^*(r_F^0(\lambda))^\top = $$
$$= (\beta^2 + \gamma_1'(\lambda) + \gamma_2'(\lambda))(F^0(\lambda)+G^0(\lambda))B_2^\top(F^0(\lambda)+G^0(\lambda)), \quad (3.130)$$

where α^2, β^2 are Lagrange multipliers, $|\gamma'(\lambda)| \leq 1$ and

$$\gamma'(\lambda) = \text{sign}\,\left\langle B_1, F^0(\lambda) - F_1(\lambda)\right\rangle : \; \left\langle B_1, F^0(\lambda) - F_1(\lambda)\right\rangle \neq 0,$$

$\gamma_1'(\lambda) \leq 0$ and $\gamma_1'(\lambda) = 0$ if $\langle B_2, G^0(\lambda)\rangle > \langle B_2, V(\lambda)\rangle$, $\gamma_2'(\lambda) \geq 0$ and $\gamma_2'(\lambda) = 0$ if $\langle B_2, G^0(\lambda)\rangle < \langle B_2, U(\lambda)\rangle$.

For the fourth pair $D_{1\delta}^4 \times D_V^{U\,4}$ we have equations

$$(r_G^0(\lambda))^*(r_G^0(\lambda))^\top = $$
$$= (F^0(\lambda)+G^0(\lambda))\left\{\alpha_{ij}\gamma_{ij}(\lambda)\right\}_{i,j=1}^T(F^0(\lambda)+G^0(\lambda)), \quad (3.131)$$

$$\frac{1}{2\pi}\int_{-\pi}^{\pi}\left|f_{ij}^0(\lambda) - f_{ij}^1(\lambda)\right| d\lambda = \delta_i^j, \; i,j = \overline{1,T}, \quad (3.132)$$

$$(r_F^0(\lambda))^*(r_F^0(\lambda))^\top = $$
$$= (F^0(\lambda)+G^0(\lambda))(\vec{\beta}\cdot\vec{\beta}^* + \Gamma_1(\lambda) + \Gamma_2(\lambda))(F^0(\lambda)+G^0(\lambda)) \quad (3.133)$$

where $\vec{\beta}, \alpha_{ij}$ are Lagrange multipliers, $\left|\gamma_{ij}(\lambda)\right| \leq 1$ and

$$\gamma_{ij}(\lambda) = \frac{f_{ij}^0(\lambda) - f_{ij}^1(\lambda)}{\left|f_{ij}^0(\lambda) - f_{ij}^1(\lambda)\right|} : \; f_{ij}^0(\lambda) - f_{ij}^1(\lambda) \neq 0, \; i,j = \overline{1,T},$$

$\Gamma_1(\lambda) \leq 0$ and $\Gamma_1(\lambda) = 0$ if $G^0(\lambda) > V(\lambda)$, $\Gamma_2(\lambda) \geq 0$ and $\Gamma_2(\lambda) = 0$ if $G^0(\lambda) < U(\lambda)$.

The following theorem and corollaries are true.

Theorem 3.10. *The least favourable spectral densities $F^0(\lambda)$, $G^0(\lambda)$ in the classes $D_{1\delta}^k \times D_V^{U^k}, k = 1,2,3,4$, for the optimal linear filtering of the functional $A\vec{\xi}$ are determined by relations (3.122) – (3.133), the minimality condition (3.93), the constrained optimization problem (3.107), and restrictions on densities from the corresponding classes $D_{1\delta}^k \times D_V^{U^k}, k = 1,2,3,4$. The minimax-robust spectral characteristic of the optimal estimate of the functional $A\vec{\xi}$ is determined by formula (3.103).*

Corollary 3.15. *Assume that the spectral density $G(\lambda)$ is known. Let the function $F^0(\lambda) + G(\lambda)$ satisfy the minimality condition (3.93). The spectral density $F^0(\lambda)$ is the least favourable in the classes $D_{1\delta}^k$, $k = \overline{1,4}$, for the optimal linear filtering of the functional $A\vec{\xi}$ if it satisfies relations (3.122) – (3.123), (3.125) – (3.126), (3.128) – (3.129), (3.131) – (3.132), respectively, and the pair $(F^0(\lambda), G(\lambda))$ determines a solution of the optimization problem (3.107). The minimax-robust spectral characteristic of the optimal estimate of the functional $A\vec{\xi}$ is determined by formula (3.103).*

Corollary 3.16. *Assume that the spectral density $F(\lambda)$ is known. Let the function $F(\lambda) + G^0(\lambda)$ satisfy the minimality condition (3.93). The spectral density $G^0(\lambda)$ is the least favourable in the classes $D_V^{U^k}$, $k = \overline{1,4}$, for the optimal linear filtering of the functional $A\vec{\xi}$ if it satisfies relations (3.124), (3.127), (3.130), (3.133), respectively, and the pair $(F(\lambda), G^0(\lambda))$ determines a solution of the optimization problem (3.107). The minimax-robust spectral characteristic of the optimal estimate of the functional $A\vec{\xi}$ is determined by formula (3.103).*

Conclusion

In this chapter, we present results of the investigation of the interpolation, extrapolation, and filtering problems for linear functionals which depend on unobserved values of multidimensional stationary stochastic sequences. We develop methods of finding the optimal estimates of the functionals in the case where some observations are missing.

We, first, investigate the problems under the condition of spectral certainty, where spectral density matrices of the involved stationary sequences are known. In this case, we propose an approach based on the Hilbert space projection method.

In the case of extrapolation problem, formulas for calculating values of the mean-square error and the spectral characteristic of the optimal estimate are derived for the functional

$$A\vec{\xi} = \sum_{j=0}^{\infty} \vec{a}(j)^{\top}\vec{\xi}(j)$$

constructed from unobserved values of a multidimensional stationary stochastic sequence $\{\vec{\xi}(j), j \in \mathbb{Z}\}$. Estimates are based on observations of the sequence $\{\vec{\xi}(j) + \vec{\eta}(j)\}$, where $\{\vec{\eta}(j)\}$ is a stationary correlated with $\xi(j)$ stochastic sequence, at points

$$j \in \mathbb{Z}_{-}\backslash S = \{\dots, -2, -1\}\backslash S,\ S = \bigcup_{l=1}^{s}\{-M_l - N_l, -M_l - N_l + 1, \dots, -M_l\}.$$

In the case of interpolation problem, we describe methods of finding solutions of the problem of mean-square optimal linear estimation of the functional

$$A_s\vec{\xi} = \sum_{l=0}^{s-1} \sum_{j=M_l}^{M_l+N_{l+1}} \vec{a}(j)^{\top}\vec{\xi}(j),$$

$$M_l = \sum_{k=0}^{l} (N_k + K_k),\ N_0 = K_0 = 0$$

constructed from unobserved values of the stationary stochastic sequence $\vec{\xi}(j) = \{\xi_k(j)\}_{k=1}^{T}$. Estimates are based on observations of the sequence $\vec{\xi}(j)$ at points $j \in \mathbb{Z}\backslash S$, where $S = \bigcup_{l=0}^{s-1}\{M_l, M_l + 1, \dots, M_l + N_{l+1}\}$.

In the case of filtering problem, we propose results of the investigation of the problem of mean-square optimal linear estimation of the functional

$$A\vec{\xi} = \sum_{j\in Z^S} \vec{a}(j)^{\top}\vec{\xi}(-j)$$

constructed from unobserved values of a multidimensional stationary sequence $\{\vec{\xi}(j), j \in \mathbb{Z}\}$. Estimates are based on observations of the sequence $\{\vec{\xi}(j) + \vec{\eta}(j)\}$ at points $j \in \mathbb{Z}_- \setminus S$, where $\{\vec{\eta}(j), j \in \mathbb{Z}\}$ is uncorrelated with $\{\vec{\xi}(j), j \in \mathbb{Z}\}$ multidimensional stationary noise sequence,

$$S = \bigcup_{l=1}^{s} \{-(M_l + N_l), \ldots, -M_l\}, \quad Z^S = \{1, 2, \ldots\} \setminus S^+,$$

$$S^+ = \bigcup_{l=1}^{s} \{M_l, \ldots, M_l + N_l\}, M_0 = 0, N_0 = 0.$$

Next, we investigate the problems in the case of spectral uncertainty, where the spectral density matrices of the multidimensional stationary stochastic sequences are not exactly known, while some special classes of admissible spectral density matrices are given. In this case, we apply the minimax robust method of estimation. This method allows us to find estimates that minimize the maximal values of the mean-square errors of the estimates for all spectral density matrices from a given class of admissible spectral densities. We derive formulas that determine the least favourable spectral densities and the minimax spectral characteristics for some special classes of admissible spectral density matrices. These results are expressed in terms of the Fourier coefficients of some functions from the spectral densities of the stochastic sequences.

These least favourable spectral density matrices are solutions of the optimization problem

$$\Delta_D(F, G) = -\Delta(h(F^0, G^0); F, G) + \delta((F, G) \,|D_F \times D_G) \to \inf,$$

which is characterized by the condition $0 \in \partial\Delta_D(F^0, G^0)$, where $\partial\Delta_D(F^0, G^0)$ is the subdifferential of the convex functional $\Delta_D(F, G)$ at point (F^0, G^0). The form of the functional $\Delta(h(F^0, G^0); F, G)$ is convenient for application of the Lagrange method of indefinite multipliers for finding a solution of the optimization problem. The complexity of the problem is determined by the complexity of calculation of the subdifferential of the convex functional $\Delta_D(F, G)$. Making use of the method of Lagrange multipliers and the form of subdifferentials of the indicator functions, we describe relations that determine the least favourable spectral densities in some special classes of spectral densities. These classes of spectral densities are as follows. The classes D_W of densities with the moment restrictions. The classes $D_{1\delta}$ of spectral densities which form the "δ-neighbourhood" models in the space L_1 of a fixed bounded spectral density matrix. The classes $D_{2\delta}$ of spectral densities which form the "δ-neighbourhood" models in the space L_2 of a fixed bounded spectral density matrix. The classes D_ε of spectral densities which form the "ε–contamination" models of spectral density matrices. The classes D_V^U of spectral densities which form the "strip" models of spectral density matrices.

It is shown that spectral densities of the autoregressive stochastic sequences are the least favourable in some classes of spectral density matrices.

Chapter 4

Estimation of Multidimensional Continuous Time Stationary Stochastic Processes with Missing Observations

1. Extrapolation Problem for Multidimensional Stationary Stochastic Processes with Missing Observations

In this section we deal with the problem of mean-square optimal linear estimation of the functional

$$A\vec{\xi} = \int_{0}^{\infty} \vec{a}(t)^{\top}\vec{\xi}(t)dt$$

constructed from unobserved values of a multidimensional stationary stochastic process $\vec{\xi}(t) = \{\xi_k(t)\}_{k=1}^{T}$, $t \in \mathbb{R}^{+}$. Estimates are based on observations of the signal process $\vec{\xi}(t)$ with additive noise process $\vec{\eta}(t)$ at points $t \in \mathbb{R}^{-}\setminus S$, $S = \bigcup_{l=1}^{s}[-M_l - N_l, -M_l]$. The problem is investigated in the case of spectral certainty, as well as in the case of spectral uncertainty. Formulas for calculating the spectral characteristic and the mean-square error of the optimal linear estimate of the functional are derived in the case of spectral certainty, where the spectral densities of the signal and noise processes are exactly known. In the case of spectral uncertainty, where the spectral densities of the signal and noise processes are not exactly known while a set of admissible spectral densities is given, the minimax method of estimation is applied. Formulas for determining the least favourable spectral densities and minimax-robust spectral characteristics of the optimal estimates of the functional are proposed for some specific classes of admissible spectral densities.

1.1. Hilbert Space Projection Method of Extrapolation

Consider two uncorrelated mean square continuous multidimensional stationary stochastic processes $\vec{\xi}(t)=\{\xi_k(t)\}_{k=1}^{T}, t\in\mathbb{R}$ and $\vec{\eta}(t)=\{\eta_k(t)\}_{k=1}^{T}, t\in\mathbb{R}$ with zero first moments, $E\vec{\xi}(t)=\vec{0}$, $E\vec{\eta}(t)=\vec{0}$, and the correlation functions $R_{\xi}(t)=E\vec{\xi}(t+s)(\vec{\xi(s)})^*$ and $R_{\eta}(t)=E\vec{\eta}(t+s)(\vec{\eta(s)})^*$ represented in the form

$$R_{\xi}(t)=\frac{1}{2\pi}\int_{-\infty}^{\infty}e^{it\lambda}F(\lambda)d\lambda,\quad R_{\eta}(t)=\frac{1}{2\pi}\int_{-\infty}^{\infty}e^{it\lambda}G(\lambda)d\lambda,$$

where $F(\lambda)=\{f_{kl}(\lambda)\}_{k,l=1}^{T}$ and $G(\lambda)=\{g_{kl}(\lambda)\}_{k,l=1}^{T}$ are the spectral densities of the processes $\vec{\xi}(t)$ and $\vec{\eta}(t)$ (see the book by I. I. Gikhman and A. V. Skorokhod [153] for more details).

Suppose that the spectral densities of the processes satisfy the minimality condition

$$\int_{-\infty}^{\infty}(b(\lambda))^{\top}(F(\lambda)+G(\lambda))^{-1}\overline{b(\lambda)}d\lambda<\infty, \tag{4.1}$$

where $b(\lambda)=\sum\limits_{l=1}^{s}\int\limits_{-M_l-N_l}^{-M_l}\vec{\alpha}(t)e^{it\lambda}dt$ is a nontrivial function of the exponential type. Under this condition the error-free extrapolation is impossible (see, for example, Yu. A. Rozanov [10]).

The multidimensional stationary stochastic processes $\vec{\xi}(t)$ and $\vec{\eta}(t)$ admit the spectral representations (see I. I. Gikhman and A. V. Skorokhod [153])

$$\vec{\xi}(t) = \int\limits_{-\infty}^{\infty} e^{it\lambda} Z_{\xi}(d\lambda), \qquad \vec{\eta}(t) = \int\limits_{-\infty}^{\infty} e^{it\lambda} Z_{\eta}(d\lambda), \tag{4.2}$$

where $Z_{\xi}(d\lambda)$ and $Z_{\eta}(d\lambda)$ are orthogonal stochastic measures defined on $(-\infty,\infty)$ such that the following relations are true

$$EZ_{\xi}(\Delta_1)(Z_{\xi}(\Delta_2))^* = \frac{1}{2\pi}\int_{\Delta_1\cap\Delta_2} F(\lambda)d\lambda,$$

$$EZ_{\eta}(\Delta_1)(Z_{\eta}(\Delta_2))^* = \frac{1}{2\pi}\int_{\Delta_1\cap\Delta_2} G(\lambda)d\lambda.$$

Consider the problem of mean-square optimal linear extrapolation of the functional

$$A\vec{\xi} = \int\limits_{0}^{\infty} \vec{a}(t)^{\top}\vec{\xi}(t)dt$$

based on observations of the process $\vec{\xi}(t)+\vec{\eta}(t)$ at points $t\in\mathbb{R}^{-}\setminus S$, where $S = \bigcup\limits_{l=1}^{s}[-M_l-N_l,-M_l]$.

It follows from the spectral representation of the process $\vec{\xi}(t)$ that the functional $A\vec{\xi}$ can be represented in the form

$$A\vec{\xi} = \int\limits_{-\infty}^{\infty} (A(\lambda))^{\top} Z_{\xi}(d\lambda), \quad A(\lambda) = \int\limits_{0}^{\infty} \vec{a}(t)e^{it\lambda}dt.$$

Consider the Hilbert space $H = L_2(\Omega,\mathcal{F},P)$ generated by random variables ξ with zero mathematical expectations, $E\xi = 0$, finite variations, $E|\xi|^2<\infty$, and the inner product $(\xi,\eta) = E\xi\overline{\eta}$. Denote by $H^s(\xi+\eta)$ the closed linear subspace generated by elements $\{\xi_k(t)+\eta_k(t): t\in\mathbb{R}\setminus S, k=1,\dots,T\}$ in the Hilbert space $H = L_2(\Omega,\mathcal{F},P)$.

Let $L_2(F+G)$ be the Hilbert space of complex-valued functions $\vec{a}(\lambda) = \{a_k(\lambda)\}_{k=1}^{T}$ such that

$$\int_{-\infty}^{\infty} \vec{a}(\lambda)^{\top}(F(\lambda)+G(\lambda))\overline{\vec{a}(\lambda)}d\lambda =$$

$$= \int_{-\infty}^{\infty} \sum_{k,l=1}^{T} a_k(\lambda)\overline{a_l(\lambda)}(f_{kl}(\lambda) + g_{kl}(\lambda))d\lambda < \infty.$$

Denote by $L_2^s(F+G)$ the subspace of $L_2(F+G)$ generated by functions $\{e^{it\lambda}\delta_k,\ \delta_k = \{\delta_{kl}\}_{l=1}^{T}, k=1,\dots,T, t\in\mathbb{R}\backslash S\}$.

The mean-square optimal linear estimate $\hat{A}\vec{\xi}$ of the functional $A\vec{\xi}$ is of the form

$$\hat{A}\vec{\xi} = \int_{-\infty}^{\infty} (h(\lambda))^{\top}(Z_{\xi}(d\lambda) + Z_{\eta}(d\lambda)),$$

where $h(\lambda) = \{h_k(\lambda)\}_{k=1}^{T} \in L_2^s(F+G)$ is the spectral characteristic of the estimate.

The mean-square error $\Delta(h;F,G)$ of the estimate $\hat{A}\vec{\xi}$ can be calculated by the formula

$$\Delta(h;F,G) = E\left|A\vec{\xi} - \hat{A}\vec{\xi}\right|^2 =$$

$$= \frac{1}{2\pi}\int_{-\infty}^{\infty} (A(\lambda)-h(\lambda))^{\top}F(\lambda)\overline{(A(\lambda)-h(\lambda))}d\lambda + \frac{1}{2\pi}\int_{-\infty}^{\infty} (h(\lambda))^{\top}G(\lambda)\overline{h(\lambda)}d\lambda.$$

We first consider the problem in the case of spectral certainty, where the spectral densities of the stationary processes $\vec{\xi}(t)$ and $\vec{\eta}(t)$ are exactly known. In this case we apply the method of orthogonal projections in the Hilbert space to find the optimal estimate. According to this method, the optimal linear estimate of the functional $A\vec{\xi}$ is a projection of the element $A\vec{\xi}$ of the space $H = L_2(\Omega, \mathcal{F}, P)$ onto the subspace $H^s(\xi+\eta)$. The estimate is determined by the following two conditions:

$$1)\hat{A}\vec{\xi} \in H^s(\xi+\eta),$$
$$2)A\vec{\xi} - \hat{A}\vec{\xi} \perp H^s(\xi+\eta).$$

It follows from the second condition, that the spectral characteristic $h(\lambda)$ of the optimal linear estimate $\hat{A}\vec{\xi}$ for any $t\in\mathbb{R}^-\backslash S$ satisfies the equation

$$\frac{1}{2\pi}\int_{-\infty}^{\infty} (A(\lambda)-h(\lambda))^{\top} F(\lambda)e^{-it\lambda}d\lambda - \frac{1}{2\pi}\int_{-\infty}^{\infty} (h(\lambda))^{\top}G(\lambda)e^{-it\lambda}d\lambda = \vec{0}.$$

The last equation can be represented in the form

$$\frac{1}{2\pi}\int_{-\infty}^{\infty}\left[(A(\lambda))^{\top}F(\lambda)-(h(\lambda))^{\top}(F(\lambda)+G(\lambda))\right]e^{-it\lambda}d\lambda=0,\ t\in\mathbb{R}^{-}\backslash S. \quad (4.3)$$

Consider the function

$$(C(\lambda))^{\top}=(A(\lambda))^{\top}F(\lambda)-(h(\lambda))^{\top}(F(\lambda)+G(\lambda))$$

and its Fourier transform

$$\vec{\mathbf{c}}(t)=\frac{1}{2\pi}\int_{-\infty}^{\infty}C(\lambda)e^{-it\lambda}d\lambda, \quad t\in\mathbb{R}.$$

According to condition (4.3) the function $\vec{\mathbf{c}}(t)$ can be nonzero only on the set $U=S\cup[0,\infty)$. Therefore the function $C(\lambda)$ is of the form

$$C(\lambda)=\sum_{l=1}^{s}\int_{-M_l-N_l}^{-M_l}\vec{\mathbf{c}}(t)e^{it\lambda}dt+\int_{0}^{\infty}\vec{\mathbf{c}}(t)e^{it\lambda}dt,$$

and the spectral characteristic of the estimate $\hat{A}\vec{\xi}$ is of the form

$$\begin{aligned}(h(\lambda))^{\top}&=\\&=(A(\lambda))^{\top}F(\lambda)(F(\lambda)+G(\lambda))^{-1}-(C(\lambda))^{\top}(F(\lambda)+G(\lambda))^{-1}.\end{aligned} \quad (4.4)$$

It follows from the first condition, $\hat{A}\vec{\xi}\in H^{s}(\xi+\eta)$, which determines the optimal linear estimate of the functional $A\vec{\xi}$, that for any $t\in U$ the following relation is true

$$\int_{-\infty}^{\infty}\left((A(\lambda))^{\top}F(\lambda)(H(\lambda))^{-1}-(C(\lambda))^{\top}(H(\lambda))^{-1}\right)e^{-it\lambda}d\lambda=0, \quad (4.5)$$

where $H(\lambda)=F(\lambda)+G(\lambda)$

In order to represent the last relation in terms of linear operators in the space $L_2(U)$, we introduce the operators

$$(\mathbf{B}\mathbf{x})(t) = \frac{1}{2\pi}\sum_{l=1}^{s}\int_{-M_l-N_l}^{-M_l}(\vec{\mathbf{x}}(u))^{\top}\int_{-\infty}^{\infty}(H(\lambda))^{-1}e^{i\lambda(u-t)}d\lambda du+$$

$$+\frac{1}{2\pi}\int_{0}^{\infty}(\vec{\mathbf{x}}(u))^{\top}\int_{-\infty}^{\infty}(H(\lambda))^{-1}e^{i\lambda(u-t)}d\lambda du,$$

$$(\mathbf{R}\mathbf{x})(t) = \frac{1}{2\pi}\sum_{l=1}^{s}\int_{-M_l-N_l}^{-M_l}(\vec{\mathbf{x}}(u))^{\top}\int_{-\infty}^{\infty}F(\lambda)(H(\lambda))^{-1}e^{i\lambda(u-t)}d\lambda du+$$

$$+\frac{1}{2\pi}\int_{0}^{\infty}(\vec{\mathbf{x}}(u))^{\top}\int_{-\infty}^{\infty}F(\lambda)(H(\lambda))^{-1}e^{i\lambda(u-t)}d\lambda du,$$

$$(\mathbf{Q}\mathbf{x})(t) = \frac{1}{2\pi}\sum_{l=1}^{s}\int_{-M_l-N_l}^{-M_l}(\vec{\mathbf{x}}(u))^{\top}\int_{-\infty}^{\infty}F(\lambda)(H(\lambda))^{-1}G(\lambda)e^{i\lambda(u-t)}d\lambda du+$$

$$+\frac{1}{2\pi}\int_{0}^{\infty}(\vec{\mathbf{x}}(u))^{\top}\int_{-\infty}^{\infty}F(\lambda)(H(\lambda))^{-1}G(\lambda)e^{i\lambda(u-t)}d\lambda du,$$

$$\vec{\mathbf{x}}(t) \in L_2(U), \quad t \in U.$$

Consider the function $\vec{\mathbf{a}}(t)$, $t \in U$, such that $\vec{\mathbf{a}}(t) = \vec{0}$, $t \in S$, $\vec{\mathbf{a}}(t) = \vec{a}(t), t \geq 0$. Making use of the introduced operators and functions relation (4.5) can be represented in the form

$$(\mathbf{R}\mathbf{a})(t) = (\mathbf{B}\mathbf{c})(t), \quad t \in U. \tag{4.6}$$

Suppose that the operator $\mathbf{B}$ is invertible. Then the function $\vec{\mathbf{c}}(t)$ can be calculated by the formula

$$\vec{\mathbf{c}}(t) = (\mathbf{B}^{-1}\mathbf{R}\mathbf{a})(t), \quad t \in U.$$

Therefore, the spectral characteristic $h(\lambda)$ of the estimate $\hat{A}\vec{\xi}$ can be calculated by the formula

$$(h(\lambda))^{\top} = (A(\lambda))^{\top}F(\lambda)(F(\lambda)+G(\lambda))^{-1} - (C(\lambda))^{\top}(F(\lambda)+G(\lambda))^{-1}, \tag{4.7}$$

$$C(\lambda) = \sum_{l=1}^{s} \int_{-M_l-N_l}^{-M_l} (\mathbf{B}^{-1}\mathbf{Ra})(t)e^{it\lambda}dt + \int_{0}^{\infty} (\mathbf{B}^{-1}\mathbf{Ra})(t)e^{it\lambda}dt.$$

The mean-square error of the estimate $\hat{A}\vec{\xi}$ can be calculated by the formula

$$\Delta(h;F,G) = \frac{1}{2\pi}\int_{-\infty}^{\infty} ((A(\lambda))^{\top}G(\lambda) + (C(\lambda))^{\top})(F(\lambda)+G(\lambda))^{-1}F(\lambda)\times$$

$$\times(F(\lambda)+G(\lambda))^{-1}((A(\lambda))^{\top}G(\lambda)+(C(\lambda))^{\top})^{*}d\lambda+$$

$$+\frac{1}{2\pi}\int_{-\infty}^{\infty} ((A(\lambda))^{\top}F(\lambda) - (C(\lambda))^{\top})(F(\lambda)+G(\lambda))^{-1}G(\lambda)\times$$

$$\times(F(\lambda)+G(\lambda))^{-1}((A(\lambda))^{\top}G(\lambda)+(C(\lambda))^{\top})^{*}d\lambda =$$

$$= \langle(\mathbf{Ra})(t), (\mathbf{B}^{-1}\mathbf{Ra})(t)\rangle + \langle(\mathbf{Qa})(t), \vec{\mathbf{a}}(t)\rangle, \qquad (4.8)$$

where $\left\langle \vec{a}(t), \vec{b}(t) \right\rangle$ is the inner product in the space $L_2(U)$.

The following theorem is true.

Theorem 4.1. *Let $\vec{\xi}(t)$ and $\vec{\eta}(t)$ be uncorrelated multidimensional stationary stochastic processes with the spectral densities $F(\lambda)$ and $G(\lambda)$ satisfying the minimality condition (4.1). The spectral characteristic $h(\lambda)$ and the mean-square error $\Delta(F,G)$ of the optimal linear estimate of the functional $A\vec{\xi}$ based on observations of the process $\vec{\xi}(t)+\vec{\eta}(t)$, $t \in \mathbb{R}^{-}\backslash S$, can be calculated by formulas (4.7), (4.8).*

Estimation of the Functional $A_N\vec{\xi}$

Consider the problem of mean-square optimal linear estimation of the functional

$$A_N\vec{\xi} = \int_{0}^{N} \vec{a}(t)^{\top}\vec{\xi}(t)dt,$$

based on observations of the process $\vec{\xi}(t)+\vec{\eta}(t)$ at points $t \in \mathbb{R}^{-}\backslash S$.

Each linear estimate $\hat{A}_N\vec{\xi}$ of the functional $A_N\vec{\xi}$ is of the form

$$\hat{A}_N\vec{\xi} = \int_{-\infty}^{\infty} (h_N(\lambda))^\top (Z_\xi(d\lambda) + Z_\eta(d\lambda)),$$

where $h_N(\lambda) = \{h_{kN}(\lambda)\}_{k=1}^T \in L_2^s(F+G)$ is the spectral characteristic of the estimate.

Introduce the function $\vec{\mathbf{a}}_N(t)$ in the following way

$$\vec{\mathbf{a}}_N(t) = \vec{a}(t), t \in [0,N], \quad \vec{\mathbf{a}}_N(t) = \vec{0}, t \in S \cup R_+ \setminus [0,N].$$

Making use of the introduced function, we can come to the conclusion that the spectral characteristic $h_N(\lambda)$ of the estimate $\hat{A}_N\vec{\xi}$ can be calculated by the formula

$$(h_N(\lambda))^\top = (A_N(\lambda))^\top F(\lambda)(H(\lambda))^{-1} - (C_N(\lambda))^\top (H(\lambda))^{-1}, \tag{4.9}$$

$$C_N(\lambda) = \sum_{l=1}^{s} \int_{-M_l-N_l}^{-M_l} (\mathbf{B}^{-1}\mathbf{R}\mathbf{a}_N)(t)e^{it\lambda}dt + \int_0^\infty (\mathbf{B}^{-1}\mathbf{R}\mathbf{a}_N)(t)e^{it\lambda}dt,$$

where

$$A_N(\lambda) = \int_0^N \vec{a}(t)e^{-it\lambda}dt.$$

The mean-square error $\Delta(h_N;F,G)$ of the estimate $\hat{A}_N\vec{\xi}$ is calculated by the formula

$$\Delta(h_N;F,G) = \langle (\mathbf{R}\mathbf{a}_N)(t), (\mathbf{B}^{-1}\mathbf{R}\mathbf{a}_N)(t)\rangle + \langle (\mathbf{Q}\mathbf{a}_N)(t), \vec{\mathbf{a}}_N(t)\rangle. \tag{4.10}$$

We obtain the following corollary.

Corollary 4.1. *Let $\vec{\xi}(t)$ and $\vec{\eta}(t)$ be uncorrelated multidimensional stationary processes with the spectral densities $F(\lambda)$ and $G(\lambda)$ satisfying the minimality condition (4.1). The spectral characteristic $h_N(\lambda)$ and the mean-square error $\Delta(h_N;F,G)$ of the optimal linear estimate of the functional $A_N\vec{\xi}$ based on observations of the process $\vec{\xi}(t)+\vec{\eta}(t)$, $t \in \mathbb{R}^-\setminus S$, can be calculated by formulas (4.9), (4.10).*

Estimation from Observations without Noise

Consider the problem of mean-square optimal estimation of the functional $A\vec{\xi}$ in the case where the stationary process $\vec{\xi}(t)$ is observed without noise. In this case the spectral characteristic of the optimal estimate $\hat{A}\vec{\xi}$ of the functional is of the form

$$(h(\lambda))^{\top} = (A(\lambda))^{\top} - (C(\lambda))^{\top}(F(\lambda))^{-1},$$
$$C(\lambda) = \sum_{l=1}^{s} \int_{-M_l-N_l}^{-M_l} \vec{\mathbf{c}}(t)e^{it\lambda}dt + \int_{0}^{\infty} \vec{\mathbf{c}}(t)e^{it\lambda}dt. \tag{4.11}$$

Relation (4.6) in this case is of the form

$$\vec{\mathbf{a}}(t) = (\mathbf{Bc})(t), \quad t \in U. \tag{4.12}$$

If the operator $\mathbf{B}$ is invertible, then the unknown function $\vec{\mathbf{c}}(t)$ can be calculated by the formula

$$\vec{\mathbf{c}}(t) = (\mathbf{B}^{-1}\mathbf{a})(t), \quad t \in U.$$

Hence, the spectral characteristic of the estimate $\hat{A}\vec{\xi}$ can be calculated by formula

$$(h(\lambda)^{\top} = (A(\lambda))^{\top} - (C(\lambda))^{\top}(F(\lambda))^{-1}, \tag{4.13}$$
$$C(\lambda) = \sum_{l=1}^{s} \int_{-M_l-N_l}^{-M_l} (\mathbf{B}^{-1}\mathbf{a})(t)e^{it\lambda}dt + \int_{0}^{\infty} (\mathbf{B}^{-1}\mathbf{a})(t)e^{it\lambda}dt.$$

The mean-square error of the estimate of the functional is calculated by formula

$$\Delta(h;F) = \langle (\mathbf{B}^{-1}\mathbf{a})(t), \vec{\mathbf{a}}(t) \rangle. \tag{4.14}$$

The following theorem is true.

Theorem 4.2. *Let $\vec{\xi}(t)$ be a multidimensional stationary stochastic process with the spectral density $F(\lambda)$ satisfying the minimality condition*

$$\int_{-\pi}^{\pi} (b(\lambda))^{\top}(F(\lambda))^{-1}\overline{b(\lambda)}d\lambda < \infty \tag{4.15}$$

for a nonzero vector-valued function of the exponential type

$$b(\lambda) = \sum_{l=1}^{s} \int_{-M_l-N_l}^{-M_l} \vec{\alpha}(t) e^{it\lambda} dt.$$

The spectral characteristic $h(\lambda)$ *and the mean-square error* $\Delta(h,F)$ *of the optimal linear estimate* $\hat{A}\vec{\xi}$ *of the functional* $A\vec{\xi}$ *based on observations of the process* $\vec{\xi}(t)$ *at points* $t \in \mathbb{R}^- \setminus S$, $S = \bigcup_{l=1}^{s} [-M_l - N_l, -M_l]$, *can be calculated by formulas (4.13),(4.14).*

1.2. Minimax Robust Method of Extrapolation

The formulas obtained in the section above can be used to find a solution of the estimation problem only in the case where the spectral densities of the processes are exactly known. In the case of spectral uncertainty, where information about the exact form of densities is not complete, while it is known that the spectral densities belong to a certain class of spectral densities, the minimax robust method of estimation of functionals may be applied. This method gives us a possibility to find an estimate that minimizes the maximal values of the mean-square errors of the estimates for all spectral densities from a given class of admissible spectral densities.

Definition 4.1. *For a given class of spectral densities* $D = D_F \times D_G$ *the spectral densities* $F^0(\lambda) \in D_F$, $G^0(\lambda) \in D_G$ *are called the least favourable in the class* D *for the optimal linear extrapolation of the functional* $A\vec{\xi}$ *if the following relation is true*

$$\Delta\left(F^0, G^0\right) = \Delta\left(h\left(F^0, G^0\right); F^0, G^0\right) = \max_{(F,G)\in D_F \times D_G} \Delta\left(h(F,G); F, G\right).$$

Definition 4.2. *For a given class of spectral densities* $D = D_F \times D_G$ *the spectral characteristic* $h^0(\lambda)$ *of the optimal linear extrapolation of the functional* $A\vec{\xi}$ *is called minimax robust if there are satisfied conditions*

$$h^0(\lambda) \in H_D = \bigcap_{(F,G)\in D_F \times D_G} L_2^s(F+G),$$

$$\min_{h\in H_D}\max_{(F,G)\in D}\Delta(h;F,G)=\max_{(F,G)\in D}\Delta\left(h^0;F,G\right).$$

Making use of the presented definitions and the obtained results, we can verify that the following statements are true.

Lemma 4.1. *The spectral densities $F^0(\lambda)\in D_F$, $G^0(\lambda)\in D_G$, satisfying the minimality condition (4.1), are the least favourable in the class $D=D_F\times D_G$ for the optimal linear extrapolation of the functional $A\vec{\xi}$ if the Fourier coefficients of the functions*

$$(F^0(\lambda)+G^0(\lambda))^{-1},\ F_0(\lambda)(F^0(\lambda)+G^0(\lambda))^{-1},\ F^0(\lambda)(F^0(\lambda)+G^0(\lambda))^{-1}G^0(\lambda)$$

define operators $\mathbf{B}^0,\mathbf{R}^0,\mathbf{Q}^0$, which determine a solution of the constrained optimization problem

$$\max_{(F,G)\in D_F\times D_G}(\langle(\mathbf{Ra})(t),(\mathbf{B}^{-1}\mathbf{Ra})(t)\rangle+\langle(\mathbf{Qa})(t),\vec{\mathbf{a}}(t)\rangle)=$$

$$=\langle(\mathbf{R}^0\mathbf{a})(t),((\mathbf{B}^0)^{-1}\mathbf{R}^0\mathbf{a})(t)\rangle+\langle(\mathbf{Q}^0\mathbf{a})(t),\vec{\mathbf{a}}(t)\rangle. \tag{4.16}$$

The minimax spectral characteristic $h^0=h(F^0,G^0)$ is determined by formula (4.7) if $h(F^0,G^0)\in H_D$.

Corollary 4.2. *Suppose that the spectral density $F^0(\lambda)\in D_F$ satisfy the minimality condition (4.15). The spectral density $F^0(\lambda)\in D_F$ is the least favourable in the class D_F for the optimal linear extrapolation of the functional $A\vec{\xi}$ from observations of the process $\vec{\xi}(t)$ at points $t\in\mathbb{R}^-\backslash S$, if the Fourier coefficients of the function $(F^0(\lambda))^{-1}$ define the operator $\mathbf{B}^0$, which determines a solution of the constrained optimization problem*

$$\max_{F\in D_F}\langle(\mathbf{B}^{-1}\mathbf{a})(t),\vec{\mathbf{a}}(t)\rangle=\langle((\mathbf{B}^0)^{-1}\mathbf{a})(t),\vec{\mathbf{a}}(t)\rangle. \tag{4.17}$$

The minimax spectral characteristic $h^0=h(F^0)$ is determined by formula (4.13) if $h(F^0)\in H_{D_F}$.

For more detailed analysis of properties of the least favourable spectral densities and the minimax-robust spectral characteristics we observe that the least favourable spectral densities $F^0(\lambda)$, $G^0(\lambda)$ and the minimax spectral characteristic $h^0 = h(F^0, G^0)$ form a saddle point of the function $\Delta(h; F, G)$ on the set $H_D \times D$. The saddle point inequalities

$$\Delta\left(h^0; F, G\right) \leq \Delta\left(h^0; F^0, G^0\right) \leq \Delta\left(h; F^0, G^0\right),$$

$$\forall h \in H_D, \forall F \in D_F, \forall G \in D_G,$$

are true if $h^0 = h(F^0, G^0)$ and $h(F^0, G^0) \in H_D$, where (F^0, G^0) is a solution of the constrained optimization problem

$$\sup_{(F,G)\in D_F\times D_G} \Delta\left(h(F^0, G^0); F, G\right) = \Delta\left(h(F^0, G^0); F^0, G^0\right). \tag{4.18}$$

The linear functional $\Delta\left(h\left(F^0, G^0\right); F, G\right)$ is calculated by the formula

$$\Delta\left(h\left(F^0, G^0\right); F, G\right) =$$

$$= \frac{1}{2\pi}\int_{-\infty}^{\infty} ((A(\lambda))^{\top} G^0(\lambda) + (C^0(\lambda))^{\top})(F^0(\lambda) + G^0(\lambda))^{-1} F(\lambda) \times$$
$$\times (F^0(\lambda) + G^0(\lambda))^{-1}((A(\lambda))^{\top} G^0(\lambda) + (C^0(\lambda))^{\top})^* d\lambda +$$
$$+ \frac{1}{2\pi}\int_{-\infty}^{\infty} ((A(\lambda))^{\top} F^0(\lambda) - (C^0(\lambda))^{\top})(F^0(\lambda) + G^0(\lambda))^{-1} G(\lambda) \times$$
$$\times (F^0(\lambda) + G^0(\lambda))^{-1}((A(\lambda))^{\top} G^0(\lambda) - (C^0(\lambda))^{\top})^* d\lambda,$$

$$C^0(\lambda) = \sum_{l=1}^{s} \int_{-M_l-N_l}^{-M_l} ((\mathbf{B}^0)^{-1}\mathbf{R}^0\mathbf{a})(t) e^{it\lambda} dt + \int_{0}^{\infty} ((\mathbf{B}^0)^{-1}\mathbf{R}^0\mathbf{a})(t) e^{it\lambda} dt.$$

The constrained optimization problem (4.18) is equivalent to the unconstrained optimization problem (see B. N. Pshenichnyj [156])

$$\Delta_D(F, G) = -\Delta(h(F^0, G^0); F, G) + \delta((F, G) \left| D_F \times D_G \right.) \to \inf, \tag{4.19}$$

where $\delta((F, G) \left| D_F \times D_G \right.)$ is the indicator function of the set $D = D_F \times D_G$. A solution of the problem (4.19) is characterized by the condition $0 \in \partial\Delta_D(F^0, G^0)$, where $\partial\Delta_D(F^0, G^0)$ is the subdifferential of the convex functional $\Delta_D(F, G)$ at point (F^0, G^0) (see the book by T. Rockafellar [158]). This condition makes it possible to find the least favourable spectral densities in some special classes of spectral densities D (see books by A. D. Ioffe and V. M. Tihomirov [161], B. N. Pshenichnyj [156], T. Rockafellar [158]).

Note, that the form of the functional $\Delta(h(F^0,G^0);F,G)$ is convenient for application of the Lagrange method of indefinite multipliers for finding a solution of the problem (4.18). Making use of the method of Lagrange multipliers and the form of subdifferentials of the indicator functions, we describe relations that determine the least favourable spectral densities in some special classes of spectral densities.

1.3. Least Favourable Spectral Densities in the Class $D = D_\varepsilon \times D_V^U$

Consider the problem of minimax extrapolation of the functional $A\vec{\xi}$ in the case where the spectral densities of the processes belong to the following classes of admissible spectral densities $D_\varepsilon^k \times D_V^{U\,k}$, $k = 1,2,3,4$,

$$D_\varepsilon^1 = \left\{ F(\lambda) \middle| \operatorname{Tr} F(\lambda) = (1-\varepsilon)\operatorname{Tr} F_1(\lambda) + \varepsilon \operatorname{Tr} W(\lambda), \right.$$
$$\left. \frac{1}{2\pi}\int_{-\infty}^{\infty} \operatorname{Tr} F(\lambda)d\lambda = p \right\},$$

$$D_V^{U\,1} = \left\{ G(\lambda) \middle| \operatorname{Tr} V(\lambda) \le \operatorname{Tr} G(\lambda) \le \operatorname{Tr} U(\lambda),\ \frac{1}{2\pi}\int_{-\infty}^{\infty} \operatorname{Tr} G(\lambda)d\lambda = q \right\},$$

$$D_\varepsilon^2 = \left\{ F(\lambda) \middle| f_{kk}(\lambda) = (1-\varepsilon) f_{kk}^1(\lambda) + \varepsilon w_{kk}(\lambda), \right.$$
$$\left. \frac{1}{2\pi}\int_{-\infty}^{\infty} f_{kk}(\lambda)d\lambda = p_k, k = \overline{1,T} \right\},$$

$$D_V^{U\,2} = \left\{ G(\lambda) \middle| v_{kk}(\lambda) \le g_{kk}(\lambda) \le u_{kk}(\lambda), \right.$$
$$\left. \frac{1}{2\pi}\int_{-\infty}^{\infty} g_{kk}(\lambda)d\lambda = q_k, k = \overline{1,T} \right\},$$

$$D_\varepsilon^3=\left\{F(\lambda)\middle|\langle B_1,F(\lambda)\rangle=(1-\varepsilon)\langle B_1,F_1(\lambda)\rangle+\varepsilon\langle B_1,W(\lambda)\rangle,\right.$$
$$\left.\frac{1}{2\pi}\int_{-\infty}^{\infty}\langle B_1,F(\lambda)\rangle\,d\lambda=p\right\},$$

$$D_V^{U\,3}=\left\{G(\lambda)\middle|\langle B_2,V(\lambda)\rangle\leq\langle B_2,G(\lambda)\rangle\leq\langle B_2,U(\lambda)\rangle,\right.$$
$$\left.\frac{1}{2\pi}\int_{-\infty}^{\infty}\langle B_2,G(\lambda)\rangle\,d\lambda=q\right\},$$

$$D_\varepsilon^4=\left\{F(\lambda)\middle|F(\lambda)=(1-\varepsilon)F_1(\lambda)+\varepsilon W(\lambda),\frac{1}{2\pi}\int_{-\infty}^{\infty}F(\lambda)d\lambda=P\right\},$$
$$D_V^{U\,4}=\left\{G(\lambda)\middle|V(\lambda)\leq G(\lambda)\leq U(\lambda),\frac{1}{2\pi}\int_{-\infty}^{\infty}G(\lambda)d\lambda=Q\right\},$$

where the spectral density matrices $V(\lambda),U(\lambda),F_1(\lambda)$ are known and fixed, $W(\lambda)$ is unknown spectral density matrix. $p,q,p_k,q_k,k=\overline{1,T}$ are given numbers, P,Q are given positive definite Hermitian matrices.
Densities from the classes $D_V^{U\,k}$, $k=1,2,3,4$, of admissible spectral densities form the "strip" models of spectral density matrices.
Densities from the classes D_ε^k, $k=1,2,3,4$, of admissible spectral densities form the "ε–contamination" models of spectral density matrices.

From the condition $0\in\partial\Delta_D(F^0,G^0)$ we find the following equations which determine the least favourable spectral densities for these given classes of admissible spectral densities.

For the first pair $D_\varepsilon^1\times D_V^{U\,1}$ we have equations

$$((A(\lambda))^\top G^0(\lambda)+(C^0(\lambda))^\top)^*((A(\lambda))^\top G^0(\lambda)+(C^0(\lambda))^\top)=$$
$$=(\alpha^2+\gamma_1(\lambda))(F^0(\lambda)+G^0(\lambda))^2, \tag{4.20}$$
$$((A(\lambda))^\top F^0(\lambda)-(C^0(\lambda))^\top)^*((A(\lambda))^\top F^0(\lambda)-(C^0(\lambda))^\top)=$$
$$=(\beta^2+\gamma_2(\lambda)+\gamma_3(\lambda))(F^0(\lambda)+G^0(\lambda))^2, \tag{4.21}$$

where α^2,β^2 are Lagrange multipliers, $\gamma_1(\lambda)\leq0$ and $\gamma_1(\lambda)=0$ if $\operatorname{Tr}F^0(\lambda)>(1-\varepsilon)\operatorname{Tr}F_1(\lambda)$, $\gamma_2(\lambda)\leq0$ and $\gamma_2(\lambda)=0$ if $\operatorname{Tr}G^0(\lambda)>\operatorname{Tr}V(\lambda)$, $\gamma_3(\lambda)\geq0$, and $\gamma_3(\lambda)=0$ if $\operatorname{Tr}G^0(\lambda)<\operatorname{Tr}U(\lambda)$.

For the second pair $D_\varepsilon^2 \times D_V^{U\,2}$ we have equations

$$((A(\lambda))^\top G^0(\lambda) + (C^0(\lambda))^\top)^*((A(\lambda))^\top G^0(\lambda) + (C^0(\lambda))^\top) =$$

$$= (F^0(\lambda) + G^0(\lambda)) \left\{(\alpha_k^2 + \gamma_{1k}(\lambda))\delta_{kl}\right\}_{k,l=1}^{T} (F^0(\lambda) + G^0(\lambda)), \tag{4.22}$$

$$((A(\lambda))^\top F^0(\lambda) - (C^0(\lambda))^\top)^*((A(\lambda))^\top F^0(\lambda) - (C^0(\lambda))^\top) =$$

$$= (F^0(\lambda) + G^0(\lambda)) \left\{(\beta_k^2 + \gamma_{2k}(\lambda) + \gamma_{3k}(\lambda))\delta_{kl}\right\}_{k,l=1}^{T} (F^0(\lambda) + G^0(\lambda)), \tag{4.23}$$

where α_k^2, β_k^2 are Lagrange multipliers, δ_{kl} are Kronecker symbols, $\gamma_{1k}(\lambda) \leq 0$ and $\gamma_{1k}(\lambda) = 0$ if $f_{kk}^0(\lambda) > (1-\varepsilon) f_{kk}^1(\lambda)$, $\gamma_{2k}(\lambda) \leq 0$ and $\gamma_{2k}(\lambda) = 0$ if $g_{kk}^0(\lambda) > v_{kk}(\lambda)$, $\gamma_{3k}(\lambda) \geq 0$, and $\gamma_{3k}(\lambda) = 0$ if $g_{kk}^0(\lambda) < u_{kk}(\lambda)$.

For the third pair $D_\varepsilon^3 \times D_V^{U\,3}$ we have equations

$$((A(\lambda))^\top G^0(\lambda) + (C^0(\lambda))^\top)^*((A(\lambda))^\top G^0(\lambda) + (C^0(\lambda))^\top) =$$

$$= (\alpha^2 + \gamma_1'(\lambda))(F^0(\lambda) + G^0(\lambda)) B_1^\top (F^0(\lambda) + G^0(\lambda)), \tag{4.24}$$

$$((A(\lambda))^\top F^0(\lambda) - (C^0(\lambda))^\top)^*((A(\lambda))^\top F^0(\lambda) - (C^0(\lambda))^\top) =$$

$$= (\beta^2 + \gamma_2'(\lambda) + \gamma_3'(\lambda))(F^0(\lambda) + G^0(\lambda)) B_2^\top (F^0(\lambda) + G^0(\lambda)), \tag{4.25}$$

where α^2, β^2 are Lagrange multipliers, $\gamma_1'(\lambda) \leq 0$ and $\gamma_1'(\lambda) = 0$ if $\langle B_1, F^0(\lambda)\rangle > (1-\varepsilon)\langle B_1, F_1(\lambda)\rangle$, $\gamma_2'(\lambda) \leq 0$ and $\gamma_2'(\lambda) = 0$ if $\langle B_2, G^0(\lambda\rangle > \langle B_2, V(\lambda)\rangle$, $\gamma_3'(\lambda) \geq 0$, and $\gamma_3'(\lambda) = 0$ if $\langle B_2, G^0(\lambda\rangle < \langle B_2, U(\lambda)\rangle$.

For the fourth pair $D_\varepsilon^4 \times D_V^{U\,4}$ we have equations

$$((A(\lambda))^\top G^0(\lambda) + (C^0(\lambda))^\top)^*((A(\lambda))^\top G^0(\lambda) + (C^0(\lambda))^\top) =$$

$$= (F^0(\lambda) + G^0(\lambda))(\vec{\alpha} \cdot \vec{\alpha}^* + \Gamma_1(\lambda))(F^0(\lambda) + G^0(\lambda)), \tag{4.26}$$

$$((A(\lambda))^\top F^0(\lambda) - (C^0(\lambda))^\top)^*((A(\lambda))^\top F^0(\lambda) - (C^0(\lambda))^\top) =$$

$$= (F^0(\lambda) + G^0(\lambda))(\vec{\beta} \cdot \vec{\beta}^* + \Gamma_2(\lambda) + \Gamma_3(\lambda))(F^0(\lambda) + G^0(\lambda)) \tag{4.27}$$

where $\vec{\alpha}, \vec{\beta}$ are Lagrange multipliers, $\Gamma_1(\lambda) \leq 0$ and $\Gamma_1(\lambda) = 0$ if $F^0(\lambda) > (1-\varepsilon)F_1(\lambda)$, $\Gamma_2(\lambda) \leq 0$ and $\Gamma_2(\lambda) = 0$ if $G^0(\lambda) > V(\lambda)$, $\Gamma_3(\lambda) \geq 0$, and $\Gamma_3(\lambda) = 0$ if $G^0(\lambda) < U(\lambda)$.

Thus, the following statement is true.

Theorem 4.3. *The least favourable spectral densities* $F^0(\lambda)$, $G^0(\lambda)$ *in the classes* $D_\varepsilon^k \times D_V^{U\,k}$, $k = 1,2,3,4$, *for the optimal linear extrapolation of the functional* $A\vec{\xi}$ *are determined by relations (4.20) – (4.27), the minimality condition (4.1), the constrained optimization problem (4.16), and restrictions on densities from the corresponding classes* $D_\varepsilon^k \times D_V^{U\,k}$, $k = 1,2,3,4$. *The minimax robust spectral characteristic of the optimal estimate of the functional* $A\vec{\xi}$ *is determined by the formula (4.7).*

Corollary 4.3. *The least favourable spectral densities* $F^0(\lambda)$ *in the classes* D_ε^k, $k = 1,2,3,4$, *for the optimal linear extrapolation of the functional* $A\vec{\xi}$ *based on observations of the process* $\vec{\xi}(t)$ *at points* $t \in \mathbb{R}^-\backslash S$, $S = \bigcup_{l=1}^{s} [-M_l - N_l, -M_l]$, *are determined by the following equations, respectively,*

$$((C^0(\lambda))^\top)^* \cdot (C^0(\lambda))^\top = (\alpha^2 + \gamma_1(\lambda))(F^0(\lambda))^2, \tag{4.28}$$

$$((C^0(\lambda))^\top)^* \cdot (C^0(\lambda))^\top = F^0(\lambda) \left\{(\alpha_k^2 + \gamma_{1k}(\lambda))\delta_{kl}\right\}_{k,l=1}^{T} F^0(\lambda), \tag{4.29}$$

$$((C^0(\lambda))^\top)^* \cdot (C^0(\lambda))^\top = (\alpha^2 + \gamma_1'(\lambda))F^0(\lambda)(B_1)^\top F^0(\lambda), \tag{4.30}$$

$$((C^0(\lambda))^\top)^* \cdot (C^0(\lambda))^\top = F^0(\lambda)(\vec{\alpha}\cdot\vec{\alpha}^* + \Gamma_1(\lambda))F^0(\lambda), \tag{4.31}$$

the minimality condition (4.15), the constrained optimization problem (4.17), and restrictions on densities from the corresponding classes D_ε^k, $k = 1,2,3,4$. *The minimax spectral characteristic of the optimal estimate of the functional* $A\vec{\xi}$ *is determined by the formula (4.13).*

Corollary 4.4. *The least favourable spectral densities* $F^0(\lambda)$ *in the classes* $D_V^{U\,k}$, $k = 1,2,3,4$, *for the optimal linear extrapolation of the functional* $A\vec{\xi}$ *based on observations of the process* $\vec{\xi}(t)$ *at points* $t \in \mathbb{R}^-\backslash S$, $S = \bigcup_{l=1}^{s} [-M_l - N_l, -M_l]$, *are determined by the following equations, respectively,*

$$((C^0(\lambda))^\top)^* \cdot (C^0(\lambda))^\top = (\beta^2 + \gamma_2(\lambda) + \gamma_3(\lambda))(F^0(\lambda))^2, \tag{4.32}$$

$$((C^0(\lambda))^\top)^* \cdot (C^0(\lambda))^\top = F^0(\lambda) \left\{(\beta_k^2 + \gamma_{2k}(\lambda) + \gamma_{3k}(\lambda))\delta_{kl}\right\}_{k,l=1}^{T} F^0(\lambda), \tag{4.33}$$

$$((C^0(\lambda))^\top)^* \cdot (C^0(\lambda))^\top = (\beta^2 + \gamma_2'(\lambda) + \gamma_3'(\lambda))F^0(\lambda)B_2^\top F^0(\lambda), \tag{4.34}$$

$$((C^0(\lambda))^\top)^* \cdot (C^0(\lambda))^\top = F^0(\lambda)(\vec{\beta}\cdot\vec{\beta}^* + \Gamma_2(\lambda) + \Gamma_3(\lambda))F^0(\lambda), \qquad (4.35)$$

the minimality condition (4.15), the constrained optimization problem (4.17), and restrictions on densities from the corresponding classes $D_V^{U\,k}$, $k = 1,2,3,4$. The minimax spectral characteristic of the optimal estimate of the functional $A\vec{\xi}$ is determined by the formula (4.13).

1.4. Least Favourable Spectral Densities in the Class $D = D_{1\delta} \times D_{2\delta}$

Consider the problem of extrapolation of the functional $A\vec{\xi}$ in the case where spectral densities of the processes belong to the following classes of admissible spectral densities $D_{1\delta}^k \times D_{2\delta}^k, k = 1,2,3,4$,

$$D_{1\delta}^1 = \left\{ F(\lambda) \left| \frac{1}{2\pi}\int_{-\infty}^{\infty} |\mathrm{Tr}(F(\lambda) - F_1(\lambda))|\, d\lambda \le \delta_1 \right.\right\},$$

$$D_{2\delta}^1 = \left\{ G(\lambda) \left| \frac{1}{2\pi}\int_{-\infty}^{\infty} |\mathrm{Tr}(G(\lambda) - G_1(\lambda))|^2\, d\lambda \le \delta_2 \right.\right\};$$

$$D_{1\delta}^2 = \left\{ F(\lambda) \left| \frac{1}{2\pi}\int_{-\infty}^{\infty} \left|f_{kk}(\lambda) - f_{kk}^1(\lambda)\right| d\lambda \le \delta_{1k}, k = \overline{1,T} \right.\right\},$$

$$D_{2\delta}^2 = \left\{ G(\lambda) \left| \frac{1}{2\pi}\int_{-\infty}^{\infty} \left|g_{kk}(\lambda) - g_{kk}^1(\lambda)\right|^2 d\lambda \le \delta_{2k}, k = \overline{1,T} \right.\right\};$$

$$D_{1\delta}^3 = \left\{ F(\lambda) \left| \frac{1}{2\pi}\int_{-\infty}^{\infty} |\langle B_1, F(\lambda) - F_1(\lambda)\rangle|\, d\lambda \le \delta_1 \right.\right\},$$

$$D_{2\delta}^3 = \left\{ G(\lambda) \left| \frac{1}{2\pi}\int_{-\infty}^{\infty} |\langle B_2, G(\lambda) - G_1(\lambda)\rangle|^2\, d\lambda \le \delta_2 \right.\right\};$$

$$D_{1\delta}^4 = \left\{ F(\lambda) \left| \frac{1}{2\pi}\int_{-\infty}^{\infty} \left|f_{ij}(\lambda) - f_{ij}^1(\lambda)\right| d\lambda \le \delta_{ij}^1, i,j = \overline{1,T} \right.\right\},$$

$$D_{2\delta}^4 = \left\{ G(\lambda) \left| \frac{1}{2\pi}\int_{-\infty}^{\infty} \left|g_{ij}(\lambda) - g_{ij}^1(\lambda)\right|^2 d\lambda \le \delta_{ij}^2, i,j = \overline{1,T} \right.\right\},$$

where $F_1(\lambda), G_1(\lambda)$ are known and fixed spectral density matrices, $\delta_1, \delta_2, \delta_{1k}, \delta_{2k}, k = 1,\ldots,T$, $\delta_{ij}^1, \delta_{ij}^2, k = 1,\ldots,T$, are given numbers, B_1, B_2 are given positive-definite Hermitian matrices.

Densities from the classes $D^k_{1\delta}k = 1,2,3,4$, of admissible spectral densities form the "δ–neighbourhood" models in the space L_1 of a given bounded spectral density $F_1(\lambda)$.
Densities from the classes $D^k_{2\delta}, k = 1,2,3,4$, of admissible spectral densities form the "δ–neighbourhood" models in the space L_2 of a given bounded spectral density $G_1(\lambda)$.

From the condition $0 \in \partial\Delta_D(F^0, G^0)$ we find the following equations which determine the least favourable spectral densities for these given sets of admissible spectral densities.
For the first pair $D^1_{1\delta} \times D^1_{2\delta}$ we have equations

$$((A(\lambda))^\top G^0(\lambda) + (C^0(\lambda))^\top)^*((A(\lambda))^\top G^0(\lambda) + (C^0(\lambda))^\top) =$$

$$= \alpha^2\gamma(\lambda)(F^0(\lambda) + G^0(\lambda))^2, \tag{4.36}$$

$$((A(\lambda))^\top F^0(\lambda) - (C^0(\lambda))^\top)^*((A(\lambda))^\top F^0(\lambda) - (C^0(\lambda))^\top) =$$

$$= \beta^2 \mathrm{Tr}\,(G^0(\lambda) - G_1(\lambda))(F^0(\lambda) + G^0(\lambda))^2, \tag{4.37}$$

$$\frac{1}{2\pi}\int_{-\infty}^{\infty} \left|\mathrm{Tr}\,(F^0(\lambda) - F_1(\lambda))\right| d\lambda = \delta_1, \tag{4.38}$$

$$\frac{1}{2\pi}\int_{-\infty}^{\infty} \left|\mathrm{Tr}\,(G^0(\lambda) - G_1(\lambda))\right|^2 d\lambda = \delta_2, \tag{4.39}$$

where α^2, β^2 are Lagrange multipliers, $|\gamma(\lambda)| \leq 1$ and

$$\gamma(\lambda) = \mathrm{sign}\,(\mathrm{Tr}\,(F^0(\lambda) - F_1(\lambda))): \ \mathrm{Tr}\,(F^0(\lambda) - F_1(\lambda)) \neq 0.$$

For the second pair $D^2_{1\delta} \times D^2_{2\delta}$ we have equations

$$((A(\lambda))^\top G^0(\lambda) + (C^0(\lambda))^\top)^*((A(\lambda))^\top G^0(\lambda) + (C^0(\lambda))^\top) =$$

$$= (F^0(\lambda) + G^0(\lambda)) \left\{\alpha_k^2\gamma_k(\lambda)\delta_{kl}\right\}_{k,l=1}^T (F^0(\lambda) + G^0(\lambda)), \tag{4.40}$$

$$((A(\lambda))^\top F^0(\lambda) - (C^0(\lambda))^\top)^*((A(\lambda))^\top F^0(\lambda) - (C^0(\lambda))^\top) =$$

$$= (F^0(\lambda) + G^0(\lambda)) \left\{\beta_k^2(g^0_{kk}(\lambda) - g^1_{kk}(\lambda))\delta_{kl}\right\}_{k,l=1}^T (F^0(\lambda) + G^0(\lambda)), \tag{4.41}$$

$$\frac{1}{2\pi}\int_{-\infty}^{\infty} \left|f^0_{kk}(\lambda) - f^1_{kk}(\lambda)\right| d\lambda = \delta_{1k}, \ k = \overline{1,T}, \tag{4.42}$$

$$\frac{1}{2\pi}\int_{-\infty}^{\infty}\left|g_{kk}^{0}(\lambda)-g_{kk}^{1}(\lambda)\right|^{2}d\lambda=\delta_{2k},\ k=\overline{1,T}, \tag{4.43}$$

where α_k^2,β_k^2 are Lagrange multipliers, $|\gamma_k(\lambda)|\leq 1$ and

$$\gamma_k(\lambda)=\text{sign}\,(f_{kk}^{0}(\lambda)-f_{kk}^{1}(\lambda)):\ f_{kk}^{0}(\lambda)-f_{kk}^{1}(\lambda)\neq 0,\ k=\overline{1,T}.$$

For the third pair $D_{1\delta}^{3}\times D_{2\delta}^{3}$ we have equations

$$((A(\lambda))^{\top}G^{0}(\lambda)+(C^{0}(\lambda))^{\top})^{*}((A(\lambda))^{\top}G^{0}(\lambda)+(C^{0}(\lambda))^{\top})=$$

$$=\alpha^{2}\gamma'(\lambda)(F^{0}(\lambda)+G^{0}(\lambda))B_{1}^{\top}(F^{0}(\lambda)+G^{0}(\lambda)), \tag{4.44}$$

$$((A(\lambda))^{\top}F^{0}(\lambda)-(C^{0}(\lambda))^{\top})^{*}((A(\lambda))^{\top}F^{0}(\lambda)-(C^{0}(\lambda))^{\top})=$$

$$=\beta^{2}\left\langle B_{2},G^{0}(\lambda)-G_{1}(\lambda)\right\rangle(F^{0}(\lambda)+G^{0}(\lambda))^{2}, \tag{4.45}$$

$$\frac{1}{2\pi}\int_{-\infty}^{\infty}\left|\left\langle B_{1},F^{0}(\lambda)-F_{1}(\lambda)\right\rangle\right|d\lambda=\delta_{1}, \tag{4.46}$$

$$\frac{1}{2\pi}\int_{-\infty}^{\infty}\left|\left\langle B_{2},G^{0}(\lambda)-G_{1}(\lambda)\right\rangle\right|^{2}d\lambda=\delta_{2}, \tag{4.47}$$

where α^2,β^2 are Lagrange multipliers, $|\gamma'(\lambda)|\leq 1$ and

$$\gamma'(\lambda)=\text{sign}\left\langle B_{1},F^{0}(\lambda)-F_{1}(\lambda)\right\rangle:\ \left\langle B_{1},F^{0}(\lambda)-F_{1}(\lambda)\right\rangle\neq 0.$$

For the fourth pair $D_{1\delta}^{4}\times D_{2\delta}^{4}$ we have equations

$$((A(\lambda))^{\top}G^{0}(\lambda)+(C^{0}(\lambda))^{\top})^{*}((A(\lambda))^{\top}G^{0}(\lambda)+(C^{0}(\lambda))^{\top})=$$

$$=(F^{0}(\lambda)+G^{0}(\lambda))\left\{\alpha_{ij}\gamma_{ij}(\lambda)\right\}_{i,j=1}^{T}(F^{0}(\lambda)+G^{0}(\lambda)), \tag{4.48}$$

$$((A(\lambda))^{\top}F^{0}(\lambda)-(C^{0}(\lambda))^{\top})^{*}((A(\lambda))^{\top}F^{0}(\lambda)-(C^{0}(\lambda))^{\top})=$$

$$=(F^{0}(\lambda)+G^{0}(\lambda))\left\{\beta_{ij}(g_{ij}^{0}(\lambda)-g_{ij}^{1}(\lambda))\right\}_{i,j=1}^{T}(F^{0}(\lambda)+G^{0}(\lambda)), \tag{4.49}$$

$$\frac{1}{2\pi}\int_{-\infty}^{\infty}\left|f_{ij}^{0}(\lambda)-f_{ij}^{1}(\lambda)\right|d\lambda=\delta_{ij}^{1},\ i,j=\overline{1,T}, \tag{4.50}$$

$$\frac{1}{2\pi}\int_{-\infty}^{\infty}\left|g_{ij}^{0}(\lambda)-g_{ij}^{1}(\lambda)\right|^{2}d\lambda=\delta_{ij}^{2},\ i,j=\overline{1,T}, \tag{4.51}$$

where $\beta ij, \alpha_{ij}$ are Lagrange multipliers, $|\gamma_{ij}(\lambda)| \leq 1$ and

$$\gamma_{ij}(\lambda) = \frac{f_{ij}^0(\lambda) - f_{ij}^1(\lambda)}{\left|f_{ij}^0(\lambda) - f_{ij}^1(\lambda)\right|} \text{ if } f_{ij}^0(\lambda) - f_{ij}^1(\lambda) \neq 0,\ i, j = \overline{1, T}.$$

Thus, the following theorem is true.

Theorem 4.4. *The least favourable spectral densities $F^0(\lambda)$, $G^0(\lambda)$ in the classes $D_{1\delta}^k \times D_{2\delta}^k, k = 1,2,3,4,$ for the optimal linear extrapolation of the functional $A\vec{\xi}$ are determined by relations (4.36) – (4.51), the minimality condition (4.1), the constrained optimization problem (4.16), and restrictions on densities from the corresponding classes $D_{1\delta}^k \times D_{2\delta}^k, k = 1,2,3,4$. The minimax robust spectral characteristic of the optimal estimate of the functional $A\vec{\xi}$ is determined by the formula (4.7).*

Corollary 4.5. *The least favourable spectral densities $F^0(\lambda)$ in the classes $D_{1\delta}^k$, $k = 1,2,3,4,$ for the optimal linear extrapolation of the functional $A\vec{\xi}$ from observations of the process $\vec{\xi}(t)$ at points $t \in \mathbb{R}^- \backslash S$, where $S = \bigcup_{l=1}^{s} [-M_l - N_l, -M_l]$, are determined by the following equations, respectively,*

$$((C^0(\lambda))^\top)^* \cdot (C^0(\lambda))^\top = \alpha^2 \gamma(\lambda)(F^0(\lambda))^2, \tag{4.52}$$

$$((C^0(\lambda))^\top)^* \cdot (C^0(\lambda))^\top = F^0(\lambda)\left\{\alpha_k^2 \gamma_k(\lambda)\delta_{kl}\right\}_{k,l=1}^T F^0(\lambda), \tag{4.53}$$

$$((C^0(\lambda))^\top)^* \cdot (C^0(\lambda))^\top = \alpha^2 \gamma'(\lambda) F^0(\lambda) B_1^\top F^0(\lambda), \tag{4.54}$$

$$((C^0(\lambda))^\top)^* \cdot (C^0(\lambda))^\top = F^0(\lambda)\left\{\alpha_{ij}(\lambda)\gamma_{ij}(\lambda)\right\}_{i,j=1}^T F^0(\lambda), \tag{4.55}$$

the minimality condition (4.15), the constrained optimization problem (4.17), and the following restrictions on densities from the corresponding classes $D_{1\delta}^k$, $k = 1,2,3,4$, respectively,

$$\frac{1}{2\pi}\int_{-\infty}^{\infty} \left|\mathrm{Tr}\,(F^0(\lambda) - F_1(\lambda))\right| d\lambda = \delta_1, \tag{4.56}$$

$$\frac{1}{2\pi}\int_{-\infty}^{\infty} \left|f_{kk}^0(\lambda) - f_{kk}^1(\lambda)\right| d\lambda = \delta_{1k}, \tag{4.57}$$

$$\frac{1}{2\pi}\int_{-\infty}^{\infty}\left|\left\langle B_1,F^0(\lambda)-F_1(\lambda)\right\rangle\right|d\lambda=\delta_1, \tag{4.58}$$

$$\frac{1}{2\pi}\int_{-\infty}^{\infty}\left|f_{ij}^0(\lambda)-f_{ij}^1(\lambda)\right|d\lambda=\delta_{ij}^1. \tag{4.59}$$

The minimax spectral characteristic of the optimal estimate of the functional $A\vec{\xi}$ *is determined by the formula (4.13).*

Corollary 4.6. *The least favourable spectral densities* $F^0(\lambda)$ *in the classes* $D_{2\delta}^k$, $k=1,2,3,4$, *for the optimal linear extrapolation of the functional* $A\vec{\xi}$ *from observations of the process* $\vec{\xi}(t)$ *at points* $t\in\mathbb{R}^-\setminus S$, *where* $S=\bigcup_{l=1}^{s}[-M_l-N_l,-M_l]$,, *are determined by the following equations, respectively,*

$$((C^0(\lambda))^\top)^*\cdot(C^0(\lambda))^\top=\beta^2\mathrm{Tr}\,(F^0(\lambda)-G_1(\lambda))(F^0(\lambda))^2, \tag{4.60}$$

$$((C^0(\lambda))^\top)^*\cdot(C^0(\lambda))^\top=F^0(\lambda)\left\{\beta_k^2(f_{kk}^0(\lambda)-g_{kk}^1(\lambda))\delta_{kl}\right\}_{k,l=1}^T F^0(\lambda), \tag{4.61}$$

$$((C^0(\lambda))^\top)^*\cdot(C^0(\lambda))^\top=\beta^2\left\langle B_2,F^0(\lambda)-G_1(\lambda)\right\rangle(F^0(\lambda))^2, \tag{4.62}$$

$$((C^0(\lambda))^\top)^*\cdot(C^0(\lambda))^\top=F^0(\lambda)\left\{\beta_{ij}(f_{ij}^0(\lambda)-g_{ij}^1(\lambda))\right\}_{i,j=1}^T F^0(\lambda) \tag{4.63}$$

the minimality condition (4.15), the constrained optimization problem (4.17), and the following restrictions on densities from the corresponding classes $D_{2\delta}^k$, $k=1,2,3,4$, *respectively,*

$$\frac{1}{2\pi}\int_{-\infty}^{\infty}\left|\mathrm{Tr}\,(F^0(\lambda)-G_1(\lambda))\right|^2d\lambda=\delta_2, \tag{4.64}$$

$$\frac{1}{2\pi}\int_{-\infty}^{\infty}\left|f_{kk}^0(\lambda)-g_{kk}^1(\lambda)\right|^2d\lambda=\delta_{2k},\ k=\overline{1,T}, \tag{4.65}$$

$$\frac{1}{2\pi}\int_{-\infty}^{\infty}\left|\left\langle B_2,F^0(\lambda)-G_1(\lambda)\right\rangle\right|^2d\lambda=\delta_2, \tag{4.66}$$

$$\frac{1}{2\pi}\int_{-\infty}^{\infty}\left|f_{ij}^0(\lambda)-g_{ij}^1(\lambda)\right|^2d\lambda=\delta_{ij}^2,\ i,j=\overline{1,T}. \tag{4.67}$$

The minimax spectral characteristic of the optimal estimate of the functional $A\vec{\xi}$ *is determined by the formula (4.13).*

2. Interpolation Problem for Multidimensional Stationary Stochastic Processes with Missing Observations

In this section, we deal with the problem of mean-square optimal linear estimation of the functional

$$A_s\vec{\xi} = \sum_{l=1}^{s} \int_{-M_l-N_l}^{-M_l} \vec{a}(t)^{\top}\vec{\xi}(t)dt$$

constructed from unobserved values of a multidimensional continuous time stationary stochastic process $\vec{\xi}(t)$. Estimates are based on observations of the process $\vec{\xi}(t)$ with additive stationary stochastic noise process $\vec{\eta}(t)$ at points $t \in \mathbb{R}\backslash S$, $S = \bigcup_{l=1}^{s}[-M_l - N_l, -M_l]$. Formulas for calculating the spectral characteristic and the mean-square error of the optimal linear estimate of the functional are derived in the case of spectral certainty, where the spectral densities of the signal and noise processes are exactly known. In the case of spectral uncertainty, where the spectral densities are not exactly known while a set of admissible spectral densities is given, the minimax method of estimation is applied. Formulas for determining the least favourable spectral densities and the minimax-robust spectral characteristics of the optimal estimates of the functional are proposed for some specific classes of admissible spectral densities.

2.1. Hilbert Space Projection Method of Interpolation

Let $\vec{\xi}(t) = \{\xi_k(t)\}_{k=1}^{T}, t \in \mathbb{R}$, and $\vec{\eta}(t) = \{\eta_k(t)\}_{k=1}^{T}, t \in \mathbb{R}$, be uncorrelated mean square continuous multidimensional stationary stochastic processes with zero first moments, $E\vec{\xi}(t) = \vec{0}$, $E\vec{\eta}(t) = \vec{0}$, and the correlation functions which admit the spectral representation (see I. I. Gikhman and A. V. Skorokhod [153])

$$R_{\xi}(t) = \frac{1}{2\pi}\int_{-\infty}^{\infty} e^{it\lambda}F(\lambda)d\lambda, \quad R_{\eta}(t) = \frac{1}{2\pi}\int_{-\infty}^{\infty} e^{it\lambda}G(\lambda)d\lambda,$$

where $F(\lambda) = \{f_{kl}(\lambda)\}_{k,l=1}^{T}$ and $G(\lambda) = \{g_{kl}(\lambda)\}_{k,l=1}^{T}$ are spectral density matrices of the processes $\vec{\xi}(t)$ and $\vec{\eta}(t)$ respectively.

Suppose that the spectral densities of the processes satisfy the minimality condition

$$\int_{-\infty}^{\infty}(b(\lambda))^{\top}(F(\lambda)+G(\lambda))^{-1}\overline{b(\lambda)}d\lambda<\infty, \tag{4.68}$$

where

$$b(\lambda)=\sum_{l=1}^{s}\int_{-M_l-N_l}^{-M_l}\vec{\alpha}(t)e^{it\lambda}dt$$

is a nontrivial function of the exponential type. Under this condition the error-free interpolation of the process $\vec{\xi}(t)+\vec{\eta}(t)$ is impossible (see, for example, Yu. A. Rozanov [10]).

The stationary processes $\vec{\xi}(t)$ and $\vec{\eta}(t)$ admit the spectral representations (see I. I. Gikhman and A. V. Skorokhod [153])

$$\vec{\xi}(t)=\int_{-\infty}^{\infty}e^{it\lambda}Z_{\xi}(d\lambda), \qquad \vec{\eta}(t)=\int_{-\infty}^{\infty}e^{it\lambda}Z_{\eta}(d\lambda), \tag{4.69}$$

where $Z_{\xi}(d\lambda)$ and $Z_{\eta}(d\lambda)$ are vector-valued orthogonal stochastic measures such that

$$EZ_{\xi}(\Delta_1)(Z_{\xi}(\Delta_2))^{*}=\frac{1}{2\pi}\int_{\Delta_1\cap\Delta_2}F(\lambda)d\lambda,$$
$$EZ_{\eta}(\Delta_1)(Z_{\eta}(\Delta_2))^{*}=\frac{1}{2\pi}\int_{\Delta_1\cap\Delta_2}G(\lambda)d\lambda.$$

Consider the problem of mean-square optimal linear estimation of the functional

$$A_s\vec{\xi}=\sum_{l=1}^{s}\int_{-M_l-N_l}^{-M_l}\vec{a}(t)^{\top}\vec{\xi}(t)dt$$

constructed from unobserved values of the process $\vec{\xi}(t)$. Estimates are based on observations of the process $\vec{\xi}(t)+\vec{\eta}(t)$ at points $t\in\mathbb{R}\backslash S$, where $S=\bigcup_{l=1}^{s}[-M_l-N_l,-M_l]$.

We assume that the function $\vec{a}(t)$ satisfies the following conditions

$$\sum_{k=1}^{T}\sum_{l=1}^{s}\int_{-M_l-N_l}^{-M_l}|a_k(t)|dt<\infty, \quad \sum_{l=1}^{s}\int_{-M_l-N_l}^{-M_l}t\|\vec{a}(t)\|^2dt<\infty. \tag{4.70}$$

These conditions ensure that the functional $A_s\vec{\xi}$ has a finite second moment.

Making use of the spectral representation (4.69) of the stationary process $\vec{\xi}(t)$, we can represent the functional $A_s\vec{\xi}$ in the following form

$$A_s\vec{\xi} = \int_{-\infty}^{\infty} (A_s(\lambda))^\top Z_\xi(d\lambda), \quad A_s(\lambda) = \sum_{l=1}^{s} \int_{-M_l-N_l}^{-M_l} \vec{a}(t)e^{it\lambda}dt.$$

Consider values $\xi_k(t), k = 1,\dots,T, t \in \mathbb{R}$ and $\eta_k(t), k = 1,\dots,T, t \in \mathbb{R}$ as elements of the Hilbert space $H = L_2(\Omega, \mathcal{F}, P)$ generated by random variables ξ with zero mathematical expectations, $E\xi = 0$, finite variations, $E|\xi|^2 < \infty$, and the inner product $(\xi, \eta) = E\xi\overline{\eta}$.

Denote by $H^s(\xi+\eta)$ the closed linear subspace generated by elements $\{\xi_k(t)+\eta_k(t) : t \in \mathbb{R}\backslash S, k = 1,\dots,T\}$ in the Hilbert space $H = L_2(\Omega, \mathcal{F}, P)$.

Denote by $L_2(F+G)$ the Hilbert space of vector-valued functions $\vec{a}(\lambda) = \{a_k(\lambda)\}_{k=1}^T$ such that

$$\int_{-\infty}^{\infty} \vec{a}(\lambda)^\top (F(\lambda)+G(\lambda))\overline{\vec{a}(\lambda)}d\lambda =$$

$$= \int_{-\infty}^{\infty} \sum_{k,l=1}^{T} a_k(\lambda)\overline{a_l(\lambda)}\,(f_{kl}(\lambda)+g_{kl}(\lambda))\,d\lambda < \infty,$$

and denote by $L_2^s(F+G)$ the subspace of the space $L_2(F+G)$ generated by the functions $\{e^{it\lambda}\delta_k,\ \delta_k = \{\delta_{kl}\}_{l=1}^T, k = 1,\dots,T, t \in \mathbb{R}\backslash S\}$.

Each linear estimate $\hat{A}_s\vec{\xi}$ of the functional $A_s\vec{\xi}$ can be represented in the form

$$\hat{A}_s\vec{\xi} = \int_{-\infty}^{\infty} (h(\lambda))^\top (Z_\xi(d\lambda)+Z_\eta(d\lambda)),$$

where $h(\lambda) \in L_2^s(F+G)$ is the spectral characteristic of the estimate.

The mean-square error $\Delta(h; F, G)$ of the estimate $\hat{A}_s\vec{\xi}$ is given by the formula

$$\Delta(h; F, G) = E\left|A_s\xi - \hat{A}_s\xi\right|^2 =$$

$$\frac{1}{2\pi}\int_{-\infty}^{\infty} (A_s(\lambda)-h(\lambda))^\top F(\lambda)\overline{(A_s(\lambda)-h(\lambda))}d\lambda + \frac{1}{2\pi}\int_{-\infty}^{\infty} (h(\lambda))^\top G(\lambda)\overline{h(\lambda)}d\lambda.$$

We first consider the problem in the case where the spectral densities of the stationary processes $\vec{\xi}(t)$ and $\vec{\eta}(t)$ are exactly known. In this case we can apply the method of orthogonal projections in the Hilbert space to find the optimal estimate. According to this method, the optimal linear estimate of the functional $A\vec{\xi}$ is a projection of the element $A\vec{\xi}$ of the space $H = L_2(\Omega, \mathcal{F}, P)$ onto the subspace $H^s(\xi+\eta)$. The estimate is determined by the following two conditions:

$$1)\hat{A}_s\vec{\xi} \in H^s(\xi+\eta),$$
$$2)A_s\vec{\xi} - \hat{A}_s\vec{\xi} \perp H^s(\xi+\eta).$$

It follows from the second condition that the spectral characteristic $h(\lambda)$ of the optimal linear estimate $\hat{A}_s\vec{\xi}$ for any $t \in \mathbb{R}\backslash S$ satisfies the equation

$$\frac{1}{2\pi}\int_{-\infty}^{\infty} (A_s(\lambda) - h(\lambda))^{\top} F(\lambda) e^{-it\lambda} d\lambda - \frac{1}{2\pi}\int_{-\infty}^{\infty} (h(\lambda))^{\top} G(\lambda) e^{-it\lambda} d\lambda = 0.$$

The last relation can be represented in the form

$$\frac{1}{2\pi}\int_{-\infty}^{\infty} \left[(A_s(\lambda))^{\top} F(\lambda) - (h(\lambda))^{\top}(F(\lambda)+G(\lambda))\right] e^{-it\lambda} d\lambda = 0, \tag{4.71}$$
$$t \in \mathbb{R}\backslash S.$$

Define the function

$$(C_s(\lambda))^{\top} = (A_s(\lambda))^{\top} F(\lambda) - (h(\lambda))^{\top}(F(\lambda)+G(\lambda))$$

and its Fourier transform

$$\vec{c}(t) = \frac{1}{2\pi}\int_{-\infty}^{\infty} C_s(\lambda) e^{-it\lambda} d\lambda, \quad t \in \mathbb{R}.$$

It follows from the relation (4.71) that the function $\vec{c}(t)$ is nonzero only on the set S. Hence, the function $C_s(\lambda)$ is of the form

$$C_s(\lambda) = \sum_{l=1}^{s} \int_{-M_l-N_l}^{-M_l} \vec{c}(t) e^{it\lambda} dt,$$

and the spectral characteristic of the estimate $\hat{A}_s\vec{\xi}$ is of the form

$$(h(\lambda))^\top = \\ = (A_s(\lambda))^\top F(\lambda)(F(\lambda)+G(\lambda))^{-1} - (C_s(\lambda))^\top (F(\lambda)+G(\lambda))^{-1}. \tag{4.72}$$

It follows from the first condition, $\hat{A}_s\vec{\xi} \in H^s(\xi+\eta)$, which determine the optimal linear estimate of the functional $A_s\vec{\xi}$, that the following relation is true

$$\int_{-\infty}^{\infty}\Big((A_s(\lambda))^\top F(\lambda)(F(\lambda)+G(\lambda))^{-1} - \\ - (C_s(\lambda))^\top (F(\lambda)+G(\lambda))^{-1}\Big)e^{-it\lambda}d\lambda = 0, t \in S. \tag{4.73}$$

The last relation can be represented in terms of linear operators in the space $L_2(S)$. We define operators

$$(\mathbf{B}_s\mathbf{a})(t) = \frac{1}{2\pi}\sum_{l=1}^{s}\int_{-M_l-N_l}^{-M_l}\int_{-\infty}^{\infty}(\vec{a}(u))^\top (F(\lambda)+G(\lambda))^{-1}e^{i\lambda(u-t)}d\lambda du,$$

$$(\mathbf{R}_s\mathbf{a})(t) = \frac{1}{2\pi}\sum_{l=1}^{s}\int_{-M_l-N_l}^{-M_l}\int_{-\infty}^{\infty}(\vec{a}(u))^\top F(\lambda)(F(\lambda)+G(\lambda))^{-1}e^{i\lambda(u-t)}d\lambda du,$$

$$(\mathbf{Q}_s\mathbf{a})(t) = \\ = \frac{1}{2\pi}\sum_{l=1}^{s}\int_{-M_l-N_l}^{-M_l}\int_{-\infty}^{\infty}(\vec{a}(u))^\top F(\lambda)(F(\lambda)+G(\lambda))^{-1}G(\lambda)e^{i\lambda(u-t)}d\lambda du,$$

$$\vec{a}(t) \in L_2(S), \quad t \in S.$$

Making use of the introduced operators, relation (4.73) can be represented in the form

$$(\mathbf{R}_s\mathbf{a})(t) = (\mathbf{B}_s\mathbf{c})(t), \quad t \in S. \tag{4.74}$$

Suppose that the operator $\mathbf{B}_s$ is invertible (see paper by H. Salehi [155] for more details). Then the function $\vec{c}(t)$ can be calculated by the formula

$$\vec{c}(t) = (\mathbf{B}_s^{-1}\mathbf{R}_s\mathbf{a})(t), \quad t \in S.$$

Consequently, the spectral characteristic $h(\lambda)$ of the estimate $\hat{A}_s\vec{\xi}$ is calculated by the formula

$$(h(\lambda))^{\top} = \\ = (A_s(\lambda))^{\top}F(\lambda)(F(\lambda)+G(\lambda))^{-1} - (C_s(\lambda))^{\top}(F(\lambda)+G(\lambda))^{-1}, \tag{4.75}$$

$$C_s(\lambda) = \sum_{l=1}^{s} \int_{-M_l-N_l}^{-M_l} (\mathbf{B}_s^{-1}\mathbf{R}_s\mathbf{a})(t)e^{it\lambda}dt, \quad t \in S.$$

The mean-square error of the estimate $\hat{A}_s\vec{\xi}$ can be calculated by the formula

$$\Delta(h;F,G) = \frac{1}{2\pi}\int_{-\infty}^{\infty}((A_s(\lambda))^{\top}G(\lambda)+(C_s(\lambda))^{\top})(F(\lambda)+G(\lambda))^{-1}F(\lambda)\times$$

$$\times(F(\lambda)+G(\lambda))^{-1}((A_s(\lambda))^{\top}G(\lambda)+(C_s(\lambda))^{\top})^{*}d\lambda+$$

$$+\frac{1}{2\pi}\int_{-\infty}^{\infty}((A_s(\lambda))^{\top}F(\lambda)-(C_s(\lambda))^{\top})(F(\lambda)+G(\lambda))^{-1}G(\lambda)\times$$

$$\times(F(\lambda)+G(\lambda))^{-1}((A_s(\lambda))^{\top}G(\lambda)+(C_s(\lambda))^{\top})^{*}d\lambda =$$

$$= \langle(\mathbf{R}_s\mathbf{a})(t),(\mathbf{B}_s^{-1}\mathbf{R}_s\mathbf{a})(t)\rangle + \langle(\mathbf{Q}_s\mathbf{a})(t),\vec{a}(t)\rangle, \tag{4.76}$$

where $\langle a(t),b(t)\rangle$ is the inner product in the space $L_2(S)$.

Let us summarize the obtained results and present them in the form of a theorem.

Theorem 4.5. *Let $\{\vec{\xi}(t), t\in\mathbb{R}\}$ and $\{\vec{\eta}(t), t\in\mathbb{R}\}$ be uncorrelated multidimensional stationary stochastic processes with the spectral density matrices $F(\lambda)$ and $G(\lambda)$ satisfying the minimality condition (4.68). The spectral characteristic $h(\lambda)$ and the mean-square error $\Delta(F,G)$ of the optimal linear estimate of the functional $A_s\vec{\xi}$ based on observations of the process $\vec{\xi}(t)+\vec{\eta}(t)$, $t\in\mathbb{R}\backslash S$ can be calculated by formulas (4.75), (4.76).*

Estimation from Observations without Noise

Consider the problem of the optimal estimation of the functional in case where the stationary process $\vec{\xi}(t)$ is observed without noise. In this case the spectral characteristic of the estimate $\hat{A}_s\vec{\xi}$ is of the form

$$(h(\lambda))^{\top} = (A_s(\lambda))^{\top} - (C_s(\lambda))^{\top}(F(\lambda))^{-1},$$
$$C_s(\lambda) = \sum_{l=1}^{s} \int_{-M_l-N_l}^{-M_l} \vec{c}(t)e^{it\lambda}dt, \tag{4.77}$$

and relation (4.74) is of the form

$$\vec{a}(t) = (\mathbf{B}_s\mathbf{c})(t), \quad t \in S. \tag{4.78}$$

Suppose that the operator $\mathbf{B}_s$ is invertible. Then the unknown function $\vec{c}(t)$ can be calculated by the formula

$$\vec{c}(t) = (\mathbf{B}_s^{-1}\mathbf{a})(t), \quad t \in S,$$

and the spectral characteristic of the estimate $\hat{A}_s\vec{\xi}$ is of the form

$$(h(\lambda))^{\top} = (A_s(\lambda))^{\top} - (C_s(\lambda))^{\top}(F(\lambda))^{-1},$$
$$C_s(\lambda) = \sum_{l=1}^{s} \int_{-M_l-N_l}^{-M_l} (\mathbf{B}_s^{-1}\mathbf{a})(t)e^{it\lambda}dt, \quad t \in S. \tag{4.79}$$

The mean-square error of the estimate can be calculated by the formula

$$\Delta(h;F) = \langle(\mathbf{B}_s^{-1})\mathbf{a}(t), \vec{a}(t)\rangle. \tag{4.80}$$

The following theorem is true.

Theorem 4.6. *Let $\{\vec{\xi}(t), t \in \mathbb{R}\}$ be a multidimensional stationary stochastic process with the spectral density matrix $F(\lambda)$, satisfying the minimality condition*

$$\int_{-\infty}^{\infty} (b(\lambda))^{\top}(F(\lambda))^{-1}\overline{b(\lambda)}d\lambda < \infty \tag{4.81}$$

for a nonzero vector-valued function of the exponential type $b(\lambda) = \sum_{l=1}^{s} \int_{-M_l-N_l}^{-M_l} \vec{\alpha}(t)e^{it\lambda}dt$. *The spectral characteristic* $h(\lambda)$ *and the mean-square error* $\Delta(F)$ *of the optimal linear estimate of the functional* $A_s\vec{\xi}$ *based on observations of the process* $\vec{\xi}(t)$ *at points* $t \in \mathbb{R}\setminus S$, $S = \bigcup_{l=1}^{s}[-M_l - N_l, -M_l]$, *can be calculated by formulas (4.79), (4.80).*

2.2. Minimax Approach to Interpolation Problem for Stationary Processes

The derived formulas can be applied to estimate the value of the functional only in the case of spectral certainty, where the spectral density matrices of the considered processes are exactly known. However, in practice, we do not have exact values of the spectral density matrices while, instead, we have a class of admissible spectral density matrices, where the spectral density matrices of the processes belong to. In this case, the minimax method of estimation can be applied to estimate the value of the functional. This method allows us to find estimates that minimize the maximal values of the mean-square errors of the estimates for all spectral density matrices from a given class of admissible spectral density matrices.

Definition 4.3. *For a given class of spectral densities* $D = D_F \times D_G$ *the spectral densities* $F^0(\lambda) \in D_F$, $G^0(\lambda) \in D_G$ *are called the least favourable in the class* D *for the optimal linear interpolation of the functional* $A_s\vec{\xi}$ *if the following relation is true*

$$\Delta\left(F^0,G^0\right) = \Delta\left(h\left(F^0,G^0\right);F^0,G^0\right) = \max_{(F,G)\in D_F\times D_G} \Delta\left(h\left(F,G\right);F,G\right).$$

Definition 4.4. *For a given class of spectral densities* $D = D_F \times D_G$ *the spectral characteristic* $h^0(\lambda)$ *of the optimal linear interpolation of the functional* $A_s\vec{\xi}$ *is called minimax-robust if there are satisfied conditions*

$$h^0(\lambda) \in H_D = \bigcap_{(F,G)\in D_F\times D_G} L_2^s(F+G),$$

$$\min_{h\in H_D}\max_{(F,G)\in D} \Delta(h;F,G) = \max_{(F,G)\in D} \Delta\left(h^0;F,G\right).$$

From the introduced definitions and formulas derived in the previous section we can obtain the following statements.

Lemma 4.2. *The spectral densities $F_0(\lambda) \in D_F$, $G_0(\lambda) \in D_G$ satisfying the minimality condition (4.68) are the least favourable in the class $D = D_F \times D_G$ for the optimal linear interpolation of the functional $A_s\vec{\xi}$, if the Fourier coefficients of the functions*

$$(F^0(\lambda)+G^0(\lambda))^{-1},\ F^0(\lambda)(F^0(\lambda)+G^0(\lambda))^{-1},\ F^0(\lambda)(F^0(\lambda)+G^0(\lambda))^{-1}G^0(\lambda)$$

define operators $\mathbf{B}_s^0, \mathbf{R}_s^0, \mathbf{Q}_s^0$, which determine a solution of the constrained optimization problem

$$\begin{aligned} \max_{(F,G)\in D_F\times D_G} & \langle(\mathbf{R}_s\mathbf{a})(t),(\mathbf{B}_s^{-1}\mathbf{R}_s\mathbf{a})(t)\rangle+\langle(\mathbf{Q}_s\mathbf{a})(t),\vec{a}(t)\rangle = \\ &= \langle(\mathbf{R}_s^0\mathbf{a})(t),((\mathbf{B}_s^0)^{-1}\mathbf{R}_s^0\mathbf{a})(t)\rangle+\langle(\mathbf{Q}_s^0\mathbf{a})(t),\vec{a}(t)\rangle. \end{aligned} \tag{4.82}$$

The minimax spectral characteristic $h^0 = h(F^0,G^0)$ is determined by the formula (4.75) if $h(F^0,G^0) \in H_D$.

Corollary 4.7. *The spectral density $F^0(\lambda) \in D_F$ satisfying the minimality condition (4.81) is the least favourable in the class D_F for the optimal linear interpolation of the functional $A_s\vec{\xi}$ from observations of the process $\vec{\xi}(t)$ at points $t \in \mathbb{R}\backslash S$, if the Fourier coefficients of the function $(F^0(\lambda))^{-1}$ define the operator $\mathbf{B}_s^0$ which determines a solution of the constrained optimization problem*

$$\max_{F\in D_F}\langle(\mathbf{B}_s^{-1}\mathbf{a})(t),\vec{a}(t)\rangle = \langle(((\mathbf{B}_s^0)^{-1}\mathbf{a})(t),\vec{a}(t)\rangle. \tag{4.83}$$

The minimax spectral characteristic $h^0 = h(F^0)$ is determined by the formula (4.79) if $h(F^0) \in H_{D_F}$.

For more detailed analysis of properties of the least favourable spectral densities and the minimax-robust spectral characteristics we observe that the least favourable spectral densities $F^0(\lambda)$, $G^0(\lambda)$ and the minimax spectral characteristic $h^0 = h(F^0,G^0)$ form a saddle point of the function $\Delta(h;F,G)$ on the set $H_D \times D$. The saddle point inequalities

$$\Delta\left(h^0;F,G\right) \leq \Delta\left(h^0;F^0,G^0\right) \leq \Delta\left(h;F^0,G^0\right),$$

$$\forall h \in H_D, \forall F \in D_F, \forall G \in D_G,$$

are true if $h^0 = h(F^0, G^0)$ and $h(F^0, G^0) \in H_D$, where (F^0, G^0) is a solution of the constrained optimization problem

$$\sup_{(F,G)\in D_F\times D_G} \Delta\left(h(F^0,G^0);F,G\right) = \Delta\left(h(F^0,G^0);F^0,G^0\right). \tag{4.84}$$

The linear functional $\Delta\left(h\left(F^0,G^0\right);F,G\right)$ is calculated by the formula

$$\Delta\left(h\left(F^0,G^0\right);F,G\right) =$$

$$= \frac{1}{2\pi}\int_{-\infty}^{\infty}((A_s(\lambda))^{\top}G^0(\lambda)+(C_s^0(\lambda))^{\top})(F^0(\lambda)+G^0(\lambda))^{-1}F(\lambda)\times$$

$$\times(F^0(\lambda)+G^0(\lambda))^{-1}((A_s(\lambda))^{\top}G^0(\lambda)+(C_s^0(\lambda))^{\top})^{*}d\lambda+$$

$$+\frac{1}{2\pi}\int_{-\infty}^{\infty}((A_s(\lambda))^{\top}F^0(\lambda)-(C_s^0(\lambda))^{\top})(F^0(\lambda)+G^0(\lambda))^{-1}G(\lambda)\times$$

$$\times(F^0(\lambda)+G^0(\lambda))^{-1}((A_s(\lambda))^{\top}G^0(\lambda)-(C_s^0(\lambda))^{\top})^{*}d\lambda,$$

$$C_s^0(\lambda) = \sum_{l=1}^{s}\int_{-M_l-N_l}^{-M_l}((\mathbf{B}_s^0)^{-1}\mathbf{R}_s^0\mathbf{a})(t)e^{it\lambda}dt, \quad t\in S.$$

The constrained optimization problem (4.84) is equivalent to the unconstrained optimization problem (see B. N. Pshenichnyj [156])

$$\Delta_D(F,G) = -\Delta(h(F^0,G^0);F,G)+\delta((F,G)\,|D_F\times D_G)\to \inf, \tag{4.85}$$

where $\delta((F,G)|D_F\times D_G)$ is the indicator function of the set $D = D_F\times D_G$. A solution of the problem (4.85) is characterized by the condition $0\in\partial\Delta_D(F^0,G^0)$, where $\partial\Delta_D(F^0,G^0)$ is the subdifferential of the convex functional $\Delta_D(F,G)$ at point (F^0,G^0). This condition makes it possible to find the least favourable spectral densities in some special classes of spectral densities D (see books by A. D. Ioffe and V. M. Tihomirov [161], B. N. Pshenichnyj [156], T. Rockafellar [158]).

Note, that the form of the functional $\Delta(h(F^0,G^0);F,G)$ is convenient for application of the Lagrange method of indefinite multipliers for finding a solution of the problem (4.85). Making use of the method of Lagrange multipliers and the form of subdifferentials of the indicator functions, we describe relations that determine the least favourable spectral densities in some special classes of spectral densities (see books by M. P. Moklyachuk [55, 163], M. P. Moklyachuk and O. Yu. Masyutka [96] for additional details).

Taking into consideration the introduced definitions and the derived relations we can verify that the following lemma is true.

Lemma 4.3. *Let (F^0, G^0) be a solution of the optimization problem (4.85). The spectral densities $F^0(\lambda)$, $G^0(\lambda)$ are the least favourable in the class $D = D_F \times D_G$ and the spectral characteristic $h^0 = h(F^0, G^0)$ is the minimax spectral characteristic of the optimal linear estimate of the functional $A_s\vec{\xi}$ if $h(F^0, G^0) \in H_D$.*

2.3. Least Favourable Spectral Densities in the Class $D = D_0 \times D_\varepsilon$

Consider the problem of mean-square optimal interpolation of the functional $A_s\vec{\xi}$ in the case where the spectral density matrices of the processes are not exactly known while the admissible spectral density matrices are from the classes $D = D_0^k \times D_\varepsilon^k, k = 1,2,3,4$, where

$$D_0^1 = \left\{ F(\lambda) \left| \frac{1}{2\pi} \int_{-\infty}^{\infty} \mathrm{Tr}\, F(\lambda) d\lambda = p \right. \right\},$$

$$D_\varepsilon^1 = \left\{ G(\lambda) \left| \mathrm{Tr}\, G(\lambda) = (1-\varepsilon)\mathrm{Tr}\, G_1(\lambda) + \varepsilon \mathrm{Tr}\, W(\lambda), \right. \right.$$
$$\left. \frac{1}{2\pi} \int_{-\infty}^{\infty} \mathrm{Tr}\, G(\lambda) d\lambda = q \right\};$$

$$D_0^2 = \left\{ F(\lambda) \left| \frac{1}{2\pi} \int_{-\infty}^{\infty} f_{kk}(\lambda) d\lambda = p_k, k = \overline{1,T} \right. \right\},$$

$$D_\varepsilon^2 = \left\{ G(\lambda) \left| g_{kk}(\lambda) = (1-\varepsilon) g_{kk}^1(\lambda) + \varepsilon w_{kk}(\lambda), \right. \right.$$
$$\left. \frac{1}{2\pi} \int_{-\infty}^{\infty} g_{kk}(\lambda) d\lambda = q_k, k = \overline{1,T} \right\};$$

$$D_0^3 = \left\{ F(\lambda) \left| \frac{1}{2\pi} \int_{-\infty}^{\infty} \langle B_1, F(\lambda) \rangle \, d\lambda = p \right. \right\},$$

$$D_\varepsilon^3 = \left\{ G(\lambda) \left| \langle B_2, G(\lambda) \rangle = (1-\varepsilon) \langle B_2, G_1(\lambda) \rangle + \varepsilon \langle B_2, W(\lambda) \rangle, \right. \right.$$
$$\left. \frac{1}{2\pi} \int_{-\infty}^{\infty} \langle B_2, G(\lambda) \rangle \, d\lambda = q \right\};$$

$$D_0^4 = \left\{ F(\lambda) \left| \frac{1}{2\pi} \int_{-\infty}^{\infty} F(\lambda) d\lambda = P \right. \right\},$$

$$D_\varepsilon^4 = \left\{ G(\lambda) \left| G(\lambda) = (1-\varepsilon) G_1(\lambda) + \varepsilon W(\lambda), \frac{1}{2\pi} \int_{-\infty}^{\infty} G(\lambda) d\lambda = Q \right. \right\}.$$

Here $G_1(\lambda)$ is a known and fixed spectral density matrix while $W(\lambda)$ is unknown spectral density matrix, $p, q, p_k, q_k, k = \overline{1,T}$ are given numbers, P, Q, B_1, B_2 are given positive definite Hermitian matrices.
Densities from the classes of $D_\varepsilon^k, k = 1,2,3,4,$ of spectral densities form the "ε–contamination" model of spectral density matrices.

From the condition $0 \in \partial\Delta_D(F^0, G^0)$ we find the following equations which determine the least favourable spectral densities for these given sets of admissible spectral densities.

For the first pair $D_0^1 \times D_\varepsilon^1$ we have equations

$$((A_s(\lambda))^\top G^0(\lambda) + (C_s^0(\lambda))^\top)^* ((A_s(\lambda))^\top G^0(\lambda) + (C_s^0(\lambda))^\top) =$$
$$= \alpha^2 (F^0(\lambda) + G^0(\lambda))^2, \quad (4.86)$$

$$((A_s(\lambda))^\top F^0(\lambda) - (C_s^0(\lambda))^\top)^* ((A_s(\lambda))^\top F^0(\lambda) - (C_s^0(\lambda))^\top) =$$
$$= (\beta^2 + \gamma(\lambda))(F^0(\lambda) + G^0(\lambda))^2, \quad (4.87)$$

where α^2, β^2 are Lagrange multipliers, $\gamma(\lambda) \leq 0$ and $\gamma(\lambda) = 0$ if $\mathrm{Tr}\, F^0(\lambda) > (1-\varepsilon)\mathrm{Tr}\, G_1(\lambda)$.

For the second pair $D_0^2 \times D_\varepsilon^2$ we have equations

$$((A_s(\lambda))^\top G^0(\lambda) + (C_s^0(\lambda))^\top)^* ((A_s(\lambda))^\top G^0(\lambda) + (C_s^0(\lambda))^\top) =$$
$$= (F^0(\lambda) + G^0(\lambda)) \left\{ \alpha_k^2 \delta_{kl} \right\}_{k,l=1}^T (F^0(\lambda) + G^0(\lambda)), \quad (4.88)$$

$$((A_s(\lambda))^\top F^0(\lambda)-(C_s^0(\lambda))^\top)^*((A_s(\lambda))^\top F^0(\lambda)-(C_s^0(\lambda))^\top)=$$
$$=(F^0(\lambda)+G^0(\lambda))\left\{(\beta_k^2+\gamma_k(\lambda))\delta_{kl}\right\}_{k,l=1}^T(F^0(\lambda)+G^0(\lambda)),\quad (4.89)$$

where α_k^2,β_k^2 are Lagrange multipliers, δ_{kl} are Kronecker symbols, $\gamma_k(\lambda)\leq 0$ and $\gamma_k(\lambda)=0$ if $g_{kk}^0(\lambda)>(1-\varepsilon)g_{kk}^1(\lambda)$.

For the third pair $D_0^3\times D_\varepsilon^3$ we have equations

$$((A_s(\lambda))^\top G^0(\lambda)+(C_s^0(\lambda))^\top)^*((A_s(\lambda))^\top G^0(\lambda)+(C_s^0(\lambda))^\top)=$$
$$=\alpha^2(F^0(\lambda)+G^0(\lambda))B_1^\top(F^0(\lambda)+G^0(\lambda)),\quad (4.90)$$

$$((A_s(\lambda))^\top F^0(\lambda)-(C_s^0(\lambda))^\top)^*((A_s(\lambda))^\top F^0(\lambda)-(C_s^0(\lambda))^\top)=$$
$$=(\beta^2+\gamma'(\lambda))(F^0(\lambda)+G^0(\lambda))B_2^\top(F^0(\lambda)+G^0(\lambda)),\quad (4.91)$$

where α^2,β^2 are Lagrange multipliers, $\gamma'(\lambda)\leq 0$ and $\gamma'(\lambda)=0$ if $\langle B_2,G^0(\lambda)\rangle>(1-\varepsilon)\langle B_2,G_1(\lambda)\rangle$.

For the fourth pair $D_0^4\times D_\varepsilon^4$ we have equations

$$((A_s(\lambda))^\top G^0(\lambda)+(C_s^0(\lambda))^\top)^*((A_s(\lambda))^\top G^0(\lambda)+(C_s^0(\lambda))^\top)=$$
$$=(F^0(\lambda)+G^0(\lambda))\vec{\alpha}\cdot\vec{\alpha}^*(F^0(\lambda)+G^0(\lambda)),\quad (4.92)$$

$$((A_s(\lambda))^\top F^0(\lambda)-(C_s^0(\lambda))^\top)^*((A_s(\lambda))^\top F^0(\lambda)-(C_s^0(\lambda))^\top)=$$
$$=(F^0(\lambda)+G^0(\lambda))(\vec{\beta}\cdot\vec{\beta}^*+\Gamma(\lambda))(F^0(\lambda)+G^0(\lambda)),\quad (4.93)$$

where $\vec{\alpha},\vec{\beta}$ are Lagrange multipliers, $\Gamma(\lambda)\leq 0$ and $\Gamma_3(\lambda)=0$ if $G^0(\lambda)>(1-\varepsilon)G_1(\lambda)$.

Thus, the following statement is true.

Theorem 4.7. *The least favourable spectral densities $F^0(\lambda)$, $G^0(\lambda)$ in the classes $D_0^k\times D_\varepsilon^k, k=1,2,3,4$, for the optimal linear interpolation of the functional $A_s\vec{\xi}$ are determined by relations (4.86) – (4.93), the minimality condition (4.68), the constrained optimization problem (4.82), and restrictions on densities from the corresponding classes $D_0^k\times D_\varepsilon^k, k=1,2,3,4$. The minimax-robust spectral characteristic of the optimal estimate of the functional $A_s\vec{\xi}$ is determined by formula (4.75).*

Corollary 4.8. *The least favourable spectral densities $F^0(\lambda)$ in the classes D_0^k, $k = 1,2,3,4$, for the optimal linear interpolation of the functional $A_s\vec{\xi}$, based on observations of the process $\vec{\xi}(t)$ at points $t \in \mathbb{R}\backslash S$, are determined by the following equations, respectively,*

$$((C_s^0(\lambda))^\top)^* \cdot (C_s^0(\lambda))^\top = \alpha^2(F^0(\lambda))^2, \tag{4.94}$$

$$((C_s^0(\lambda))^\top)^* \cdot (C_s^0(\lambda))^\top = F^0(\lambda)\left\{\alpha_k^2\delta_{kl}\right\}_{k,l=1}^T F^0(\lambda), \tag{4.95}$$

$$((C_s^0(\lambda))^\top)^* \cdot (C_s^0(\lambda))^\top = \alpha^2 F^0(\lambda)(B_2)^\top F^0(\lambda), \tag{4.96}$$

$$((C_s^0(\lambda))^\top)^* \cdot (C_s^0(\lambda))^\top = F^0(\lambda)\vec{\alpha}\cdot\vec{\alpha}^* F^0(\lambda), \tag{4.97}$$

the minimality condition (4.81), the constrained optimization problem (4.83), and restrictions on densities from the corresponding classes D_0^k, $k = 1,2,3,4$. The minimax spectral characteristic of the optimal estimate of the functional $A_s\vec{\xi}$ is determined by formula (4.79).

Corollary 4.9. *The least favourable spectral densities $F^0(\lambda)$ in the classes D_ε^k, $k = 1,2,3,4$, for the optimal linear interpolation of the functional $A_s\vec{\xi}$, based on observations of the process $\vec{\xi}(t)$ at points $t \in \mathbb{R}\backslash S$, are determined by the following equations, respectively,*

$$((C_s^0(\lambda))^\top)^* \cdot (C_s^0(\lambda))^\top = (\beta^2 + \gamma(\lambda))(F^0(\lambda))^2, \tag{4.98}$$

$$((C_s^0(\lambda))^\top)^* \cdot (C_s^0(\lambda))^\top = F^0(\lambda)\left\{(\beta_k^2 + \gamma_k(\lambda))\delta_{kl}\right\}_{k,l=1}^T F^0(\lambda), \tag{4.99}$$

$$((C_s^0(\lambda))^\top)^* \cdot (C_s^0(\lambda))^\top = (\beta^2 + \gamma'(\lambda))F^0(\lambda)(B_2)^\top F^0(\lambda), \tag{4.100}$$

$$((C_s^0(\lambda))^\top)^* \cdot (C_s^0(\lambda))^\top = F^0(\lambda)(\vec{\beta}\cdot\vec{\beta}^* + \Gamma(\lambda))F^0(\lambda), \tag{4.101}$$

the minimality condition (4.81), the constrained optimization problem (4.83), and restrictions on densities from the corresponding classes D_ε^k, $k = 1,2,3,4$. The minimax spectral characteristic of the optimal estimate of the functional $A_s\vec{\xi}$ is determined by formula (4.79).

2.4. Least Favourable Spectral Densities in the Class $D = D_V^U \times D_{2\delta}$

Consider the problem of mean-square optimal interpolation of the functional $A_s\vec{\xi}$ in the case where the spectral density matrices of the processes are not exactly known while the admissible spectral density matrices are from the classes $D_V^{U^k} \times D_{2\delta}^k, k = 1,2,3,4$, where

$$D_V^{U^1} = \left\{ F(\lambda) \middle| \operatorname{Tr} V(\lambda) \leq \operatorname{Tr} F(\lambda) \leq \operatorname{Tr} U(\lambda), \quad \frac{1}{2\pi}\int_{-\infty}^{\infty} \operatorname{Tr} F(\lambda) d\lambda = p \right\},$$

$$D_{2\delta}^1 = \left\{ G(\lambda) \middle| \frac{1}{2\pi}\int_{-\infty}^{\infty} |\operatorname{Tr}(G(\lambda) - G_1(\lambda))|^2 d\lambda \leq \delta \right\};$$

$$D_V^{U^2} = \left\{ F(\lambda) \middle| v_{kk}(\lambda) \leq f_{kk}(\lambda) \leq u_{kk}(\lambda), \right.$$
$$\left. \frac{1}{2\pi}\int_{-\infty}^{\infty} f_{kk}(\lambda) d\lambda = p_k, k = \overline{1,T} \right\},$$

$$D_{2\delta}^2 = \left\{ G(\lambda) \middle| \frac{1}{2\pi}\int_{-\infty}^{\infty} \left|g_{kk}(\lambda) - g_{kk}^1(\lambda)\right|^2 d\lambda \leq \delta_k, k = \overline{1,T} \right\};$$

$$D_V^{U^3} = \left\{ F(\lambda) \middle| \langle B_1, V(\lambda)\rangle \leq \langle B_1, F(\lambda)\rangle \leq \langle B_1, U(\lambda)\rangle, \right.$$
$$\left. \frac{1}{2\pi}\int_{-\infty}^{\infty} \langle B_1, F(\lambda)\rangle d\lambda = p \right\},$$

$$D_{2\delta}^3 = \left\{ G(\lambda) \middle| \frac{1}{2\pi}\int_{-\infty}^{\infty} |\langle B_2, G(\lambda) - G_1(\lambda)\rangle|^2 d\lambda \leq \delta \right\};$$

$$D_V^{U^4} = \left\{ F(\lambda) \middle| V(\lambda) \leq F(\lambda) \leq U(\lambda), \frac{1}{2\pi}\int_{-\infty}^{\infty} F(\lambda) d\lambda = P \right\}.$$

$$D_{2\delta}^4 = \left\{ G(\lambda) \middle| \frac{1}{2\pi}\int_{-\infty}^{\infty} \left|g_{ij}(\lambda) - g_{ij}^1(\lambda)\right|^2 d\lambda \leq \delta_i^j, i,j = \overline{1,T} \right\},$$

Here the spectral density matrices $V(\lambda), U(\lambda), G_1(\lambda)$ are known and fixed, $\delta, p, \delta_k, p_k, k = \overline{1,T}$, $\delta_i^j, i,j = \overline{1,T}$, are given numbers, P, B_1, B_2 are given positive-definite Hermitian matrices.

Spectral densities from the classes $D_V^{U\,k}, k = 1, 2, 3, 4$, form the "strip" models of spectral density matrices.
Densities from the classes $D_{2\delta}^k, k = 1, 2, 3, 4$, of admissible spectral densities form models of "δ–neighbourhood"in the space L_2 of a given bounded spectral density $G_1(\lambda)$.

From the condition $0 \in \partial\Delta_D(F^0, G^0)$ we find the following equations which determine the least favourable spectral densities for these given sets of admissible spectral densities.

For the first pair $D_V^{U\,1} \times D_{2\delta}^1$ we have equations

$$((A_s(\lambda))^\top G^0(\lambda) + (C_s^0(\lambda))^\top)^*((A_s(\lambda))^\top G^0(\lambda) + (C_s^0(\lambda))^\top) =$$
$$= (\alpha^2 + \gamma_1(\lambda) + \gamma_2(\lambda))(F^0(\lambda) + G^0(\lambda))^2, \quad (4.102)$$

$$((A_s(\lambda))^\top F^0(\lambda) - (C_s^0(\lambda))^\top)^*((A_s(\lambda))^\top F^0(\lambda) - (C_s^0(\lambda))^\top) =$$
$$= \beta^2 \mathrm{Tr}\,(G^0(\lambda) - G_1(\lambda))(F^0(\lambda) + G^0(\lambda))^2, \quad (4.103)$$

$$\frac{1}{2\pi}\int_{-\infty}^{\infty} |\mathrm{Tr}\,(G(\lambda) - G_1(\lambda))|^2\, d\lambda = \delta, \quad (4.104)$$

where α^2, β^2 are Lagrange multipliers, $\gamma_1(\lambda) \leq 0$ and $\gamma_1(\lambda) = 0$ if $\mathrm{Tr}\,F^0(\lambda) > \mathrm{Tr}\,V(\lambda)$, $\gamma_2(\lambda) \geq 0$ and $\gamma_2(\lambda) = 0$ if $\mathrm{Tr}\,F^0(\lambda) < \mathrm{Tr}\,U(\lambda)$.

For the second pair $D_V^{U\,2} \times D_{2\delta}^2$ we have equations

$$((A_s(\lambda))^\top G^0(\lambda) + (C_s^0(\lambda))^\top)^*((A_s(\lambda))^\top G^0(\lambda) + (C_s^0(\lambda))^\top) =$$
$$= (F^0(\lambda) + G^0(\lambda))\left\{(\alpha_k^2 + \gamma_{1k}(\lambda) + \gamma_{2k}(\lambda))\delta_{kl}\right\}_{k,l=1}^T (F^0(\lambda) + G^0(\lambda)),$$
$$(4.105)$$

$$((A_s(\lambda))^\top F^0(\lambda) - (C_s^0(\lambda))^\top)^*((A_s(\lambda))^\top F^0(\lambda) - (C_s^0(\lambda))^\top) =$$
$$= (F^0(\lambda) + G^0(\lambda))\left\{\beta_k^2(g_{kk}^0(\lambda) - g_{kk}^1(\lambda))\delta_{kl}\right\}_{k,l=1}^T (F^0(\lambda) + G^0(\lambda)), \quad (4.106)$$

$$\frac{1}{2\pi}\int_{-\infty}^{\infty} \left|g_{kk}(\lambda) - g_{kk}^1(\lambda)\right|^2 d\lambda = \delta_k,\ k = \overline{1, T}, \quad (4.107)$$

where α_k^2, β_k^2 are Lagrange multipliers, $\gamma_{1k}(\lambda) \leq 0$ and $\gamma_{1k}(\lambda) = 0$ if $f_{kk}^0(\lambda) > v_{kk}(\lambda)$, $\gamma_{2k}(\lambda) \geq 0$ and $\gamma_{2k}(\lambda) = 0$ if $f_{kk}^0(\lambda) < u_{kk}(\lambda)$.

For the third pair $D_V^{U3} \times D_{2\delta}^3$ we have equations

$$((A_s(\lambda))^\top G^0(\lambda) + (C_s^0(\lambda))^\top)^*((A_s(\lambda))^\top G^0(\lambda) + (C_s^0(\lambda))^\top) =$$

$$= (\alpha^2 + \gamma_1'(\lambda) + \gamma_2'(\lambda))(F^0(\lambda) + G^0(\lambda))(B_1)^\top (F^0(\lambda) + G^0(\lambda)), \quad (4.108)$$

$$((A_s(\lambda))^\top F^0(\lambda) - (C_s^0(\lambda))^\top)^*((A_s(\lambda))^\top F^0(\lambda) - (C_s^0(\lambda))^\top) =$$

$$= \beta^2 \left\langle B_2, G^0(\lambda) - G_1(\lambda) \right\rangle (F^0(\lambda) + G^0(\lambda))^2, \quad (4.109)$$

$$\frac{1}{2\pi} \int_{-\infty}^{\infty} |\langle B_2, G(\lambda) - G_1(\lambda) \rangle|^2 d\lambda = \delta, \quad (4.110)$$

where α^2, β^2 are Lagrange multipliers, $\gamma_1'(\lambda) \leq 0$ and $\gamma_1'(\lambda) = 0$ if $\langle B_1, F^0(\lambda) \rangle > \langle B_1, V(\lambda) \rangle$, $\gamma_2'(\lambda) \geq 0$ and $\gamma_2'(\lambda) = 0$ if $\langle B_1, F^0(\lambda) \rangle < \langle B_1, U(\lambda) \rangle$.

For the fourth pair $D_V^{U4} \times D_{2\delta}^4$ we have equations

$$((A_s(\lambda))^\top G^0(\lambda) + (C_s^0(\lambda))^\top)^*((A_s(\lambda))^\top G^0(\lambda) + (C_s^0(\lambda))^\top) =$$

$$= (F^0(\lambda) + G^0(\lambda))(\vec{\alpha} \cdot \vec{\alpha}^* + \Gamma_1(\lambda) + \Gamma_2(\lambda))(F^0(\lambda) + G^0(\lambda)) \quad (4.111)$$

$$((A_s(\lambda))^\top F^0(\lambda) - (C_s^0(\lambda))^\top)^*((A_s(\lambda))^\top F^0(\lambda) - (C_s^0(\lambda))^\top) =$$

$$= (F^0(\lambda) + G^0(\lambda)) \left\{ \beta_{ij}(g_{ij}^0(\lambda) - g_{ij}^1(\lambda)) \right\}_{i,j=1}^T (F^0(\lambda) + G^0(\lambda)), \quad (4.112)$$

$$\frac{1}{2\pi} \int_{-\infty}^{\infty} \left| g_{ij}(\lambda) - g_{ij}^1(\lambda) \right|^2 d\lambda = \delta_i^j, \ i, j = \overline{1, T}. \quad (4.113)$$

where $\vec{\alpha}, \beta_{ij}$ are Lagrange multipliers, $\Gamma_1(\lambda) \leq 0$ and $\Gamma_1(\lambda) = 0$ if $F^0(\lambda) > V(\lambda)$, $\Gamma_2(\lambda) \geq 0$ and $\Gamma_2(\lambda) = 0$ if $F^0(\lambda) < U(\lambda)$.

The following theorem and corollaries are true.

Theorem 4.8. *The least favourable spectral densities $F^0(\lambda)$, $G^0(\lambda)$ in the classes $D_0^k \times D_\varepsilon^k, k = 1,2,3,4$, for the optimal linear interpolation of the functional $A_s\xi$ are determined by relations (4.102) – (4.113), the minimality condition (4.68), the constrained optimization problem (4.82), and restrictions on densities from the corresponding classes $D_0^k \times D_\varepsilon^k, k = 1,2,3,4$. The minimax-robust spectral characteristic of the optimal estimate of the functional $A_s\vec{\xi}$ is determined by the formula (4.75).*

Corollary 4.10. *The least favourable spectral densities $F^0(\lambda)$ in the classes $D_V^{U^k}$, $k = 1,2,3,4$, for the optimal linear estimation of the functional $A_s\vec{\xi}$, based on observations of the process $\vec{\xi}(t)$ at points $t \in \mathbb{R}\backslash S$, are determined by the following equations, respectively,*

$$((C_s^0(\lambda))^\top)^* \cdot (C_s^0(\lambda))^\top = (\alpha^2 + \gamma_1(\lambda) + \gamma_2(\lambda))(F^0(\lambda))^2, \tag{4.114}$$

$$((C_s^0(\lambda))^\top)^* \cdot (C_s^0(\lambda))^\top = F^0(\lambda)\left\{\beta_k^2(g_{kk}^0(\lambda) - g_{kk}^1(\lambda))\delta_{kl}\right\}_{k,l=1}^T F^0(\lambda), \tag{4.115}$$

$$((C_s^0(\lambda))^\top)^* \cdot (C_s^0(\lambda))^\top = (\alpha^2 + \gamma_1'(\lambda) + \gamma_2'(\lambda))F^0(\lambda)(B_1)^\top F^0(\lambda), \tag{4.116}$$

$$((C_s^0(\lambda))^\top)^* \cdot (C_s^0(\lambda))^\top = F^0(\lambda)(\vec{\alpha} \cdot \vec{\alpha}^* + \Gamma_1(\lambda) + \Gamma_2(\lambda))F^0(\lambda), \tag{4.117}$$

the minimality condition (4.81), the constrained optimization problem (4.83), and restrictions on densities from the corresponding classes $D_V^{U^k}$, $k = 1,2,3,4$. The minimax spectral characteristic of the optimal estimate of the functional $A_s\vec{\xi}$ is determined by the formula (4.79).

Corollary 4.11. *The least favourable spectral densities $F^0(\lambda)$ in the classes $D_{2\delta}^k$, $k = 1,2,3,4$, for the optimal linear estimation of the functional $A_s\vec{\xi}$, based on observations of the process $\vec{\xi}(t)$ at points $t \in \mathbb{R}\backslash S$, are determined by the following equations, respectively,*

$$((C_s^0(\lambda))^\top)^* \cdot (C_s^0(\lambda))^\top = \beta^2 \mathrm{Tr}\,(F^0(\lambda) - G_1(\lambda))(F^0(\lambda))^2, \tag{4.118}$$

$$((C_s^0(\lambda))^\top)^* \cdot (C_s^0(\lambda))^\top = F^0(\lambda)\left\{\beta_k^2(f_{kk}^0(\lambda) - g_{kk}^1(\lambda))\delta_{kl}\right\}_{k,l=1}^T F^0(\lambda), \tag{4.119}$$

$$((C_s^0(\lambda))^\top)^* \cdot (C_s^0(\lambda))^\top = \beta^2 \left\langle B_2, F^0(\lambda) - G_1(\lambda)\right\rangle (F^0(\lambda))^2, \tag{4.120}$$

$$((C_s^0(\lambda))^\top)^* \cdot (C_s^0(\lambda))^\top = F^0(\lambda)\left\{\beta_{ij}(f_{ij}^0(\lambda) - g_{ij}^1(\lambda))\right\}_{i,j=1}^T F^0(\lambda) \tag{4.121}$$

the minimality condition (4.81), the constrained optimization problem (4.83), and the following restrictions on densities from the corresponding classes $D_{2\delta}^k$, $k = 1,2,3,4$, respectively,

$$\frac{1}{2\pi}\int_{-\infty}^{\infty} |\mathrm{Tr}\,(G(\lambda) - G_1(\lambda))|^2 \, d\lambda = \delta, \tag{4.122}$$

$$\frac{1}{2\pi}\int_{-\infty}^{\infty} \left|g_{kk}(\lambda) - g_{kk}^1(\lambda)\right|^2 d\lambda = \delta_k, \; k = \overline{1,T}, \tag{4.123}$$

$$\frac{1}{2\pi}\int_{-\infty}^{\infty} |\langle B_2, G(\lambda) - G_1(\lambda)\rangle|^2 d\lambda = \delta, \tag{4.124}$$

$$\frac{1}{2\pi}\int_{-\infty}^{\infty} \left|g_{ij}(\lambda) - g^1_{ij}(\lambda)\right|^2 d\lambda = \delta^j_i,\ i,j=\overline{1,T}. \tag{4.125}$$

The minimax spectral characteristic of the optimal estimate of the functional $A_s\vec{\xi}$ is determined by the formula (4.79).

3. Filtering of Multidimensional Stationary Stochastic Processes with Missing Observations

In this section we deal with the problem of mean-square optimal linear estimation of the functional

$$A\vec{\xi} = \int_{R^s} \vec{a}(t)^\top \vec{\xi}(-t)dt$$

constructed from unobserved values of a multidimensional stationary stochastic process $\vec{\xi}(t)$. Estimates are based on observations of the process $\vec{\xi}(t)+\vec{\eta}(t)$ at points $t \in \mathbb{R}^-\backslash S$, $S = \bigcup_{l=1}^{s}[-M_l - N_l, -M_l]$, $R^s = [0,\infty)\backslash S^+$, $S^+ = \bigcup_{l=1}^{s}[M_l,\ M_l + N_l]$. The problem is investigated in the case of spectral certainty, as well as in the case of spectral uncertainty. Formulas for calculating the spectral characteristic and the mean-square error of the optimal linear estimate of the functional are derived in the case of spectral certainty, where the spectral densities of the processes are exactly known. In the case of spectral uncertainty, where the spectral densities are not exactly known while a set of admissible spectral densities is given, the minimax method of estimation is applied. Formulas for determining the least favourable spectral densities and the minimax-robust spectral characteristics of the optimal estimates of the functional are proposed for some specific classes of admissible spectral densities.

3.1. Hilbert Space Projection Method of Filtering

Let $\vec{\xi}(t) = \{\xi_k(t)\}_{k=1}^T, t \in \mathbb{R}$ and $\vec{\eta}(t) = \{\eta_k(t)\}_{k=1}^T, t \in \mathbb{R}$ be uncorrelated mean square continuous multidimensional stationary stochastic processes with zero first moments, $E\vec{\xi}(t) = \vec{0}$, $E\vec{\eta}(t) = \vec{0}$, and the correlation functions which admit the spectral representation

$$R_\xi(t) = \frac{1}{2\pi}\int\limits_{-\infty}^{\infty} e^{it\lambda}F(\lambda)d\lambda, \quad R_\eta(t) = \frac{1}{2\pi}\int\limits_{-\infty}^{\infty} e^{it\lambda}G(\lambda)d\lambda,$$

where $F(\lambda) = \{f_{kl}(\lambda)\}_{k,l=1}^{T}$ and $G(\lambda) = \{g_{kl}(\lambda)\}_{k,l=1}^{T}$ are spectral density matrices of the processes $\vec{\xi}(t)$ and $\vec{\eta}(t)$ respectively.

Suppose that the spectral densities of the processes satisfy the minimality condition

$$\int_{-\infty}^{\infty}(b(\lambda))^\top(F(\lambda)+G(\lambda))^{-1}\overline{b(\lambda)}d\lambda < \infty, \tag{4.126}$$

where

$$b(\lambda) = \sum_{l=1}^{s}\int\limits_{-M_l-N_l}^{-M_l} \vec{\alpha}(t)e^{it\lambda}dt$$

is a nontrivial function of the exponential type.

Under this condition the error-free estimate of the process $\vec{\xi}(t)+\vec{\eta}(t)$ is impossible (see, for example, Yu. A. Rozanov [10]).

The stationary processes $\vec{\xi}(t)$ and $\vec{\eta}(t)$ admit the spectral representations

$$\vec{\xi}(t) = \int\limits_{-\infty}^{\infty} e^{it\lambda}Z_\xi(d\lambda), \qquad \vec{\eta}(t) = \int\limits_{-\infty}^{\infty} e^{it\lambda}Z_\eta(d\lambda), \tag{4.127}$$

where $Z_\xi(d\lambda)$ and $Z_\eta(d\lambda)$ are vector-valued orthogonal stochastic measures such that

$$EZ_\xi(\Delta_1)(Z_\xi(\Delta_2))^* = \frac{1}{2\pi}\int_{\Delta_1\cap\Delta_2} F(\lambda)d\lambda,$$

$$EZ_\eta(\Delta_1)(Z_\eta(\Delta_2))^* = \frac{1}{2\pi}\int_{\Delta_1\cap\Delta_2} G(\lambda)d\lambda.$$

Consider the problem of mean-square optimal linear estimatation of the functional

$$A\vec{\xi} = \int\limits_{R^s} \vec{a}(t)^\top\vec{\xi}(-t)dt$$

constructed from unobserved values of the process $\vec{\xi}(t)$. Estimates are based on the observed values of the process $\vec{\xi}(t)+\vec{\eta}(t)$ at points $t\in\mathbb{R}^-\backslash S$, where

$$S=\bigcup_{l=1}^{s}[-M_l-N_l,-M_l],\ R^s=[0,\infty)\backslash S^+,\ S^+=\bigcup_{l=1}^{s}[M_l,\ M_l+N_l]$$

. We assume that the function $\vec{a}(t)$ satisfies the following conditions

$$\sum_{k=1}^{T}\sum_{l=1}^{s}\int_{M_l}^{M_l+N_l}|a_k(t)|dt<\infty,\quad \sum_{l=1}^{s}\int_{M_l}^{M_l+N_l}t\|\vec{a}(t)\|^2dt<\infty. \tag{4.128}$$

These conditions ensure that the functional $A_s\vec{\xi}$ has a finite second moment.

Making use of the spectral representation (4.127) of the stationary process $\vec{\xi}(t)$, we can represent the functional $A\vec{\xi}$ in the following form

$$A\vec{\xi}=\int_{-\infty}^{\infty}(A(\lambda))^{\top}Z_{\xi}(d\lambda),\quad A(\lambda)=\int_{R^s}\vec{a}(t)e^{-it\lambda}dt.$$

Consider the Hilbert space $H=L_2(\Omega,\mathcal{F},P)$ generated by random variables ξ with zero mathematical expectations, $E\xi=0$, finite variations, $E|\xi|^2<\infty$, and the inner product $(\xi,\eta)=E\xi\overline{\eta}$. Denote by $H^s(\xi+\eta)$ the closed linear subspace generated by elements $\{\xi_k(t)+\eta_k(t):t\in\mathbb{R}^-\backslash S,k=\overline{1,T}\}$ in the Hilbert space $H=L_2(\Omega,\mathcal{F},P)$.

Let $L_2(F+G)$ be the Hilbert space of complex-valued functions $\vec{a}(\lambda)=\{a_k(\lambda)\}_{k=1}^{T}$ such that

$$\int_{-\infty}^{\infty}\vec{a}(\lambda)^{\top}(F(\lambda)+G(\lambda))\overline{\vec{a}(\lambda)}d\lambda=$$

$$=\int_{-\infty}^{\infty}\sum_{k,l=1}^{T}a_k(\lambda)\overline{a_l(\lambda)}(f_{kl}(\lambda)+g_{kl}(\lambda))d\lambda<\infty.$$

Denote by $L_2^s(F+G)$ the subspace of $L_2(F+G)$ generated by functions

$$e^{it\lambda}\delta_k,\ \delta_k=\{\delta_{kl}\}_{l=1}^{T},\ k=\overline{1,T},\ t\in\mathbb{R}^-\backslash S.$$

Each linear estimate $\hat{A}_s\vec{\xi}$ of the functional $A_s\vec{\xi}$ is of the form

$$\hat{A}\vec{\xi} = \int_{-\infty}^{\infty} (h(\lambda))^{\top}(Z_{\xi}(d\lambda) + Z_{\eta}(d\lambda)),$$

where $h(\lambda) = \{h_k(\lambda))\}_{k=1}^{T} \in L_2^s(F+G)$ is the spectral characteristic of the estimate, and the mean-square error $\Delta(h;F,G)$ of the estimate is determined by formula

$$\Delta(h;F,G) = E\left|A\vec{\xi} - \hat{A}\vec{\xi}\right|^2 =$$

$$= \frac{1}{2\pi}\int_{-\infty}^{\infty} (A(\lambda) - h(\lambda))^{\top} F(\lambda)\overline{(A(\lambda) - h(\lambda))} d\lambda + \frac{1}{2\pi}\int_{-\infty}^{\infty} (h(\lambda))^{\top} G(\lambda)\overline{h(\lambda)} d\lambda.$$

We first consider the problem in the case where spectral densities of the stationary processes $\vec{\xi}(t)$ and $\vec{\eta}(t)$ are exactly known. In this case we can apply the method of orthogonal projections in the Hilbert space to find the optimal estimate. According to this method, the optimal linear estimate of the functional $A\vec{\xi}$ is a projection of the element $A\vec{\xi}$ of the space $H = L_2(\Omega, \mathcal{F}, P)$ onto the subspace $H^s(\xi+\eta)$. The estimate is determined by the following two conditions:

$$1) \hat{A}\vec{\xi} \in H^s(\xi+\eta),$$
$$2) A\vec{\xi} - \hat{A}\vec{\xi} \perp H^s(\xi+\eta).$$

It follows from the second condition that the spectral characteristic $h(\lambda)$ of the optimal linear estimate $\hat{A}\vec{\xi}$ for any $t \in \mathbb{R}^- \setminus S$ satisfies the relation

$$\frac{1}{2\pi}\int_{-\infty}^{\infty} (A(\lambda) - h(\lambda))^{\top} F(\lambda) e^{-it\lambda} d\lambda - \frac{1}{2\pi}\int_{-\infty}^{\infty} (h(\lambda))^{\top} G(\lambda) e^{-it\lambda} d\lambda = \vec{0}.$$

This relation can be written in the following way

$$\frac{1}{2\pi}\int_{-\infty}^{\infty} \left[(A(\lambda))^{\top} F(\lambda) - (h(\lambda))^{\top}(F(\lambda) + G(\lambda)))\right] e^{-it\lambda} d\lambda = 0, \quad t \in \mathbb{R}^- \setminus S. \tag{4.129}$$

Consider the function $(C(\lambda))^{\top}=(A(\lambda))^{\top}F(\lambda)-(h(\lambda))^{\top}(F(\lambda)+G(\lambda))$ and its Fourier transform

$$\vec{\mathbf{c}}(t)=\frac{1}{2\pi}\int\limits_{-\infty}^{\infty}C(\lambda)e^{-it\lambda}d\lambda,\quad t\in\mathbb{R}.$$

It follows from relation (4.129), that the function $\mathbf{c}(t)$ can be nonzero only on the set $U=S\cup[0,\infty)$. Hence, the function $C(\lambda)$ is of the form

$$C(\lambda)=\sum_{l=1}^{s}\int\limits_{-M_l-N_l}^{-M_l}\vec{\mathbf{c}}(t)e^{it\lambda}dt+\int\limits_{0}^{\infty}\vec{\mathbf{c}}(t)e^{it\lambda}dt,$$

and the spectral characteristic of the estimate $\hat{A}\vec{\xi}$ is of the form

$$(h(\lambda))^{\top}=$$
$$=(A(\lambda))^{\top}F(\lambda)(F(\lambda)+G(\lambda))^{-1}-(C(\lambda))^{\top}(F(\lambda)+G(\lambda))^{-1}.\quad(4.130)$$

It follows from the first condition, $\hat{A}\vec{\xi}\in H^{s}(\xi+\eta)$, which determines the optimal estimate of the functional $A\vec{\xi}$, that for any $t\in U$ the following relation is true

$$\int\limits_{-\infty}^{\infty}\Big((A(\lambda))^{\top}F(\lambda)(F(\lambda)+G(\lambda))^{-1}-$$
$$-(C(\lambda))^{\top}(F(\lambda)+G(\lambda))^{-1}\Big)e^{-it\lambda}d\lambda=0.\quad(4.131)$$

In order to represent the last relation in terms of linear operators in the space

$L_2(U)$, we introduce the following operators

$$(\mathbf{Bx})(t) = \frac{1}{2\pi}\sum_{l=1}^{s}\int_{-M_l-N_l}^{-M_l}(\vec{\mathbf{x}}(u))^\top\int_{-\infty}^{\infty}(F(\lambda)+G(\lambda))^{-1}e^{i\lambda(u-t)}d\lambda du+$$

$$+\frac{1}{2\pi}\int_{0}^{\infty}(\vec{\mathbf{x}}(u))^\top\int_{-\infty}^{\infty}(F(\lambda)+G(\lambda))^{-1}e^{i\lambda(u-t)}d\lambda du,$$

$$(\mathbf{Rx})(t) = \frac{1}{2\pi}\sum_{l=1}^{s}\int_{-M_l-N_l}^{-M_l}(\vec{\mathbf{x}}(u))^\top\int_{-\infty}^{\infty}F(\lambda)(F(\lambda)+G(\lambda))^{-1}e^{i\lambda(u+t)}d\lambda du+$$

$$+\frac{1}{2\pi}\int_{0}^{\infty}(\vec{\mathbf{x}}(u))^\top\int_{-\infty}^{\infty}F(\lambda)(F(\lambda)+G(\lambda))^{-1}e^{i\lambda(u-t)}d\lambda du,$$

$$(\mathbf{Qx})(t) = \frac{1}{2\pi}\sum_{l=1}^{s}\int_{-M_l-N_l}^{-M_l}(\vec{\mathbf{x}}(u))^\top\int_{-\infty}^{\infty}F(\lambda)(F(\lambda)+G(\lambda))^{-1}G(\lambda)e^{i\lambda(u-t)}d\lambda du+$$

$$+\frac{1}{2\pi}\int_{0}^{\infty}(\vec{\mathbf{x}}(u))^\top\int_{-\infty}^{\infty}F(\lambda)(F(\lambda)+G(\lambda))^{-1}G(\lambda)e^{i\lambda(u-t)}d\lambda du,$$

$$\vec{\mathbf{x}}(t)\in L_2(U),\quad t\in U.$$

The equality (4.131) can be represented in the form

$$\int_{-\infty}^{\infty}\int_{R^s}\vec{a}(u)^\top F(\lambda)(F(\lambda)+G(\lambda))^{-1}e^{i(u-t)}dud\lambda-$$

$$-\int_{-\infty}^{\infty}\left(\sum_{l=1}^{s}\int_{-M_l-N_l}^{-M_l}\vec{\mathbf{c}}(t)^\top(F(\lambda)+G(\lambda))^{-1}e^{i(u-t)\lambda}du\right)d\lambda-$$

$$-\int_{-\infty}^{\infty}\int_{0}^{\infty}\vec{\mathbf{c}}(t)^\top(F(\lambda)+G(\lambda))^{-1}e^{i(u-t)\lambda}dud\lambda=0,\quad t\in U. \tag{4.132}$$

Consider the function $\vec{\mathbf{a}}(t)$ such that

$$\vec{\mathbf{a}}(t)=\vec{0},\ t\in S,\quad \vec{\mathbf{a}}(t)=\vec{a}(t),t\in R^s\quad \vec{\mathbf{a}}(t)=\vec{0},\ t\in S^+.$$

Making use of the introduced operators and functions, we can represent equality (4.132) in terms of the linear operators in the space $L_2(U)$

$$(\mathbf{Ra})(t) = (\mathbf{Bc})(t), \quad t \in U. \tag{4.133}$$

Assume that the operator $\mathbf{B}$ is invertible. Then the function $\vec{\mathbf{c}}(t)$ is calculated by the formula

$$\vec{\mathbf{c}}(t) = (\mathbf{B}^{-1}\mathbf{Ra})(t), \quad t \in U.$$

The spectral characteristic $h(\lambda)$ of the estimate $\hat{A}\vec{\xi}$ can be calculated by the formula

$$\begin{gathered}
(h(\lambda))^{\top} = \\
= (A(\lambda))^{\top} F(\lambda)(F(\lambda)+G(\lambda))^{-1} - (C(\lambda))^{\top}(F(\lambda)+G(\lambda))^{-1}, \\
C(\lambda) = \sum_{l=1}^{s} \int_{-M_l-N_l}^{-M_l} (\mathbf{B}^{-1}\mathbf{Ra})(t)e^{it\lambda}dt + \int_{0}^{\infty} (\mathbf{B}^{-1}\mathbf{Ra})(t)e^{it\lambda}dt.
\end{gathered} \tag{4.134}$$

The mean-square error of the estimate $\hat{A}\vec{\xi}$ is calculated by the formula

$$\begin{aligned}
\Delta(h;F,G) = \frac{1}{2\pi}\int_{-\infty}^{\infty} &((A(\lambda))^{\top}G(\lambda) + (C(\lambda))^{\top})(F(\lambda)+G(\lambda))^{-1}F(\lambda)\times \\
&\times(F(\lambda)+G(\lambda))^{-1}((A(\lambda))^{\top}G(\lambda)+(C(\lambda))^{\top})^{*}d\lambda+ \\
+\frac{1}{2\pi}\int_{-\infty}^{\infty} &((A(\lambda))^{\top}F(\lambda) - (C(\lambda))^{\top})(F(\lambda)+G(\lambda))^{-1}G(\lambda)\times \\
&\times(F(\lambda)+G(\lambda))^{-1}((A(\lambda))^{\top}G(\lambda)+(C(\lambda))^{\top})^{*}d\lambda = \\
&= \langle(\mathbf{Ra})(t),(\mathbf{B}^{-1}\mathbf{Ra})(t)\rangle + \langle(\mathbf{Qa})(t),\vec{\mathbf{a}}(t)\rangle,
\end{aligned} \tag{4.135}$$

where

$$\langle\vec{a}(t),\vec{b}(t)\rangle = \sum_{l=1}^{s}\int_{-M_l-N_l}^{-M_l}\sum_{k=1}^{T} a_k(t)\overline{b_k(t)}dt + \int_{0}^{\infty}\sum_{k=1}^{T} a_k(t)\overline{b_k(t)}dt$$

is the inner product in the space $L_2(U)$.

The obtained results can be represented in the form of a theorem.

Theorem 4.9. *Let $\vec{\xi}(t)$ and $\vec{\eta}(t)$ be uncorrelated multidimensional stationary stochastic processes with the spectral densities $F(\lambda)$ and $G(\lambda)$ satisfying the minimality condition (4.126). The spectral characteristic $h(\lambda)$ and the mean-square error $\Delta(h;F,G)$ of the optimal linear estimate of the functional $A\vec{\xi}$ based on observations of the process $\vec{\xi}(t)+\vec{\eta}(t)$, $t \in \mathbb{R}^-\backslash S$ can be calculated by formulas (4.134), (4.135).*

Estimation of the Functional $A_N\vec{\xi}$

Consider the filtering problem for the functional

$$A_N\vec{\xi} = \int_{N^s} \vec{a}(t)^\top \vec{\xi}(-t)dt,$$

$$N^s = [0,N] \cap R^s,\ R^s = [0,\infty)\backslash S^+,\ S^+ = \bigcup_{l=1}^{s}[M_l,\ M_l+N_l],$$

constructed from unobserved values of the process $\vec{\xi}(t)$. Estimates are based on observations of the process $\vec{\xi}(t)+\vec{\eta}(t)$ at points $t \in \mathbb{R}^-\backslash S$.

The optimal linear estimate $\hat{A}_N\vec{\xi}$ of the functional $A_N\vec{\xi}$ is of the form

$$\hat{A}_N\vec{\xi} = \int_{-\infty}^{\infty} (h_N(\lambda))^\top (Z_\xi(d\lambda)+Z_\eta(d\lambda)),$$

where $h_N(\lambda) = \{h_{kN}(\lambda)\}_{k=1}^{T} \in L_2^s(F+G)$ is the spectral characteristic of the estimate.

Consider the function $\vec{\mathbf{a}}_N(t)$ such that

$$\vec{\mathbf{a}}_N(t) = \vec{0}, t \in S, \quad \vec{\mathbf{a}}_N(t) = \vec{a}(t), t \in N^s, \quad \vec{\mathbf{a}}_N(t) = \vec{0}, t \in U\backslash N^s.$$

Making use of the introduced notations, we can conclude that the spectral characteristic $h_N(\lambda)$ and the mean-square error $\Delta(h_N;F,G)$ of the optimal estimate $\hat{A}_N\vec{\xi}$ can be calculated by formulas

$$(h_N(\lambda))^\top = (A_N(\lambda))^\top F(\lambda)(F(\lambda)+G(\lambda))^{-1} - (C_N(\lambda))^\top (F(\lambda)+G(\lambda))^{-1},$$

$$C_N(\lambda) = \sum_{l=1}^{s} \int_{-M_l-N_l}^{-M_l} (\mathbf{B}^{-1}\mathbf{R}\mathbf{a}_N)(t)e^{it\lambda}dt + \int_0^{\infty} (\mathbf{B}^{-1}\mathbf{R}\mathbf{a}_N)(t)e^{it\lambda}dt,$$

(4.136)

$$A_N(\lambda) = \int_{N^s} \vec{a}(t) e^{-it\lambda} dt,$$

$$\begin{aligned} \Delta(h_N; F, G) =& E\left|A_N\vec{\xi} - \hat{A}_N\vec{\xi}\right|^2 = \\ =& \langle(\mathbf{Ra}_N)(t), (\mathbf{B}^{-1}\mathbf{Ra}_N)(t)\rangle + \langle(\mathbf{Qa}_N)(t), \vec{\mathbf{a}}_N(t)\rangle. \end{aligned} \tag{4.137}$$

Thus, we obtain the following corollary.

Corollary 4.12. *Let $\vec{\xi}(t)$ and $\vec{\eta}(t)$ be uncorrelated multidimensional stationary stochastic processes with the spectral densities $F(\lambda)$ and $G(\lambda)$ satisfying the minimality condition (4.126). The spectral characteristic $h_N(\lambda)$ and the mean-square error $\Delta(h_N; F, G)$ of the optimal linear estimate of the functional $A_N\vec{\xi}$ based on observations of the process $\vec{\xi}(t) + \vec{\eta}(t)$, $t \in \mathbb{R}^- \backslash S$, can be calculated by formulas (4.136), (4.137).*

3.2. Minimax Robust Method of Filtering

In the previous section we deal with the filtering problem under the condition that we know the spectral densities of the processes. In this case we derived formulas for calculating the spectral characteristics and the mean-square errors of the optimal estimates of the introduced functionals. In the case of spectral uncertainty, where the full information on spectral densities are impossible, while it is known that the spectral densities belong to some specified classes of admissible densities, the minimax method of filtering is reasonable. This method gives us a procedure of finding estimates which minimize the maximal values of the mean-square errors of the estimates for all spectral densities from a given class of admissible spectral densities. For the description of the minimax method of filtering we propose the following definitions.

Definition 4.5. *For a given class of spectral densities $D = D_F \times D_G$ the spectral densities $F^0(\lambda) \in D_F$, $G^0(\lambda) \in D_G$ are called the least favourable in the class D for the optimal linear filtering of the functional $A\vec{\xi}$ if the following relation is true*

$$\Delta\left(F^0, G^0\right) = \Delta\left(h\left(F^0, G^0\right); F^0, G^0\right) = \max_{(F,G) \in D_F \times D_G} \Delta\left(h\left(F, G\right); F, G\right).$$

Definition 4.6. *For a given class of spectral densities $D = D_F \times D_G$ the spectral characteristic $h^0(\lambda)$ of the optimal linear filtering of the functional $A\vec{\xi}$ is called the minimax robust if there are satisfied conditions*

$$h^0(\lambda) \in H_D = \bigcap_{(F,G)\in D_F\times D_G} L_2^s(F+G),$$

$$\min_{h\in H_D}\max_{(F,G)\in D}\Delta(h;F,G) = \max_{(F,G)\in D}\Delta\left(h^0;F,G\right).$$

From the introduced definitions and derived formulas we can obtain the following statement.

Lemma 4.4. *The spectral densities $F^0(\lambda) \in D_F$, $G^0(\lambda) \in D_G$ satisfying the minimality condition (4.126) are the least favourable in the class $D = D_F \times D_G$ for the optimal linear filtering of the functional $A\vec{\xi}$, if the Fourier transforms of the functions*

$$(F^0(\lambda)+G^0(\lambda))^{-1},\ F^0(\lambda)(F^0(\lambda)+G^0(\lambda))^{-1},\ F^0(\lambda)(F^0(\lambda)+G^0(\lambda))^{-1}G^0(\lambda)$$

define the operators $\mathbf{B}^0, \mathbf{R}^0, \mathbf{Q}^0$, which determine a solution of the constrain optimization problem

$$\max_{(F,G)\in D_F\times D_G}\left(\langle(\mathbf{Ra})(t),(\mathbf{B}^{-1}\mathbf{Ra})(t)\rangle + \langle(\mathbf{Qa})(t),\vec{\mathbf{a}}(t)\rangle\right) =$$

$$= \langle(\mathbf{R}^0\mathbf{a})(t),((\mathbf{B}^0)^{-1}\mathbf{R}^0\mathbf{a})(t)\rangle + \langle(\mathbf{Q}^0\mathbf{a})(t),\vec{\mathbf{a}}(t)\rangle. \tag{4.138}$$

The minimax spectral characteristic $h^0 = h(F^0,G^0)$ is calculated by formula (4.134) if $h(F^0,G^0) \in H_D$.

For more detailed analysis of properties of the least favourable spectral densities and the minimax-robust spectral characteristics we observe that the least favourable spectral densities $F^0(\lambda)$, $G^0(\lambda)$ and the minimax spectral characteristic $h^0 = h(F^0,G^0)$ form a saddle point of the function $\Delta(h;F,G)$ on the set $H_D \times D$. The saddle point inequalities

$$\Delta\left(h^0;F,G\right) \le \Delta\left(h^0;F^0,G^0\right) \le \Delta\left(h;F^0,G^0\right),$$

$$\forall h \in H_D, \forall F \in D_F, \forall G \in D_G,$$

are true if $h^0 = h(F^0,G^0)$, $h(F^0,G^0) \in H_D$, where (F^0,G^0) is a solution of the constrained optimization problem

$$\sup_{(F,G)\in D_F\times D_G} \Delta\left(h(F^0,G^0);F,G\right) = \Delta\left(h(F^0,G^0);F^0,G^0\right). \tag{4.139}$$

The linear functional $\Delta\left(h\left(F^0,G^0\right);F,G\right)$ is calculated by the formula

$$\Delta\left(h\left(F^0,G^0\right);F,G\right) =$$

$$= \frac{1}{2\pi}\int_{-\infty}^{\infty}((A(\lambda))^{\top}G^0(\lambda)+(C^0(\lambda))^{\top})(F^0(\lambda)+G^0(\lambda))^{-1}F(\lambda)\times$$
$$\times(F^0(\lambda)+G^0(\lambda))^{-1}((A(\lambda))^{\top}G^0(\lambda)+(C^0(\lambda))^{\top})^{*}d\lambda+$$
$$+\frac{1}{2\pi}\int_{-\infty}^{\infty}((A(\lambda))^{\top}F^0(\lambda)-(C^0(\lambda))^{\top})(F^0(\lambda)+G^0(\lambda))^{-1}G(\lambda)\times$$
$$\times(F^0(\lambda)+G^0(\lambda))^{-1}((A(\lambda))^{\top}G^0(\lambda)-(C^0(\lambda))^{\top})^{*}d\lambda,$$

$$C^0(\lambda) = \sum_{l=1}^{s}\int_{-M_l-N_l}^{-M_l}((\mathbf{B}^0)^{-1}\mathbf{R}^0\mathbf{a})(t)e^{it\lambda}dt + \int_{0}^{\infty}((\mathbf{B}^0)^{-1}\mathbf{R}^0\mathbf{a})(t)e^{it\lambda}dt.$$

The constrained optimization problem (4.139) is equivalent to the unconstrained optimization problem

$$\Delta_D(F,G) = -\Delta(h(F^0,G^0);F,G) + \delta((F,G)\,|D_F\times D_G\,) \to \inf, \tag{4.140}$$

where $\delta((F,G)\,|D_F\times D_G\,)$ is the indicator function of the set $D = D_F\times D_G$. A solution of the problem (4.140) is characterized by the condition $0 \in \partial\Delta_D(F^0,G^0)$, where $\partial\Delta_D(F^0,G^0)$ is the subdifferential of the convex functional $\Delta_D(F,G)$ at point (F^0,G^0). This condition makes it possible to find the least favourable spectral densities in some special classes of spectral densities D (see books by A. D. Ioffe and V. M. Tihomirov [161], B. N. Pshenichnyj [156], T. Rockafellar [158]).

Note, that the form of the functional $\Delta(h(F^0,G^0);F,G)$ is convenient for application of the Lagrange method of indefinite multipliers for finding a solution of the problem (4.139). Making use of the method of Lagrange multipliers and the form of subdifferentials of the indicator functions, we describe relations that determine the least favourable spectral densities in some special classes of spectral densities.

3.3. Least Favourable Spectral Densities in the Class $D = D_0 \times D_{1\delta}$

Consider the problem of minimax filtering of the functional $A\vec{\xi}$ in the case where spectral densities of the processes belong to the following classes of admissible spectral densities $D = D_0^k \times D_{1\delta}^k, k = 1,2,3,,$

$$D_0^1 = \left\{ F(\lambda) \left| \frac{1}{2\pi} \int_{-\infty}^{\infty} \mathrm{Tr}\, F(\lambda) d\lambda = p \right. \right\},$$

$$D_{1\delta}^1 = \left\{ G(\lambda) \left| \frac{1}{2\pi} \int_{-\infty}^{\infty} |\mathrm{Tr}(G(\lambda) - G_1(\lambda))| d\lambda \leq \delta \right. \right\};$$

$$D_0^2 = \left\{ F(\lambda) \left| \frac{1}{2\pi} \int_{-\infty}^{\infty} f_{kk}(\lambda) d\lambda = p_k, k = \overline{1,T} \right. \right\},$$

$$D_{1\delta}^2 = \left\{ G(\lambda) \left| \frac{1}{2\pi} \int_{-\infty}^{\infty} \left| g_{kk}(\lambda) - g_{kk}^1(\lambda) \right| d\lambda \leq \delta_k, k = \overline{1,T} \right. \right\};$$

$$D_0^3 = \left\{ F(\lambda) \left| \frac{1}{2\pi} \int_{-\infty}^{\infty} \langle B_1, F(\lambda) \rangle \, d\lambda = p \right. \right\},$$

$$D_{1\delta}^3 = \left\{ G(\lambda) \left| \frac{1}{2\pi} \int_{-\infty}^{\infty} |\langle B_2, G(\lambda) - G_1(\lambda) \rangle| d\lambda \leq \delta \right. \right\},$$

$$D_0^4 = \left\{ F(\lambda) \left| \frac{1}{2\pi} \int_{-\infty}^{\infty} F(\lambda) d\lambda = P \right. \right\},$$

$$D_{1\delta}^4 = \left\{ G(\lambda) \left| \frac{1}{2\pi} \int_{-\infty}^{\infty} \left| g_{ij}(\lambda) - g_{ij}^1(\lambda) \right| d\lambda \leq \delta_i^j, i, j = \overline{1,T} \right. \right\},$$

where $G_1(\lambda)$ is a known and fixed spectral density, $\delta, p, \delta_k, p_k, k = \overline{1,T}, \delta_i^j, i, j = \overline{1,T}$, are given numbers, P, B_1, B_2 are given positive-definite Hermitian matrices. Densities from the classes of admissible spectral densities $D_{1\delta}^k, k = 1,2,3,4,$ form the "δ–neighbourhood" models in the space L_1 of a given bounded spectral density $G_1(\lambda)$.

From the condition $0 \in \partial\Delta_D(F^0, G^0)$ we find the following equations which determine the least favourable spectral densities for these given sets of admissible spectral densities.

For the first pair $D_0^1 \times D_{1\delta}^1$ we have equations

$$((A(\lambda))^\top G^0(\lambda) + (C^0(\lambda))^\top)^*((A(\lambda))^\top G^0(\lambda) + (C^0(\lambda))^\top) =$$
$$= \alpha^2 (F^0(\lambda) + G^0(\lambda))^2, \quad (4.141)$$

$$((A(\lambda))^\top F^0(\lambda) - (C^0(\lambda))^\top)^*((A(\lambda))^\top F^0(\lambda) - (C^0(\lambda))^\top) =$$
$$= \beta^2 \gamma(\lambda)(F^0(\lambda) + G^0(\lambda))^2, \quad (4.142)$$

$$\frac{1}{2\pi}\int_{-\infty}^{\infty} \left|\mathrm{Tr}\,(G^0(\lambda) - G_1(\lambda))\right| d\lambda = \delta, \quad (4.143)$$

where α^2, β^2 are Lagrange multipliers, $|\gamma(\lambda)| \leq 1$ and

$$\gamma(\lambda) = \mathrm{sign}\,(\mathrm{Tr}\,(G^0(\lambda) - G_1(\lambda)))\,\mathrm{when}\,\mathrm{Tr}\,(G^0(\lambda) - G_1(\lambda)) \neq 0.$$

For the second pair $D_0^2 \times D_{1\delta}^2$ we have equations

$$((A(\lambda))^\top G^0(\lambda) + (C^0(\lambda))^\top)^*((A(\lambda))^\top G^0(\lambda) + (C^0(\lambda))^\top) =$$
$$= (F^0(\lambda) + G^0(\lambda))\left\{\alpha_k^2 \delta_{kl}\right\}_{k,l=1}^{T} (F^0(\lambda) + G^0(\lambda)), \quad (4.144)$$

$$((A(\lambda))^\top F^0(\lambda) - (C^0(\lambda))^\top)^*((A(\lambda))^\top F^0(\lambda) - (C^0(\lambda))^\top) =$$
$$= (F^0(\lambda) + G^0(\lambda))\left\{\beta_k^2 \gamma_k(\lambda) \delta_{kl}\right\}_{k,l=1}^{T} (F^0(\lambda) + G^0(\lambda)), \quad (4.145)$$

$$\frac{1}{2\pi}\int_{-\infty}^{\infty} \left|g_{kk}^0(\lambda) - g_{kk}^1(\lambda)\right| d\lambda = \delta_k, \; k = \overline{1,T}, \quad (4.146)$$

where α_k^2, β_k^2 are Lagrange multipliers, δ_{kl} are Kronecker symbols, $|\gamma_k(\lambda)| \leq 1$ and

$$\gamma_k(\lambda) = \mathrm{sign}\,(g_{kk}^0(\lambda) - g_{kk}^1(\lambda))\,\text{when}\; g_{kk}^0(\lambda) - g_{kk}^1(\lambda) \neq 0, \; k = \overline{1,T}.$$

For the third pair $D_0^3 \times D_{1\delta}^3$ we have equations

$$((A(\lambda))^\top G^0(\lambda) + (C^0(\lambda))^\top)^*((A(\lambda))^\top G^0(\lambda) + (C^0(\lambda))^\top) =$$
$$= \alpha^2 (F^0(\lambda) + G^0(\lambda)) B_1^\top (F^0(\lambda) + G^0(\lambda)), \quad (4.147)$$

$$((A(\lambda))^\top F^0(\lambda) - (C^0(\lambda))^\top)^*((A(\lambda))^\top F^0(\lambda) - (C^0(\lambda))^\top) =$$

$$= \beta^2\gamma'(\lambda)(F^0(\lambda) + G^0(\lambda))B_2^\top(F^0(\lambda) + G^0(\lambda)), \tag{4.148}$$

$$\frac{1}{2\pi}\int_{-\infty}^{\infty} \left|\left\langle B_2, G^0(\lambda) - G_1(\lambda)\right\rangle\right| d\lambda = \delta, \tag{4.149}$$

where α^2, β^2 are Lagrange multipliers, $|\gamma'(\lambda)| \leq 1$ and

$$\gamma'(\lambda) = \text{sign}\,\left\langle B_2, G^0(\lambda) - G_1(\lambda)\right\rangle \text{ when } \left\langle B_2, G^0(\lambda) - G_1(\lambda)\right\rangle \neq 0.$$

For the fourth pair $D_0^4 \times D_{1\delta}^4$ we have equations

$$((A(\lambda))^\top G^0(\lambda) + (C^0(\lambda))^\top)^*((A(\lambda))^\top G^0(\lambda) - (C^0(\lambda))^\top) =$$

$$= (F^0(\lambda) + G^0(\lambda))\vec{\alpha}\cdot\vec{\alpha}^*(F^0(\lambda) + G^0(\lambda)), \tag{4.150}$$

$$((A(\lambda))^\top F^0(\lambda) - (C^0(\lambda))^\top)^*((A(\lambda))^\top F^0(\lambda) + (C^0(\lambda))^\top) =$$

$$= (F^0(\lambda) + G^0(\lambda))\left\{\beta_{ij}\gamma_{ij}(\lambda)\right\}_{i,j=1}^{T}(F^0(\lambda) + G^0(\lambda)), \tag{4.151}$$

$$\frac{1}{2\pi}\int_{-\infty}^{\infty} \left|g_{ij}^0(\lambda) - g_{ij}^1(\lambda)\right| d\lambda = \delta_i^j,\ i,j = \overline{1,T}, \tag{4.152}$$

where $\vec{\alpha}, \beta_{ij}$ are Lagrange multipliers, $\left|\gamma_{ij}(\lambda)\right| \leq 1$ and

$$\gamma_{ij}(\lambda) = \frac{g_{ij}^0(\lambda) - g_{ij}^1(\lambda)}{\left|g_{ij}^0(\lambda) - g_{ij}^1(\lambda)\right|} \text{ when } g_{ij}^0(\lambda) - g_{ij}^1(\lambda) \neq 0,\ i,j = \overline{1,T}.$$

Thus, the following statements are true.

Theorem 4.10. *The least favourable spectral densities $F^0(\lambda)$, $G^0(\lambda)$ in the classes $D = D_0^k \times D_{1\delta}^k, k = 1,2,3,,$ for the optimal linear filtering of the functional $A\vec{\xi}$ are determined by relations (4.141) – (4.152), the minimality condition (4.126), the constrained optimization problem (4.138), and restrictions on densities from the corresponding classes $D = D_0^k \times D_{1\delta}^k, k = 1,2,3,.$ The minimax-robust spectral characteristic of the optimal estimate of the functional $A\vec{\xi}$ is determined by formula (4.134).*

Corollary 4.13. *Suppose that the spectral density $G(\lambda)$ is known, and the unknown spectral density $F^0(\lambda) \in D_0^k$. Suppose that the function $F^0(\lambda)+G(\lambda)$ satisfies the minimality condition (4.126). The spectral density $F^0(\lambda)$ is the least favourable in the classes D_0^k, $k=\overline{1,4}$ for the optimal linear filtering of the functional $A\vec{\xi}$ if it satisfies relations (4.141), (4.144), (4.147), (4.150), respectively, and the pair $(F^0(\lambda), G(\lambda))$ determines a solution of the optimization problem (4.138). The minimax-robust spectral characteristic of the optimal estimate of the functional $A\vec{\xi}$ is determined by formula (4.134).*

Corollary 4.14. *Suppose that the spectral density $F(\lambda)$ is known, and the unknown spectral density $G^0(\lambda) \in D_0^k$. Suppose that the function $F(\lambda)+G^0(\lambda)$ satisfies the minimality condition (4.126). The spectral density $G^0(\lambda)$ is the least favourable in the classes $D_{1\delta}^k$, $k=\overline{1,4}$ for the optimal linear filtering of the functional $A\vec{\xi}$ if it satisfies relations (4.142) – (4.143), (4.145) – (4.146), (4.148) – (4.149), (4.151) – (4.152), respectively, and the pair $(F(\lambda), G^0(\lambda))$ determines a solution of the optimization problem (4.138). The minimax-robust spectral characteristic of the optimal estimate of the functional $A\vec{\xi}$ is determined by formula (4.134).*

3.4. Least Favourable Spectral Densities in the Class $D = D_{2\delta} \times D_\varepsilon$

Consider the problem of filtering of the functional $A\vec{\xi}$ in the case where the spectral densities of the processes belong to the classes of admissible spectral densities $D_{2\delta}^k \times D_\varepsilon^k, k=1,2,3,4$,

$$D_{2\delta}^1 = \left\{ F(\lambda) \left| \frac{1}{2\pi} \int_{-\infty}^{\infty} |\mathrm{Tr}(F(\lambda)-F_1(\lambda))|^2 \, d\lambda \leq \delta \right.\right\};$$

$$D_\varepsilon^1 = \left\{ G(\lambda) \left| \mathrm{Tr}\, G(\lambda) = (1-\varepsilon)\mathrm{Tr}\, G_1(\lambda) + \varepsilon \mathrm{Tr}\, W(\lambda), \right.\right.$$
$$\left. \frac{1}{2\pi} \int_{-\infty}^{\infty} \mathrm{Tr}\, G(\lambda) d\lambda = q \right\};$$

$$D_{2\delta}^2 = \left\{ F(\lambda) \left| \frac{1}{2\pi} \int_{-\infty}^{\infty} \left| f_{kk}(\lambda) - f_{kk}^1(\lambda) \right|^2 d\lambda \leq \delta_k, k=\overline{1,T} \right.\right\};$$

$$D_\varepsilon^2 = \left\{ G(\lambda) \middle| g_{kk}(\lambda) = (1-\varepsilon) g_{kk}^1(\lambda) + \varepsilon w_{kk}(\lambda), \right.$$
$$\left. \frac{1}{2\pi} \int_{-\infty}^{\infty} g_{kk}(\lambda) d\lambda = q_k, k = \overline{1,T} \right\};$$

$$D_{2\delta}^3 = \left\{ F(\lambda) \middle| \frac{1}{2\pi} \int_{-\infty}^{\infty} |\langle B_1, F(\lambda) - F_1(\lambda) \rangle|^2 d\lambda \leq \delta \right\};$$

$$D_\varepsilon^3 = \left\{ G(\lambda) \middle| \langle B_2, G(\lambda) \rangle = (1-\varepsilon) \langle B_2, G_1(\lambda) \rangle + \varepsilon \langle B_2, W(\lambda) \rangle, \right.$$
$$\left. \frac{1}{2\pi} \int_{-\infty}^{\infty} \langle B_2, G(\lambda) \rangle d\lambda = q \right\};$$

$$D_{2\delta}^4 = \left\{ F(\lambda) \middle| \frac{1}{2\pi} \int_{-\infty}^{\infty} \left| f_{ij}(\lambda) - f_{ij}^1(\lambda) \right|^2 d\lambda \leq \delta_i^j, i,j = \overline{1,T} \right\},$$
$$D_\varepsilon^4 = \left\{ G(\lambda) \middle| G(\lambda) = (1-\varepsilon) G_1(\lambda) + \varepsilon W(\lambda), \frac{1}{2\pi} \int_{-\infty}^{\infty} G(\lambda) d\lambda = Q \right\},$$

where $F_1(\lambda), G_1(\lambda)$ are known and fixed spectral densities, $W(\lambda)$ is unknown spectral density, $q, \delta, q_k, \delta_k, k = \overline{1,T}$, $\delta_i^j, i,j = \overline{1,T}$, are given numbers, Q, B_1, B_2 are given positive definite Hermitian matrices.

Densities from the classes of admissible spectral densities $D_{2\delta}^k, k = 1,2,3,4$, form the "$\delta$–neighbourhood" models in the space L_2 of a given bounded spectral density $F_1(\lambda)$.

Densities from the classes of admissible spectral densities $D_\varepsilon^k, k = 1,2,3,4$, form the "$\varepsilon$–contamination" models of spectral densities.

From the condition $0 \in \partial\Delta_D(F^0, G^0)$ we find the following equations which determine the least favourable spectral densities for these given sets of admissible spectral densities.

For the first pair $D_{2\delta} \times D_\varepsilon^1$ we have equations

$$((A(\lambda))^\top G^0(\lambda) + (C^0(\lambda))^\top)^* ((A(\lambda))^\top G^0(\lambda) + (C^0(\lambda))^\top) =$$
$$= \alpha^2 \mathrm{Tr}\,(F^0(\lambda) - F_1(\lambda))(F^0(\lambda) + G^0(\lambda))^2, \quad (4.153)$$

$$\frac{1}{2\pi} \int_{-\infty}^{\infty} \left| \mathrm{Tr}\,(F^0(\lambda) - F_1(\lambda)) \right|^2 d\lambda = \delta, \quad (4.154)$$

$$((A(\lambda))^\top F^0(\lambda)-(C^0(\lambda))^\top)^*((A(\lambda))^\top F^0(\lambda)-(C^0(\lambda))^\top)=$$
$$=(\beta^2+\gamma(\lambda))(F^0(\lambda)+G^0(\lambda))^2, \quad (4.155)$$

where α^2,β^2 are Lagrange multipliers, $\gamma(\lambda)\leq 0$ and $\gamma(\lambda)=0$ if $\mathrm{Tr}\,F^0(\lambda)>(1-\varepsilon)\mathrm{Tr}\,G_1(\lambda)$.

For the second pair $D_{2\delta}^2\times D_\varepsilon^2$ we have equations

$$((A(\lambda))^\top G^0(\lambda)+(C^0(\lambda))^\top)^*((A(\lambda))^\top G^0(\lambda)+(C^0(\lambda))^\top)=$$
$$=(F^0(\lambda)+G^0(\lambda))\left\{\alpha_k^2(f_{kk}^0(\lambda)-f_{kk}^1(\lambda))\delta_{kl}\right\}_{k,l=1}^T(F^0(\lambda)+G^0(\lambda)), \quad (4.156)$$

$$\frac{1}{2\pi}\int_{-\infty}^{\infty}\left|f_{kk}^0(\lambda)-f_{kk}^1(\lambda)\right|^2 d\lambda=\delta_k,\ k=\overline{1,T}, \quad (4.157)$$

$$((A(\lambda))^\top F^0(\lambda)-(C^0(\lambda))^\top)^*((A(\lambda))^\top F^0(\lambda)-(C^0(\lambda))^\top)=$$
$$=(F^0(\lambda)+G^0(\lambda))\left\{(\beta_k^2+\gamma_k(\lambda))\delta_{kl}\right\}_{k,l=1}^T(F^0(\lambda)+G^0(\lambda)), \quad (4.158)$$

where α_k^2,β_k^2 are Lagrange multipliers, $\gamma_k(\lambda)\leq 0$ and $\gamma_k(\lambda)=0$ if $g_{kk}^0(\lambda)>(1-\varepsilon)g_{kk}^1(\lambda)$.

For the third pair $D_{2\delta}^3\times D_\varepsilon^3$ we have equations

$$((A(\lambda))^\top G^0(\lambda)+(C^0(\lambda))^\top)^*((A(\lambda))^\top G^0(\lambda)+(C^0(\lambda))^\top)=$$
$$=\alpha^2\left\langle B_1,F^0(\lambda)-F_1(\lambda)\right\rangle(F^0(\lambda)+G^0(\lambda))^2, \quad (4.159)$$

$$\frac{1}{2\pi}\int_{-\infty}^{\infty}\left|\left\langle B_1,F^0(\lambda)-F_1(\lambda)\right\rangle\right|^2 d\lambda=\delta, \quad (4.160)$$

$$((A(\lambda))^\top F^0(\lambda)-(C^0(\lambda))^\top)^*((A(\lambda))^\top F^0(\lambda)-(C^0(\lambda))^\top)=$$
$$=(\beta^2+\gamma'(\lambda))(F^0(\lambda)+G^0(\lambda))B_2^\top(F^0(\lambda)+G^0(\lambda)), \quad (4.161)$$

where α^2,β^2 are Lagrange multipliers, $\gamma'(\lambda)\leq 0$ and $\gamma'(\lambda)=0$ if $\langle B_2,G^0(\lambda)\rangle>(1-\varepsilon)\langle B_2,G_1(\lambda)\rangle$.

For the fourth pair $D_{2\delta}^4\times D_\varepsilon^4$ we have equations

$$((A(\lambda))^\top G^0(\lambda)+(C^0(\lambda))^\top)^*((A(\lambda))^\top G^0(\lambda)+(C^0(\lambda))^\top)=$$
$$=(F^0(\lambda)+G^0(\lambda))\left\{\alpha_{ij}(f_{ij}^0(\lambda)-f_{ij}^1(\lambda))\right\}_{i,j=1}^T(F^0(\lambda)+G^0(\lambda)), \quad (4.162)$$

$$\frac{1}{2\pi}\int_{-\infty}^{\infty}\left|f_{ij}^{0}(\lambda)-f_{ij}^{1}(\lambda)\right|^{2}d\lambda=\delta_{i}^{j},\ i,j=\overline{1,T}, \tag{4.163}$$

$$((A(\lambda))^{\top}F^{0}(\lambda)-(C^{0}(\lambda))^{\top})^{*}((A(\lambda))^{\top}F^{0}(\lambda)-(C^{0}(\lambda))^{\top})=$$

$$=(F^{0}(\lambda)+G^{0}(\lambda))(\vec{\beta}\cdot\vec{\beta}^{*}+\Gamma(\lambda))(F^{0}(\lambda)+G^{0}(\lambda)), \tag{4.164}$$

where $\vec{\beta},\alpha_{ij}$ are Lagrange multipliers, $\Gamma(\lambda)\leq 0$ and $\Gamma_3(\lambda)=0$ if $G^0(\lambda)>(1-\varepsilon)G_1(\lambda)$.

Thus, the following statements are true.

Theorem 4.11. *The least favourable spectral densities $F^0(\lambda)$, $G^0(\lambda)$ in the classes $D_{2\delta}^k\times D_{\varepsilon}^k, k=1,2,3,4,$ for the optimal linear filtering of the functional $A_s\vec{\xi}$ are determined by relations (4.153) – (4.164), the minimality condition (4.126), the constrained optimization problem (4.138), and restrictions on densities from the corresponding classes $D_{2\delta}^k\times D_{\varepsilon}^k, k=1,2,3,4.$ The minimax-robust spectral characteristic of the optimal estimate of the functional $A_s\vec{\xi}$ is determined by formula (4.134).*

Corollary 4.15. *Suppose that the spectral density $G(\lambda)$ is known, and the unknown spectral density $F^0(\lambda)\in D_{2\delta}$. Suppose that the function $F^0(\lambda)+G(\lambda)$ satisfies the minimality condition (4.126). The spectral density $F^0(\lambda)$ is the least favourable in the classes $D_{2\delta}^k$, $k=\overline{1,4}$ for the optimal linear filtering of the functional $A\vec{\xi}$ if it satisfies relations (4.153) – (4.154), (4.156) – (4.157), (4.159) – (4.160), (4.162) – (4.163), respectively, and the pair $(F^0(\lambda),G(\lambda))$ determines a solution of the optimization problem (4.138). The minimax-robust spectral characteristic of the optimal estimate of the functional $A\vec{\xi}$ is determined by formula (4.134).*

Corollary 4.16. *Suppose that the spectral density $F(\lambda)$ is known, and the unknown spectral density $G^0(\lambda)\in D_{\varepsilon}^k$. Suppose that the function $F(\lambda)+G^0(\lambda)$ satisfies the minimality condition (4.126). The spectral density $G^0(\lambda)$ is the least favourable in the classes D_{ε}^k, $k=\overline{1,4}$ for the optimal linear filtering of the functional $A\vec{\xi}$ if it satisfies relations (4.155), (4.158), (4.161), (4.164), respectively, and the pair $(F(\lambda),G^0(\lambda))$ determines a solution of the optimization problem (4.138). The minimax-robust spectral characteristic of the optimal estimate of the functional $A\vec{\xi}$ is determined by formula (4.134).*

Conclusion

In this chapter, we present results of the investigation of the interpolation, extrapolation, and filtering problems for linear functionals constructed from unobserved values of multidimensional continuous time stationary stochastic processes. We develop methods of finding the optimal estimates of the functionals in the case where some observations are missing.

We, first, investigate the problems under the condition of spectral certainty, where the spectral density matrices of the involved stationary processes are exactly known. In this case, we developed an approach based on the Hilbert space projection method of estimation.

In the case of extrapolation problem, formulas for calculating values of the mean-square error and the spectral characteristic of the optimal estimate are derived for the functional

$$A\vec{\xi}=\int_{0}^{\infty}\vec{a}(t)^{\top}\vec{\xi}(t)dt$$

constructed from unobserved values of a multidimensional continuous time stationary stochastic process $\vec{\xi}(t)=\{\xi_k(t)\}_{k=1}^{T}$, $t\in\mathbb{R}^+$. Estimates are based on observations of the process with additive stationary stochastic noise process $\vec{\xi}(t)+\vec{\eta}(t)$ at points $t\in\mathbb{R}^-\backslash S$, $S=\bigcup_{l=1}^{s}[-M_l-N_l,-M_l]$.

In the case of interpolation problem, we describe methods of finding solutions of the problem of mean-square optimal linear estimation of the functional

$$A_s\vec{\xi}=\sum_{l=1}^{s}\int_{-M_l-N_l}^{-M_l}\vec{a}(t)^{\top}\vec{\xi}(t)dt$$

constructed from unobserved values of a multidimensional continuous time stationary stochastic process $\vec{\xi}(t)$. Estimates are based on observations of the process with additive stationary stochastic noise process $\vec{\xi}(t)+\vec{\eta}(t)$ at points $t\in\mathbb{R}\backslash S$, $S=\bigcup_{l=1}^{s}[-M_l-N_l,-M_l]$.

In the case of filtering problem, we propose results of the investigation of the problem of mean-square optimal linear estimation of the functional

$$A\vec{\xi} = \int_{R^s} \vec{a}(t)^{\top}\vec{\xi}(-t)dt$$

constructed from unobserved values of a multidimensional stationary stochastic process $\vec{\xi}(t)$. Estimates are based on observations of the process with additive stationary stochastic noise process $\vec{\xi}(t)+\vec{\eta}(t)$ at points $t \in \mathbb{R}^{-}\backslash S$, $S = \bigcup_{l=1}^{s}[-M_l - N_l, -M_l]$, $R^s = [0,\infty)\backslash S^{+}$, $S^{+} = \bigcup_{l=1}^{s}[M_l,\, M_l + N_l]$.

Next, we investigate the problems in the case of spectral uncertainty, where the spectral density matrices of the multidimensional stationary stochastic processes are not exactly known, while some special sets of admissible spectral density matrices are given. In this case, we apply the minimax robust method of estimation. This method allows us to find estimates which minimize the maximal values of the mean-square errors of estimates for all spectral density matrices from a given class of admissible spectral densities. We derive formulas that determine the least favourable spectral density matrices and the minimax spectral characteristics for some special classes of admissible spectral density matrices.

These least favourable spectral density matrices are determined by solutions of the optimization problem

$$\Delta_D(F,G) = -\Delta(h(F^0,G^0);F,G) + \delta((F,G)\,|D_F \times D_G) \to \inf,$$

which are characterized by the condition $0 \in \partial\Delta_D(F^0,G^0)$. Here $\partial\Delta_D(F^0,G^0)$ is the subdifferential of the convex functional $\Delta_D(F,G)$ at point (F^0,G^0). The form of the functional $\Delta(h(F^0,G^0);F,G)$ is convenient for application of the Lagrange method of indefinite multipliers for finding solutions of the optimization problem. The complexity of the problem is determined by the complexity of calculation the subdifferential of the convex functional $\Delta_D(F,G)$. Making use of the method of Lagrange multipliers and the form of subdifferentials of the indicator functions, we describe relations that determine the least favourable spectral density matrices in some special classes of spectral density matrices.

These classes of admissible spectral densities are the following. The classes $D_{1\delta}$ of spectral densities which form the "δ-neighbourhood" models in the space L_1 of a fixed bounded spectral density matrix. The classes $D_{2\delta}$ of spectral densities which form the "δ-neighbourhood" models in the space L_2 of a fixed bounded spectral density matrix. The classes D_ε of spectral densities which form the "ε–contamination" model of spectral density matrices. The classes D_V^U of spectral densities which form the "strip" models of spectral density matrices.

References

[1] A. N. Kolmogorov. *Selected works by A. N. Kolmogorov. Vol. II: Probability theory and mathematical statistics*. Ed. by A. N. Shiryayev. Mathematics and its Applications. Soviet Series. 26. Kluwer, Dordrecht, 1992.

[2] H. Cramér. On the theory of stationary random processes. *Annals of Mathematics*, 41:215–230, 1940.

[3] M. H. Stone. *Linear transformations in Hilbert space*. AMS Colloquium Publications, Vol. 15, 1932.

[4] H. Wold. *A study in the analysis of stationary time series*. Almqvist & Wiksell, Uppsala, 2-d edition, 1954.

[5] N. Wiener. *Extrapolation, interpolation and smoothing of stationary time series: With engineering applications*. The M. I. T. Press, Massachusetts Institute of Technology, Cambridge, Mass., 1966.

[6] N. Wiener and P. Masani. The prediction theory of multivariate stochastic processes. Part 1: The regularity condition. *Acta Mathematica*, 98:111–150, 1957.

[7] N. Wiener and P. Masani. The prediction theory of multivariate stochastic processes. Part 2: The linear predictor. *Acta Mathematica*, 99:93–137, 1957.

[8] A. M. Yaglom. *Correlation theory of stationary and related random functions. Vol. 1: Basic results*. Springer Series in Statistics, Springer-Verlag, New York etc, 1987.

[9] A. M. Yaglom. *Correlation theory of stationary and related random functions. Vol. 2: Supplementary notes and references.* Springer Series in Statistics, Springer-Verlag, New York etc, 1987.

[10] Yu. A. Rozanov. *Stationary stochastic processes.* Holden-Day, San Francisco-Cambridge-London-Amsterdam, 1967.

[11] E. J. Hannan. *Multiple time series.* John Wiley & Sons, Inc. XI, New York etc., 1970.

[12] P. J. Brockwell and R. A. Davis. *Time series: Theory and methods.* Springer, New York, 2nd edition, 1998.

[13] D. Z. Arov and H. Dym. *Multivariate prediction, de Branges spaces, and related extension and inverse problems.* Birkhäuser, Cham, Switzerland, 2018.

[14] L. Aggoun and R. J. Elliott. *Measure theory and filtering: introduction and applications.* Cambridge University Press, Cambridge, New York etc, 2004.

[15] I. V. Basawa and B. L. S. Prakasa Rao. *Statistical inference for stochastic processes.* Academic Press, London etc, 1980.

[16] S. N. Cohen and R. J. Elliott. *Stochastic Calculus and Applications.* Birkhäuser, Basel, 2015.

[17] M. S. Grewal and A. P. Andrews. *Kalman filtering. Theory and practice with MATLAB.* John Wiley & Sons, Hoboken, NJ, 4th edition, 2015.

[18] G. Kallianpur. *Stochastic filtering theory.* Springer-Verlag, New York - Heidelberg - Berlin, 1980.

[19] Yu. A. Kutoyants. *Parameter estimation for stochastic processes.* Herdermann, Berlin, 1984.

[20] Yu. A. Kutoyants. *Statistical inference for ergodic diffusion processes.* Springer, London, 2004.

[21] R. S. Liptser and A. N. Shiryaev. *Statistics of random processes I. General theory*. Springer, Berlin, Heidelberg, 2001.

[22] R. S. Liptser and A. N. Shiryaev. *Statistics of random processes II. Applications theory*. Springer, Berlin, Heidelberg, 2001.

[23] B.L.S. Prakasa Rao. *Statistical inference for fractional diffusion processes*. John Wiley & Sons, Chichester, 2010.

[24] B.L.S. Prakasa Rao. *Associated sequences, demimartingales and nonparametric inference*. Birkhäuser, Basel, 2012.

[25] B. L. Rozovsky and S. V. Lototsky. *Stochastic evolution systems. Linear theory and applications to non-linear filtering*. Springer, Cham, Switzerland, 2018.

[26] M. B. Rajarshi. *Statistical inference for discrete time stochastic processes*. Springer, New Dehli, 2012.

[27] W. A. Woodward, H. L. Gray, and A. C. Elliott. *Applied time series analysis with R*. CRC Press, Taylor & Francis Group, Boca Raton, FL, 2017.

[28] M. I. Yadrenko. *Spectral theory of random fields*. Optimization Software Inc. Publications Division, New York, 1983.

[29] G. E. P. Box, G. M. Jenkins, G. C. Reinsel, and G. M. Ljung. *Time series analysis. Forecasting and control*. Wiley, Hoboken, New Jersey, 5th edition, 2016.

[30] S. Arvanitis and T. Magdalinos. Mildly explosive autoregression under stationary conditional heteroskedasticity. *Journal of Time Series Analysis*, 39(6):892–908, 2018.

[31] W. R. Bell and D. E. K. Martin. Computation of asymmetric signal extraction filters and meansquared error for ARIMA component models. *Journal of Time Series Analysis*, 25:603–625, 2004.

[32] M. J. Chambers and M. A. Thornton. Discrete time representation of continuoustime arma processes. *Econometric Theory*, 28:219–238, 2012.

[33] M. J. Chambers. The estimation of continuous time models with mixed frequency data. *Journal of Econometrics*, 193:390–404, 2016.

[34] D. Findley and D. Martin. Frequency domain analyses of SEATS and X-11/12-ARIMA seasonal adjustment filters for short and moderate-length time series. *Journal of Official Statistics*, 22:1–34, 2006.

[35] B. Spangl and R. Dutter. Robustness in time series: Robust frequency domain analysis. In: Becker C., Fried R., Kuhnt S. (eds) *Robustness and Complex Data Structures*. Springer, Berlin, Heidelberg, 2013.

[36] T. Proietti, M. Marczak, and G. Mazzi. A class of periodic trend models for seasonal time series. *Journal of Forecasting*, 38(2):106–121, 2018.

[37] B. Seong, S. K. Ahn, and P. A. Zadrozny. Estimation of vector error correction models with mixed-frequency data. *Journal of Time Series Analysis*, 34:194–205, 2013.

[38] M. Zhong, Y. Song, and S. Ding. Parity spacebased fault detection for linear discrete time-varying systems with unknown input. *Automatica*, 59:120–126, 2015.

[39] S. Koopman and A. Harvey. Computing observation weights for signal extraction and filtering. *Journal of Economic and Dynamic Control*, 127:1317–1333, 2003.

[40] T. McElroy and A. Sutcliffe. Signal extraction for non-stationary multivariate time series with illustrations for trend inflation. *Journal of Time Series Analysis*, 36(2):209–227, 2015.

[41] T. McElroy. Signal extraction for non-stationary multivariate time series with illustrations for trend inflation. *Journal of Time Series Analysis*, 36(2):209–227, 2015.

[42] M. Wildi. Real-time signal extraction: a shift of perspective. *Estudios de Economia Aplicada*, 28(3):497–518, 2010.

[43] M. G. Krein. On a fundamental approximation problem in the theory of extrapolation and filtration of stationary random processes. *Selected Transl. in Math. Stat. and Prob.*, 4:127–132, 1963.

[44] H. Dym and H. P. McKean. Extrapolation and interpolation of stationary gaussian processes. *Annals of Mathematical Statistics*, 41:1817–1844, 1970.

[45] H. Dym and H. P. McKean. *Gaussian processes, function theory, and the inverse spectral problem*. Academic Press, New York, 1976.

[46] D. V. Koroliouk. Stationary statistical experiments and the optimal estimator for a predictable component. *Journal of Mathematical Sciences*, 214(2):220–228, 2016.

[47] D. V. Koroliouk and V. S. Koroliuk. Filtration of stationary Gaussian statistical experiments. *Journal of Mathematical Sciences*, 229(1):30–35, 2018.

[48] S. K. Vastola and H. V. Poor. An analysis of the effects of spectral uncertainty on wiener filtering. *Automatica*, 19(3):289–293, 1983.

[49] U. Grenander. A prediction problem in game theory. *Arkiv för Matematik*, 3:371–379, 1957.

[50] M. P. Moklyachuk. Estimation of linear functionals of stationary stochastic processes and a two-person zero-sum game. *Stanford University Technical Report No, 169:1–87*, 1981.

[51] M. P. Moklyachuk. On a problem of game theory and the extrapolation of stochastic processes with values in a hilbert space. *Theory of Probability and Mathematical Statistics*, 24:107–114, 1981.

[52] M. P. Moklyachuk. On an antagonistic game and prediction of stationary random sequences in a hilbert space. *Theory of Probability and Mathematical Statistics*, 25:107–113, 1982.

[53] M. P. Moklyachuk. Minimax extrapolation and autoregressive-moving average processes. *Theory of Probability and Mathematical Statistics*, 41:77–84, 1990.

[54] M. P. Moklyachuk. Minimax filtration of linear transformations of stationary sequences. *Ukrainian Mathematical Journal*, 43(1):92–99, 1991.

[55] M. P. Moklyachuk. *Robust estimations of functionals of stochastic processes*. Kyiv University, Kyiv (In Ukrainian), 2008.

[56] M. P. Moklyachuk. Minimax-robust estimation problems for stationary stochastic sequences. *Statistics, Optimization & Information Computing*, 3(4):348–419, 2015.

[57] P. J. Huber. Robust estimation of a location parameter. *The Annals of Mathematical Statistics*, 35(1):73–101, 1964.

[58] P. J. Huber. A robust version of the probability ratio test. *The Annals of Mathematical Statistics*, 36(6):1753–1758, 1964.

[59] S. A. Kassam and H. V. Poor. Robust techniques for signal processing: A survey. *Proceedings of the IEEE*, 73(3):433–481, 1985.

[60] L. Breiman. A note of minimax filtering. *The Annals of Probability*, 1(1):175–179, 1973.

[61] Y. Hosoya. Robust linear extrapolations of second-order stationary processes. *The Annals of Probability*, 6(4):574–584, 1978.

[62] M. Taniguchi. Robust regression and interpolation for time series. *Journal of Time Series Analysis*, 2(1):53–62, 1981.

[63] S. A. Kassam. Robust hypothesis testing for bounded classes of probability densities. *IEEE Transaction on Information Theory*, 27:242–247, 1981.

[64] S. A. Kassam. Robust hypothesis testing and robust time series interpolation and regression. *Journal of Time Series Analysis*, 3(3):85–194, 1982.

[65] J. Franke. Minimax robust prediction of discrete time series. *Z. Wahrscheinlichkeitstheor. Verw. Gebiete*, 68:337–364, 1985.

[66] J. Franke. On the robust prediction and interpolation of time series in the presence of correlated noise. *Journal of Time Series Analysis*, 5(4):227–244, 1984.

[67] J. Franke and H. V. Poor. Minimax-robust filtering and finite-length robust predictors. *Robust and Nonlinear Time Series Analysis. Lecture Notes in Statistics*, Vol. 26, Springer-Verlag, pp. 87–126, 1984.

[68] H. V. Poor. On robust wiener filtering. *IEEE Transactions on Automatic Control*, AC-25:531–536, 1980.

[69] H. V. Poor. Robust matched filters. *IEEE Transactions on Information Theory*, 29(5):677–687, 1983.

[70] S. K. Vastola and H. V. Poor. Robust wiener-kolmogorov theory. *IEEE Transactions on Information Theory*, 30(2):316–327, 1984.

[71] S. Verdu and H. V. Poor. On minimax robustness: a general approach and applications. *IEEE Transactions on Information Theory*, 30(20):328–340, 1984.

[72] G. K. Golubev and M. S. Pinsker. Minimax extrapolation of sequences. *Problems Inform. Transmission*, 19:275–283, 1984.

[73] G. K. Golubev and M. S. Pinsker. Extremal problems in minimax estimation of sequences. *Problems Inform. Transmission*, 21:192–206, 1985.

[74] O. M. Kurkin, Yu. B. Korobochkin, and S. D. Shatalov. *Minimax information processing*. Energoatomizdat, Moskow (in Russian), 1990.

[75] M. P. Moklyachuk. On a filtering problem for vector-valued sequences. *Theory of Probability and Mathematical Statistics*, 47:107–118, 1993.

[76] M. P. Moklyachuk. Stochastic autoregressive sequences and minimax interpolation. *Theory of Probability and Mathematical Statistics*, 48:95–103, 1994.

[77] M. P. Moklyachuk. Estimation of stochastic processes from observations with noise. *Theory of Stochastic Processes*, 3(3-4):330–338, 1997.

[78] M. P. Moklyachuk. Extrapolation of stationary sequences from observations with noise. *Theory of Probability and Mathematical Statistics*, 57:133–141, 1998.

[79] M. P. Moklyachuk. Robust procedures in time series analysis. *Theory of Stochastic Processes*, 6(3-4):127–147, 2000.

[80] M. P. Moklyachuk. Game theory and convex optimization methods in robust estimation problems. *Theory of Stochastic Processes*, 7(1-2):253–264, 2001.

[81] D. Bertsimas, X. V. Doan, K. Natarajan, and C. P. Teo. Models for minimax stochastic linear optimization problems with risk aversion. *Mathematics of Operations Research*, 35(3):580–602, 2010.

[82] P. S. Bhattacharya and D. D. Thomakos. Robust model rankings of forecasting performance. *Journal of Forecasting*, 37(6):676–690, 2018.

[83] E. Carrizosaa, A. Olivares-Nadal, and P. Ramirez-Cobob. Robust newsvendor problem with autoregressive demand. *Computers and Operations Research*, 68:123–133, 2016.

[84] E. Delage and Y. Ye. Distributionally robust optimization under moment uncertainty with application to data-driven problems. *Operations Research*, 58(3):596–612, 2010.

[85] J. Ishihara, M. Terra, and J. Campos. Robust kalman filter for descriptor systems. *IEEE Transactions on Automatic Control*, 51(8):1354–1358, 2006.

[86] T. L. Molloy and J. J. Ford. Asymptotic minimax robust quickest change detection for dependent stochastic processes with parametric uncertainty. *IEEE Transaction on Information Theory*, 62(11):6594–6608, 2016.

[87] P. Ruckdeschel, B. Spangl, and D. Pupashenko. Robust kalman tracking and smoothing with propagating and non-propagating outliers. *Statistical Papers*, 55(1):93–123, 2014.

[88] S. Song, Y. Gong, Y. Zhang, G. Huang, and G.-B. Huang. Dimension reduction by minimum error minimax probability machine. *IEEE Trans. Syst., Man, and Cybern., Syst.*, 47(1):58–69, 2017.

[89] Yu. Kharin. *Robustness in statistical forecasting*. Springer, Berlin, Heidelberg, 2013.

[90] M. P. Moklyachuk and O. Yu. Masyutka. Interpolation of multidimensional stationary sequences. *Theory of Probability and Mathematical Statistics*, 73:125–133, 2006.

[91] M. P. Moklyachuk and O. Yu. Masyutka. Robust estimation problems for stochastic processes. *Theory of Stochastic Processes*, 12(3-4):88–113, 2006.

[92] M. P. Moklyachuk and O. Yu. Masyutka. On the problem of filtration of vector stationary sequences. *Theory of Probability and Mathematical Statistics*, 75:109–119, 2007.

[93] M. P. Moklyachuk and O. Yu. Masyutka. Robust filtering of stochastic processes. *Theory of Stochastic Processes*, 13(166-181):1–2, 2008.

[94] M. P. Moklyachuk and O. Yu. Masyutka. Minimax prediction problem for multidimentional stationary stochastic sequence. *Theory of Stochastic Processes*, 14(3-4):89–103, 2008.

[95] M. P. Moklyachuk and O. Yu. Masyutka. Minimax prediction problem for multidimensional stationary stochastic processes. *Communications in Statistics – Theory and Methods*, 40(19-20):3700–3710, 2011.

[96] M. P. Moklyachuk and O. Yu. Masyutka. *Minimax-robust estimation technique for stationary stochastic processes*. LAP Lambert Academic Publishing, 2012.

[97] I. I. Dubovets'ka, O. Yu. Masyutka, and M. P. Moklyachuk. Interpolation of periodically correlated stochastic sequences. *Theory of Probability and Mathematical Statistics*, 84:43–56, 2012.

[98] I. I. Dubovets'ka and M. P. Moklyachuk. Filtering of periodically correlated processes. *Prykl. Stat., Aktuarna Finans. Mat.*, 2:149–158, 2012.

[99] I. I. Dubovets'ka and M. P. Moklyachuk. Filtration of linear functionals of periodically correlated sequences. *Theory of Probability and Mathematical Statistics*, 86:51–64, 2013.

[100] I. I. Dubovets'ka and M. P. Moklyachuk. Extrapolation of periodically correlated processes from observations with noise. *Theory of Probability and Mathematical Statistics*, 88:43–55, 2013.

[101] I. I. Dubovets'ka and M. P. Moklyachuk. Minimax estimation problem for periodically correlated stochastic processes. *Journal of Mathematics and System Science*, 3(1):26–30, 2013.

[102] I. I. Dubovets'ka and M. P. Moklyachuk. On minimax estimation problems for periodically correlated stochastic processes. *Contemporary Mathematics and Statistics*, 2(1):123–150, 2014.

[103] I. I. Dubovets'ka, O. Yu. Masyutka, and M. P. Moklyachuk. Estimation problems for periodically correlated isotropic random fields. *Methodology and Computing in Applied Probability*, 17(1):41–57, 2015.

[104] M. P. Moklyachuk and I. I. Golichenko. *Periodically correlated processes estimates*. LAP Lambert Academic Publishing, 2016.

[105] M. P. Moklyachuk and V. I. Ostapenko. Minimax interpolation problem for harmonizable stable sequences with noise observations. *Journal of Applied Mathematics and Statistics*, 2(1):21–42, 2015.

[106] M. P. Moklyachuk and V. I. Ostapenko. Minimax interpolation of harmonizable sequences. *Theory of Probability and Mathematical Statistics*, 92:135–146, 2016.

[107] M. P. Moklyachuk and V. I. Ostapenko. Minimax extrapolation problem for harmonizable stable processes with noise observations. *Bulletin of Taras Shevchenko National University of Kyiv Series: Physics and Mathematics*, 2:15–23, 2016.

[108] M. P. Moklyachuk and V. I. Ostapenko. Minimax filtering of harmonizable stochastic sequences. *Scientific Bulletin of Uzhhorod State University, Mathematics and Informatics*, 28(1):80–89, 2016.

[109] M. P. Moklyachuk and V. I. Ostapenko. Minimax filtering of harmonizable stochastic processes. *Scientific Bulletin of Uzhhorod State University, Mathematics and Informatics*, 29(2):130–139, 2016.

[110] M. P. Moklyachuk and V. I. Ostapenko. Interpolation of harmonizable processes from observations with noise. *Bukovinian Mathematical Journal*, 5(1-2):112–122, 2017.

[111] M. P. Moklyachuk and V. I. Ostapenko. Minimax extrapolation problem for harmonizable stable sequences with noise observations. *Stochastic Modelling and Applications*, 21(1):1–22, 2017.

[112] M. M. Luz and M. P. Moklyachuk. Interpolation of functionals from stochastic sequences with stationary increments for observations with noise. *Prykl. Stat., Aktuarna Finans. Mat.*, 2:131–148, 2012.

[113] M. M. Luz and M. P. Moklyachuk. Interpolation of functional of stochastic sequences with stationary increments. *Theory of Probability and Mathematical Statistics*, 87:117–133, 2013.

[114] M. M. Luz and M. P. Moklyachuk. Minimax-robust filtering problem for stochastic sequences with stationary increments. *Theory of Probability and Mathematical Statistics*, 89:127–142, 2014.

[115] M. M. Luz and M. P. Moklyachuk. Robust extrapolation problem for stochastic processes with stationary increments. *Mathematics and Statistics*, 2(2):78–88, 2014.

[116] M. M. Luz and M. P. Moklyachuk. Minimax-robust filtering problem for stochastic sequences with stationary increments and cointegrated sequences. *Statistics, Optimization & Information Computing*, 2(3):176–199, 2014.

[117] M. M. Luz and M. P. Moklyachuk. Minimax interpolation problem for stochastic processes with stationary increments. *Statistics, Optimization & Information Computing*, 3(1):30–41, 2015.

[118] M. M. Luz and M. P. Moklyachuk. Minimax-robust prediction problem for stochastic sequences with stationary increments and cointegrated sequences. *Statistics, Optimization & Information Computing*, 3(2):160–188, 2015.

[119] M. M. Luz and M. P. Moklyachuk. Filtering problem for stochastic processes with stationary increments. *Contemporary Mathematics and Statistics*, 3(1):8–27, 2015.

[120] M. M. Luz and M. P. Moklyachuk. Minimax prediction of stochastic processes with stationary increments from observations with stationary noise. *Cogent Mathematics*, 3(11332):19, 2016.

[121] M. M. Luz and M. P. Moklyachuk. Minimax prediction of stochastic processes with stationary increments from observations with stationary noise. *Cogent Mathematics*, 3(1133219):1–17, 2016.

[122] M. M. Luz and M. P. Moklyachuk. Filtering problem for functionals of stationary sequences. *Statistics, Optimization and Information Computing*, 4(1):68–83, 2016.

[123] M. M. Luz and M. P. Moklyachuk. Minimax-robust filtering problem for stochastic sequences with stationary increments and cointegrated sequences. *Cogent Mathematics*, 3(11678):11, 2016.

[124] M. M. Luz and M. P. Moklyachuk. Minimax interpolation of sequences with stationary increments and cointegrated sequences. *Modern Stochastics: Theory and Applications*, 3(1):59–87, 2016.

[125] M. M. Luz and M. P. Moklyachuk. *Estimates of functionals from processes with stationary increments and cointegrated sequences*. NVP "Interservis" (in Ukrainian), 2016.

[126] M. P. Moklyachuk and M. M. Luz. Robust extrapolation problem for stochastic sequences with stationary increments. *Contemporary Mathematics and Statistics*, 1(3):123–150, 2013.

[127] P. S. Kozak and M. P. Moklyachuk. Estimates of functionals constructed from random sequences with periodically stationary increments. *Theory of Probability and Mathematical Statistics*, 97:85–98, 2019.

[128] P. Bondon. Prediction with incomplete past of a stationary process. *Stochastic Process and their Applications*, 98:67–76, 2002.

[129] P. Bondon. Influence of missing values on the prediction of a stationary time series. *Journal of Time Series Analysis*, 26(4):519–525, 2005.

[130] R. Cheng, A. G. Miamee, and M. Pourahmadi. Some extremal problems in $l^p(w)$. *Proceedings of the American Mathematical Society*, 126:2333–2340, 1998.

[131] R. Cheng and M. Pourahmadi. Prediction with incomplete past and interpolation of missing values. *Statistics & Probability Letters*, 33:341–346, 1996.

[132] Y. Kasahara, M. Pourahmadi, and A. Inoue. Duals of random vectors and processes with applications to prediction problems with missing values. *Statistics & Probability Letters*, 79(14):1637–1646, 2009.

[133] M. Pourahmadi, A. Inoue, and Y. Kasahara. A prediction problem in $l^2(w)$. *Proceedings of the American Mathematical Society*, 135(4):1233–1239, 2007.

[134] B. Abraham. Missing observations in time series. *Communications in Statistics – Theory and Methods*, 10:1643–1653, 1981.

[135] M. J. Daniels and J. W. Hogan. *Missing data in longitudinal studies: strategies for Bayesian modeling and sensitivity analysis*. Taylor & Francis Group, Boca Raton, 2016.

[136] C. K. Enders. *Applied missing data analysis (Methodology in the social sciences)*. Guilford Press, New York, 2010.

[137] R. J. A. Little and D. B. Rubin. *Statistical analysis with missing data*. Wiley, Chichester, 2002.

[138] P. E. McKnight, K. M. McKnight, S. Sidani, and A. J. Figueredo. *Missing data: A gentle introduction*. Guilford Press, New York, 2007.

[139] M. M. Pelagatti. *Time series modelling with unobserved components*. CRC Press, New York, 2015.

[140] M. P. Moklyachuk and M. I. Sidei. Interpolation of stationary sequences observed with the noise. *Theory of Probability and Mathematical Statistics*, 93:143–156, 2015.

[141] M. P. Moklyachuk and M. I. Sidei. Interpolation problem for stationary sequences with missing observations. *Statistics, Optimization & Information Computing*, 3(3):259–275, 2015.

[142] M. P. Moklyachuk and M. I. Sidei. Extrapolation problem for functionals of stationary processes with missing observations. *Bukovinian Mathematical Journal*, 4(1-2):122–129, 2016.

[143] M. P. Moklyachuk and M. I. Sidei. Interpolation of functionals of stationary processes with missing observations. *Bulletin of Taras Shevchenko National University of Kyiv, no*, 1:24–30, 2016.

[144] M. P. Moklyachuk and M. I. Sidei. Filtering problem for stationary sequences with missing observations. *Statistics, Optimization & Information Computing*, 4(4):308–325, 2016.

[145] M. P. Moklyachuk and M. I. Sidei. Filtering problem for functionals of stationary processes with missing observations. *Communications in Optimization Theory*, (21):1–18, 2016.

[146] M. P. Moklyachuk and M. I. Sidei. Extrapolation problem for stationary sequences with missing observations. *Statistics, Optimization & Information Computing*, 5(3):212–233, 2017.

[147] O. Yu. Masyutka, M. P. Moklyachuk, and M. I. Sidei. Minimax extrapolation of multidimensional stationary processes with missing observations. *International Journal of Mathematical Models and Methods in Applied Sciences*, 12:94–105, 2018.

[148] O. Yu. Masyutka, M. P. Moklyachu, and M. I. Sidei. Interpolation problem for multidimensional stationary sequences with missing observations. *Stochastic Modeling and Applications*, 22(2):85–103, 2018.

[149] O. Yu. Masyutka, M. P. Moklyachuk, and M. I. Sidei. Extrapolation problem for multidimensional stationary sequences with missing observations. *Statistics, Optimization and Information Computing*, 7(1):97–117, 2019.

[150] O. Yu. Masyutka, M. P. Moklyachuk, and M. I. Sidei. Interpolation problem for multidimensional stationary processes with missing observations. *Statistics, Optimization and Information Computing*, 7(1):118–132, 2019.

[151] O. Yu. Masyutka, M. P. Moklyachuk, and M. I. Sidei. Filtering of multidimensional stationary processes with missing observations. *Universal Journal of Mathematics and Applications*, 2(1):24–35, 2019.

[152] M. P. Moklyachuk, O. Yu. Masyutka, and I. I. Golichenko. *Estimates of periodically correlated isotropic random fields*. Nova Science Publishers, Inc. New York, 2018.

[153] I. I. Gikhman and A. V. Skorokhod. *The theory of stochastic processes. I.* Springer, Berlin, 2004.

[154] M. Pourahmadi. Estimation and interpolation of missing values of a stationary time series. *Journal of Time Series Analysis*, 10(2):149–169, 1989.

[155] H. Salehi. Algorithms for linear interpolator and interpolation error for minimal stationary stochastic processes. *The Annals of Probability*, 7(5):840–846, 1979.

[156] B. N. Pshenichnyj. *Necessary conditions of an extremum*. Marcel Dekker, Pure and Applied Mathematics, 4. New York, 1971.

[157] M. G. Krein and A. A. Nudelman. *The Markov moment problem and extremal problems*. American Mathematical Society, Translations of Mathematical Monographs, Vol. 50, Providence, R.I., 1977.

[158] R. T. Rockafellar. *Convex analysis*. Princeton University Press, 1997.

[159] N. I. Akhiezer and I. M. Glazman. *Theory of linear operators in Hilbert space*. Dover Publications, New York, 1993.

[160] F. R. Gantmaher. *The theory of matrices.* AMS Chelsea Publishing, AMS, 1959.

[161] A. D. Ioffe and V. M. Tihomirov. Theory of extremal problems. North-Holland Publishing Company. XII, *Studies in Mathematics and its Applications*, Vol. 6, Amsterdam, New York, Oxford, 1979.

[162] N. I. Akhiezer and M. G. Krein. Some questions in the theory of moments. *AMS Translations of Mathematical Monographs*, Vol. 2, Providence, R.I., 1962.

[163] M. P. Moklyachuk. *Nonsmooth analysis and optimization.* Kyiv University, Kyiv (In Ukrainian), 2008.

Author Contact Information

Mikhail Moklyachuk
Professor
Department of Probability Theory, Statistics and Actuarial Mathematics,
Taras Shevchenko National University of Kyiv, Kyiv, Ukraine
Email: mmp@univ.kiev.ua; Moklyachuk@gmail.com

Maria Sidei
Department of Probability Theory, Statistics and Actuarial Mathematics,
Taras Shevchenko National University of Kyiv, Kyiv, Ukraine
Email: marysidei4@gmail.com

Oleksandr Masyutka
Department of Mathematics and Theoretical Radiophysics,
Taras Shevchenko National University of Kyiv, Kyiv, Ukraine
Email: masutkaAU@bigmir.net

Index

F

G

H

I

L

M

N

O

P

R

S

T

U

V